Epiphytic Plants: Perspective on Their Diversity, Distribution, Systematics and Conservation in the Changing Environment

Epiphytic Plants: Perspective on Their Diversity, Distribution, Systematics and Conservation in the Changing Environment

Guest Editors

Thorsten Krömer
Sven Peter Batke

MDPI

Basel • Beijing • Wuhan • Barcelona • Belgrade • Novi Sad • Cluj • Manchester

Guest Editors
Thorsten Krömer
Centro de Investigaciones Tropicales
Universidad Veracruzana
Xalapa
Mexico

Sven Peter Batke
Biology Department
Edge Hill University
Ormskirk
UK

Editorial Office
MDPI AG
Grosspeteranlage 5
4052 Basel, Switzerland

This is a reprint of the Special Issue, published open access by the journal *Plants* (ISSN 2223-7747), freely accessible at: https://www.mdpi.com/journal/plants/special_issues/P317F0G352.

For citation purposes, cite each article independently as indicated on the article page online and as indicated below:

Lastname, A.A.; Lastname, B.B. Article Title. *Journal Name* **Year**, *Volume Number*, Page Range.

ISBN 978-3-7258-4937-6 (Hbk)
ISBN 978-3-7258-4938-3 (PDF)
https://doi.org/10.3390/books978-3-7258-4938-3

Cover image courtesy of Thorsten Krömer

Contents

About the Editors

Thorsten Krömer

Thorsten Krömer has an MSc in biology (1997) and a PhD in natural sciences (2003) from the University of Göttingen, Germany. He conducted a postdoctoral stay (2005–2007) at the Estación de Biología Tropical "Los Tuxtlas" of the UNAM in Mexico. Currently, he is a senior researcher at the Centro de Investigaciones Tropicales (CITRO) of the Universidad Veracruzana in Xalapa, Mexico, where he is also a professor of the postgraduate program in "Tropical Ecology". He has been the director or advisor of 40 BSc, MSc, and PhD theses, and has co-authored more than 110 scientific papers. His research interest focuses on the diversity, distribution, ecology, and conservation of vascular epiphytes. His current projects include studies on the diversity patterns of these plants along gradients of elevation and human disturbance, IUCN conservation assessments and floristic inventories for different epiphyte groups, as well as the floral biology of aroids and bromeliads.

Sven Peter Batke

Sven Peter Batke is a Reader of Plant Science at Edge Hill University (EHU) and currently serves as the Associate Head of Research and Knowledge Exchange. He obtained his PhD in Plant Science from Trinity College Dublin in 2015, conducting fieldwork on epiphytes in Honduras as part of his doctoral research. Since then, he has focused extensively on the ecology, physiology, and taxonomy of epiphytes. He is currently leading the development of the Fern Flora of Honduras (a first for the country) and holds a research associate position at the Universidad Zamorano in Honduras. He also chairs the UK's Greenhouse Innovation Consortium, promoting sustainable horticultural technologies. In addition to publishing widely in peer-reviewed journals, he has led and contributed to numerous conservation management plans and whitepapers aimed at informing environmental policy. In 2018, he co-founded the BSc Plant Science program at EHU alongside Professor Paul Ashton.

plants

MDPI

Editorial

Epiphytic Plants: Perspective on Their Diversity, Distribution, Systematics and Conservation in the Changing Environment

Thorsten Krömer [1,*] and Sven Peter Batke [2]

[1] Centro de Investigaciones Tropicales, Universidad Veracruzana, Xalapa 91000, Mexico
[2] Biology Department, Edge Hill University, Ormskirk L39 4QP, UK; sven.batke@edgehill.ac.uk
* Correspondence: tkromer@uv.mx

Epiphytic plants are vital components of tropical and subtropical forests, contributing significantly to biodiversity, ecosystem function, and structural complexity. Despite their importance, they remain understudied and face increasing threats from environmental change. This Special Issue, entitled "Epiphytic Plants: Perspective on Their Diversity, Distribution, Systematics and Conservation in the Changing Environment", is a compilation of nine articles, whose authors deepen our understanding of the divergent roles of epiphytes, i.e., plants that germinate and root non-parasitically on other plants, mainly on trees. These are a conspicuous and integral component of tropical forest ecosystems in regard to local and regional plant diversity, especially in the Neotropics [1]. Roughly 10% of all vascular plants in the world are epiphytes, which are distributed in about 900 genera and 80 families. However, 85% of the more than 27,600 species recorded globally belong to only five taxonomic groups: Araceae, Bromeliaceae, Orchidaceae, *Peperomia* Ruiz & Pav., and Pteridophytes. Vascular epiphytes fulfill diverse ecological functions in tropical ecosystems, such as the accumulation of water and nutrients, in addition to amplifying biodiversity by providing microhabitats and food to different taxa (e.g., invertebrates, amphibians, reptiles, birds, and mammals). Due to their dependence on host trees and atmospheric water sources, epiphytes are considered one of the most endangered plant groups. Landscape fragmentation and forest degradation resulting from human activity are among the major threats to their survival.

This vulnerability is further illustrated by findings from a literature review by Krömer et al. (2025) [2], which showed that none of the human-modified ecosystems examined were able to fully replicate the epiphyte diversity found in undisturbed primary forests. Human disturbance in tropical forests typically leads to substantial diversity loss, reduced ecosystem services, and decreased soil health. However, over time, secondary forests may assist in the recovery of a substantial portion of species diversity and functional roles and complement primary forest remnants by enhancing landscape connectivity. Even so, primary forests are irreplaceable for epiphyte diversity conservation, and efforts to mitigate land-use impacts should focus on preserving these areas and implementing sustainable land-management practices.

Beyond secondary forests, urban environments represent an even more extreme form of habitat modification, where structural simplification and altered microclimates often lead to a further decline in epiphyte diversity. In a second paper by Landeros-López et al. (2025) [3], it was shown that the structural alterations of urban forests create stressful microclimatic conditions that can influence the morphology of sensitive plants, such as ferns. Sites with greater modifications in vegetation structure exhibited increased canopy openness, solar radiation, and temperature, but a lower relative humidity. Here, the fern leaves showed an increase in dry matter content and vein density, indicating a greater investment in resource storage and structural resistance. In the less-disturbed sites, terrestrial ferns demonstrated larger leaf area and specific leaf area, whereas epiphytes generally had smaller leaves, which could represent an adaptive advantage in more stressful environments.

These kinds of morphological shifts reflect broader ecological responses among epiphyte taxa to human disturbance. While certain patterns emerge, such as changes in abundance or adaptive traits, there is considerable variation across studies and species groups, cautioning against overly broad generalisations [2]. However, in a third paper, Siaz Torres et al. (2024) [4] found a higher abundance and species richness of epiphytic bromeliads in a disturbed forest than in a primary forest. Atmospheric species like *Tillandsia baileyi* Rose ex Small, *T. ionantha* Planch., and *T. usneoides* (L.) L. were most abundant on *Taxodium mucronatum* Ten., a dominant tree in the disturbed gallery forest. These bromeliads have morphological adaptations such as narrow leaves and abundant trichomes, which enhance their ability to capture atmospheric water and nutrients, facilitating their survival in disturbed or drier environments.

While some epiphyte species exhibit resilience in altered landscapes, broader environmental pressures continue to threaten the group as a whole. In addition to land-use change, the impacts of climate change, particularly rising temperatures and increased drought frequency, are emerging as major drivers of epiphyte decline. This is illustrated in a study on *Pleione formosana* Hayata, an endemic orchid once widespread across the mid-elevations of Taiwan, where Hsu et al. (2024) [5] found that populations have steadily declined, primarily due to orchid poaching and increasing climatic stress. Therefore, orchid plantlets were reintroduced to an old-growth cloud forest site in 2022, but the seedlings failed to survive the summer of 2023. The rising temperatures and frequent drought events threatened orchid growth, potentially increasing susceptibility to pathogens.

Alongside climatic stressors, direct human exploitation also poses a serious threat to many epiphyte populations. In particular, the mostly illegal harvest of species for commercial, ornamental, and cultural purposes has led to local declines across various taxa [2]. Due to their horticultural and ceremonial value, epiphytic orchids are often sold in local markets, but are also used in traditional medicine or as food supplements in tropical countries. *Prosthechea karwinskii* (Mart.) J.M.H. Shaw is an epiphytic orchid endemic to Mexico, threatened by the destruction of its habitat and the extraction of specimens to meet its demand for ornamental and religious use. Most of its populations are found in Oaxaca state, where variations in certain floral traits have been observed. A morphometric analysis by Santos-Escamilla et al. (2024) [6] identified the most significant floral characters as potential taxonomic markers for *P. karwinskii*, demonstrating the species' value in linking morphological variation to geographic origin.

Taxonomic challenges associated with floral diversity are also evident in the orchid genus *Cymbidium* Sw., which displays intricate floral structures, strong endemicity, and a fragmented distribution. These features have led to a remarkable range of morphological forms, but they have also complicated efforts to classify the genus with consistency. To clarify the phylogenetic relationships within the genus *Cymbidium*, Peng et al. (2025) [7] used DNA barcoding and found a consistent separation between epiphytic and terrestrial species. Both genetic and morphological analyses supported this distinction, with epiphytic orchids forming a clearly defined clade.

Understanding the evolutionary relationships of epiphytes also requires examining other structurally dependent plant groups that share similar ecological strategies. Epiphytes are one of four such types, alongside climbing plants like lianas, nomadic vines, and hemiepiphytes, all of which rely on host structures for support at some point in their life cycle [1]. The latter have captured the attention of biologists since they seemingly hold clues to the evolution of epiphytes themselves. During fieldwork conducted in Papua New Guinea, Sundue and Maraia (2024) [8] documented seven hemiepiphytic species of ferns that all started growth as low-trunk epiphytes, and later, as larger climbing plants, exhibited long feeding roots that entered the soil. These new records expand the geographic and taxonomic breadth of hemiepiphytic ferns and offer morphological and phylogenetic clues to uncover additional records.

Studying structurally dependent plant types in a comparative way may produce important insights into possible positive or negative interactions between them. Thus, another study by Ceballos et al. (2025) [9] aimed to evaluate the co-occurrence of vascular epiphytes and lianas in northwestern Argentina. Both plant groups were found together on 20% of the sampled trees; they colonized the same canopy tree species with larger diameters, whereas smaller trees were typically colonized by either lianas or epiphytes, but not both. Epiphyte species were more likely to co-occur with liana species with specialized climbing mechanisms. Tree size and forest type (mature vs. successional) emerged as key factors influencing their co-occurrence.

Several studies on richness patterns of vascular epiphytes suggested that abundance and diversity are highest in humid montane forests at intermediate elevations [1]. Although some studies have analyzed epiphytes alongside other plant life forms, comparisons of elevational patterns in both epiphyte and tree diversity and biomass in tropical dry forests remain scarce. In southwestern Ecuador, Werner and Homeier (2024) [10] observed that tree species density and total species richness increased with elevation, while basal area and biomass showed no consistent trends. In contrast, epiphyte species density, richness, and biomass all increased markedly with elevation. The authors attribute these patterns, particularly for epiphytes, to rising humidity at higher elevations, reflecting increasingly mesic conditions.

Taken together, the contributions to this Special Issue offer a timely and comprehensive perspective on the ecology, evolution, and conservation of epiphytic plants. By spanning a wide range of taxa, habitats, and methodological approaches, these studies not only deepen our scientific understanding but also highlight the urgent need to protect this vulnerable and ecologically vital plant group. We hope this collection will serve as a valuable resource for researchers, educators, and conservation practitioners alike, and that it will help stimulate further work to explore, document, and conserve the remarkable diversity of epiphytes worldwide.

Author Contributions: T.K. and S.P.B. contributed equally to this article. All authors have read and agreed to the published version of the manuscript.

Funding: This research received no external funding.

Conflicts of Interest: The authors declare no conflicts of interest.

References

1. Zotz, G. *Plants on Plants—The Biology of Vascular Epiphytes;* Springer: Cham, Switzerland, 2016; p. 282.
2. Krömer, T.; Einzmann, H.J.R.; Mendieta-Leiva, G.; Zotz, G. Impact of Land-Use Change on Vascular Epiphytes: A Review. *Plants* **2025**, *14*, 1188. [CrossRef] [PubMed]
3. Landeros-López, J.G.; Krömer, T.; Gómez-Díaz, J.A.; Velázquez-Rosas, N.; Carvajal-Hernández, C.I. Influence of Microclimatic Variations on Morphological Traits of Ferns in Urban Forests of Central Veracruz, Mexico. *Plants* **2025**, *14*, 1732. [CrossRef] [PubMed]

4. Siaz Torres, S.S.; de la Rosa-Manzano, E.; Arellano-Méndez, L.U.; Aguilar-Dorantes, K.M.; Martínez Ávalos, J.G.; Juárez Aragón, M.C. Species Richness, Abundance, and Vertical Distribution of Epiphytic Bromeliads in Primary Forest and Disturbed Forest. *Plants* **2024**, *13*, 2754. [CrossRef] [PubMed]
5. Hsu, R.C.-C.; Chen, Y.-C.; Lin, C. The Impact of Changing Climate on an Endangered Epiphytic Orchid (*Pleione formosana*) in a Montane Cloud Forest and the Conservation Challenge Ahead. *Plants* **2024**, *13*, 2414. [CrossRef] [PubMed]
6. Santos-Escamilla, M.H.; Cruz-Lustre, G.; Cuéllar-Martínez, M.; Lagunez-Rivera, L.; Solano, R. Variation in the Floral Morphology of *Prosthechea karwinskii* (Orchidaceae), a Mexican Endemic Orchid at Risk. *Plants* **2024**, *13*, 1984. [CrossRef] [PubMed]
7. Peng, Y.; Chen, Y.; Ding, H.; Liu, X.; Cao, F.; Xu, L. From Phenotypes to Genotypes: Enhancing the Identification of *Cymbidium* Species with DNA Barcoding. *Plants* **2025**, *14*, 619. [CrossRef] [PubMed]
8. Sundue, M.; Maraia, H. Field Work in Papua New Guinea Documents Seven New Records of a Hemiepiphytic Habit in Ferns. *Plants* **2024**, *13*, 1104. [CrossRef] [PubMed]
9. Ceballos, S.J.; Aráoz, E.; Rojas, T.N. Exploring Co-Occurrence Patterns to Understand Epiphyte–Liana Interactions. *Plants* **2025**, *14*, 140. [CrossRef] [PubMed]
10. Werner, F.A.; Homeier, J. Diverging Elevational Patterns of Tree vs. Epiphyte Species Density, Beta Diversity, and Biomass in a Tropical Dry Forest. *Plants* **2024**, *13*, 2555. [CrossRef] [PubMed]

plants

MDPI

Review

Impact of Land-Use Change on Vascular Epiphytes: A Review

Thorsten Krömer [1,*], Helena J. R. Einzmann [2], Glenda Mendieta-Leiva [3,4] and Gerhard Zotz [2,5]

[1] Centro de Investigaciones Tropicales, Universidad Veracruzana, Xalapa 91000, Mexico
[2] Institute of Biology and Environmental Sciences, Functional Ecology, Carl von Ossietzky Universität Oldenburg, D-26111 Oldenburg, Germany; helena.einzmann@uni-oldenburg.de (H.J.R.E.); gerhard.zotz@uni-oldenburg.de (G.Z.)
[3] Faculty of Biology, Philipps University of Marburg, D-35037 Marburg, Germany; glendamendieta@gmail.com
[4] Plant Ecology Division-Centro de Ornitología y Biodiversidad (CORBIDI), Lima 00051, Peru
[5] Smithsonian Tropical Research Institute, Balboa, Ancon, Panama City 0843-03092, Panama
* Correspondence: tkromer@uv.mx

Abstract: Human-caused habitat conversion, degradation, and climate change threaten global biodiversity, particularly in tropical forests where vascular epiphytes—non-parasitic plants growing on other plants—may be especially vulnerable. Epiphytes play vital ecological roles, in nutrient cycling and by providing habitat, but are disproportionately affected by land-use changes due to their reliance on host trees and specific microclimatic conditions. While tree species in secondary forests recover relatively quickly, epiphyte recolonization is slower, especially in humid montane regions, where species richness may decline by up to 96% compared to primary or old-growth forests. A review of nearly 300 pertinent studies has revealed a geographic bias toward the Neotropics, with limited research from tropical Asia, Africa, and temperate regions. The studies can be grouped into four main areas: 1. trade, use and conservation, 2. ecological effects of climate and land-use change, 3. diversity in human-modified habitats, and 4. responses to disturbance. In agricultural and timber plantations, particularly those using exotic species like pine and eucalyptus, epiphyte diversity is significantly reduced. In contrast, most native tree species and shade-grown agroforestry systems support higher species richness. Traditional polycultures with dense canopy cover maintain up to 88% of epiphyte diversity, while intensive management practices, such as epiphyte removal in coffee and cacao plantations, cause substantial biodiversity losses. Conservation strategies should prioritize preserving old-growth forests, maintaining forest fragments, and minimizing intensive land management. Active restoration, including the translocation of fallen epiphytes and planting vegetation nuclei, is more effective than passive approaches. Future research should include long-term monitoring to understand epiphyte dynamics and assess the broader impacts of epiphyte loss on biodiversity and ecosystem functioning.

Keywords: deforestation; disturbed forests; fragmentation; human disturbance; isolated remnant trees; literature search; tree plantations; secondary forests; selective logging; urban vegetation

1. Introduction

Human activities, especially the conversion, degradation, and fragmentation of habitats, as well as climate change, are major threats to global biodiversity, especially in the tropics [1,2]. Habitat loss, either caused by ongoing deforestation or changes in land-use, is the largest single driver of the decline in terrestrial biodiversity [1,3]. The magnitude and negative impact of the human disturbance on tropical forests and their biodiversity are well documented (e.g., [4–6]). In recent decades, increases in agricultural demand, logging activities and urban growth have led to unprecedented losses of tropical forests, with annual deforestation rates of around 0.5% since the 1990s, resulting in an annual loss of tropical forests of about 7.8 million hectares until the 2000s [7], and an annual loss of 6.4 million hectares between 2010 and 2015 [8]. At present, few truly undisturbed tropical forests exist, whereas degraded and regrowing forests after the abandonment of agricultural lands are rapidly expanding, currently comprising about half of the world's tropical forests [9,10]. Secondary forests may therefore constitute an important role as reservoirs for biodiversity [11,12]; however, the majority of relevant vegetation studies focuses on trees (e.g., [13,14]).

A smaller portion of studies are on lianas (woody vines; e.g., [15,16]), a structurally dependent component of the vegetation, which seems to be rather favored by disturbance, i.e., liana density and diversity peak following disturbance at local and forest scales [17,18]. This highlights how different components of the forests respond differently to disturbance, but research on other vegetative components of biodiversity is still very limited. For instance, despite the large-scale clearing of tropical forests since the 1970s, it took until the mid-1990s for Turner et al. [19] to show in a study from Singapore that epiphytic species, as a component of diversity, appear to be particularly threatened by habitat loss. While the species richness of the most studied life forms, such as trees (19%), shrubs (34%), herbs (17%), and climbers (23%), was much less affected, epiphyte species richness exceeded 60% loss.

Vascular epiphytes, plants that grow on other plants [20], are a conspicuous and integral component of tropical forests regarding local, regional, and continental plant diversity [21–23]. This very heterogeneous and taxonomically diverse group of plants contributes substantially to plant species richness, accounting for roughly 10% of all vascular plant species globally [24] and for up to 50% locally [25]. Epiphytes fulfil important ecosystem functions related to forest nutrients and water cycling [26–29]. Moreover, they provide food and habitat for canopy-dwelling fauna such as arthropods, birds or bats [30–33]. Because of these ecological functions, epiphytes are considered secondary foundation species [34]. These are species that provide structurally complex biogenic habitats that alter environmental conditions within the matrix of the primary foundation species, e.g., in this case, the host trees or shrubs [34]. Consequently, vascular epiphytes add structural complexity to forest canopies [20]. Therefore, any adverse effects on epiphytes cannot be viewed in isolation, as land-use change effects on epiphytes would have cascading consequences and similarly affect other inhabitants, such as associated fauna [34,35].

Without root contact to ground soil and dependent on the stability of their host, epiphytes might differ substantially from ground-rooted plants regarding their response to human-driven degradation and the conversion of tropical forests for agriculture and other land-use purposes. For instance, tree species richness increases very rapidly during secondary succession in Neotropical forests, with 80% recovery of old-growth values after only 20 years [36]. Moreover, in many secondary forests, tree species richness has surpassed that of old-growth forests [15], while epiphytes as canopy-dwelling organisms are specifically threatened by disturbance and deforestation [19,37,38], and their diversity is generally low in secondary forests [39–41].

However, little is known about how many and which epiphyte species (genera and families) may be prone to extinction or are lost, which depends on the type and magnitude of disturbance and the type of vegetation that replaces the original forest [42]. Several earlier studies (e.g., [39,40,43]) showed that compared with old-growth forests, epiphyte diversity tends to be markedly reduced following disturbance in most of the investigated human-modified habitats, but these studies were mostly carried out in humid lowland or montane forests. A growing number of later studies revealed exceptions to this trend, indicating that changes in epiphyte diversity and species composition vary significantly with the location and climate regime (e.g., rainfall and seasonality) of the study site, as well as the type and degree of habitat transformation (e.g., [41,44–46]). Diverging patterns, however, are actually unsurprising considering that disturbed and secondary habitats differ greatly in age and structure, and in the degree of structural differences from the original vegetation [47]. Considering this mixed picture, it is important to determine if epiphytes as a group really are particularly threatened by land-use change [48].

In this review, we summarize our current knowledge and try to find general patterns of the effects of land-use changes on epiphytes by analyzing all available information gathered through an intensive literature revision. As a basis for future work, we identify open questions within the larger context of human disturbance activities and conservation strategies for epiphytes in general. By recognizing the main aspects of research, we want to achieve the following: 1. Weigh into the mixed picture of the subject to determine if epiphytes as a group really are particularly threatened by land-use change. 2. Suggest plausible future directions of the field, identifying open questions in the larger context of land-use change. 3. Provide suggestions for possible conservation strategies for vascular epiphytes.

2. Materials and Methods

2.1. Literature Search

To gather relevant information on the impact of land-use change on epiphyte diversity and related topics, we conducted an extensive literature search for research articles, reviews, book chapters, and grey literature (e.g., theses, notes) with a publication date up to October 2024 utilizing common scientific databases including Clarivate Web of Science and the Scientific Electronic Library Online (SciELO) and internet search engines such as Google Scholar, using operators such as 'epiphyt*' AND 'land-use' OR 'disturbance' OR 'deforestation' OR 'logging' OR 'remnant tree' OR 'plantation' OR 'urban' OR 'fragmentation' OR 'edge effect' OR 'succession' OR 'recolonization' OR 'harvest' OR 'trade' OR 'translocation' in the titles, abstracts, and keywords. In addition, we revised the comprehensive vascular epiphyte literature database of Gerhard Zotz, which includes ca. 12,000 references, many of which are not included in the other databases. After carefully revising the abstracts of all references obtained, we excluded all publications not dealing with the study topics (e.g., focused on non-vascular epiphytes or epiphytic algae on seagrasses). We finally identified 285 relevant references in English, Portuguese, and Spanish (EndNote and Excel database; Supplementary Material, Table S1), mostly journal articles (254), but also several book chapters (19) and theses (14). A total of 156 publications focused on vascular epiphytes in general (some including hemiepiphytes, nomadic vines or accidental epiphytes; [49–51]), whereas 58 studies focused on orchids, 41 on bromeliads, and 30 on pteridophytes. Some of the latter research considered two of these three groups and a few also included terrestrial or saxicolous taxa.

2.2. Identifying the Main Aspects of Research

To obtain the main areas of research when evaluating the impact of land-use change on vascular epiphytes, we carried out a bibliometric analysis on the current state of knowledge about the impact of land-use change on vascular epiphytes. We used a text mining approach with a cooperative cluster-ordination and k-means to create groups representing the most consistent subjects [52] on the abstracts of the 285 references considered in this review. The process involved exporting references from EndNote® TM 21.5 to Microsoft Word®. The 285 references in EndNote were exported (annotated to include the title and abstract) as an RTF extension and organized in Microsoft Word. The structure was homogenized so that each paragraph contained the title and the abstract. Spanish or Portuguese content was translated into English using DeepL translator version 24.11.4.144242. After translating and homogenizing the abstracts, the RTF file was read into R programming language [53], where we carried out cleaning and stemming procedures, to finally proceed with the analyses (ordination and clustering techniques). The abstract list was cleaned by eliminating "stop words" (such as connectors and abbreviations) and special characters (such as punctuation, numbers, and symbols) and homogenized by stemming words (e.g., removing or reducing derived words, epiphytism --> epiphyt).

Following the cleaning and stemming procedures, a word matrix was created, with 285 rows representing the publications and 4080 columns representing the stemmed words. The cells of the word matrix represented the frequency with which each word appeared in the abstract of each publication. Among the most frequent terms we found two that referred to the same concept, which were therefore combined, as follows: "epiphyte" was combined with "holoepiphyte" and "host" with "phorophyte" and "tree". The packages qdap [54], stringr [55], striprtf [56], tm [57], tidyverse [58], and data.table [59] were used for text cleaning and stemming. The resulting word matrix was used in a detrended correspondence analysis (DCA; [60]) using the "decorana" command from the vegan package [61] to down-weigh uncommon occurrences (of words), thus lessening the influence of rare "terms" (words). This ordination technique is used to obtain a low-dimensional representation of the word matrix and reduce dimensionality (2–3 axes generally).

Then, to identify homogeneous groups (clusters) within the word matrix, we applied k-means clustering to the ordination scores. To find the best compromise between a low number of clusters and a high variance of the word matrix, we use the command "fviz" from the package NbClust [62]. This method identifies groups by minimizing within-cluster variation by computing the sum of all within-cluster squares (wss) for each run, which it does for a total of 10 self-defined clusters [63]. The visual assessment of the "wws" as a function of clusters was used to determine the number of suitable clusters; we also performed the same analysis using a different clustering technique (the average silhouette method), and the results were comparable (Supplementary Material, Figure S1). Once the clusters were derived from the word matrix, we assigned each abstract to one of the four resulting clusters; this involved tallying the number of words in the abstracts matching with the words defining each cluster, and once we had identified the cluster with the largest number of words matched to an abstract, this publication was assigned to the cluster (Figure 1; Supplementary Material, Table S1, list of the papers assigned to each group). We found the abstracts were grouped into four clusters representing the most common research carried out in the last two decades (Figure 1). These were plotted using the command "geom_label_repel", which allows the text labels to shift and reconfigure away from the data points for readability, but indicates with lines where the exact point was, using the packages ggrepel version 0.9.6 and ggplot2 version 3.5.1 [64–66].

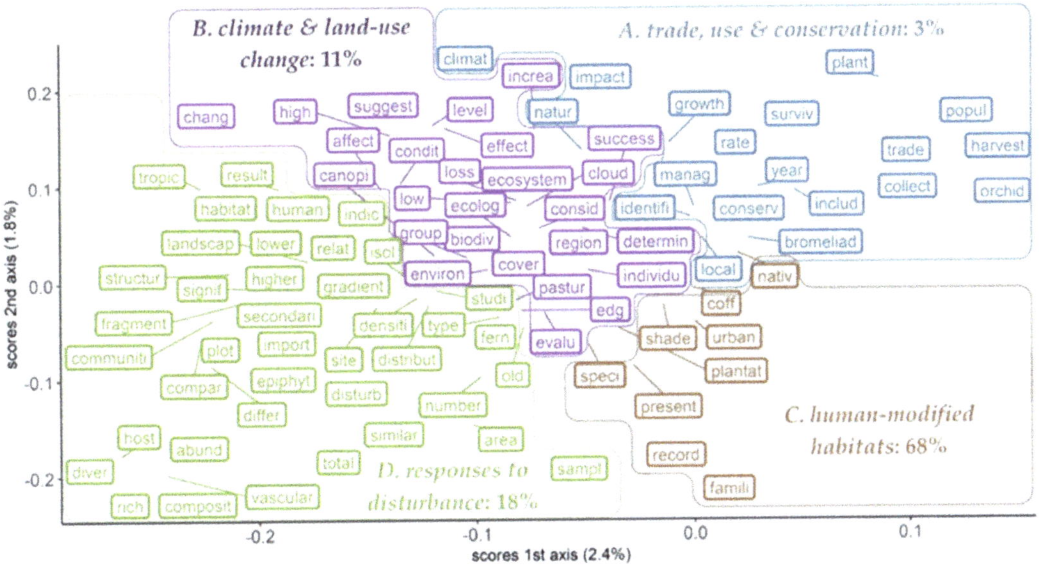

Figure 1. Four aspects of the research. A: *Trade, use and conservation*. B: *Climate and land-use change*. C: *Human-modified habitats*. D: *Responses to disturbance*. Here, we show the most frequent words per aspect (cluster) in relation to their frequency (those with a frequency larger than the highest 3rd quartile frequency of all clusters). The percentages of abstracts assigned to each cluster are shown (next to the cluster name). For the sake of comparison, we include a similar graph including the most common 20 words per cluster (Supplementary Material, Figure S2, the limit in words was set for visibility's sake). Words varied in their frequency per cluster, i.e., some clusters were more consistent than others (Supplementary Material, Figure S3).

3. Results and Discussion

3.1. General Results

Over the last two decades, many relevant studies related to the effects of land-use changes on vascular epiphytes have been conducted, although mainly at a local scale and with a strong bias towards the Neotropics compared to tropical Africa and Asia, or temperate regions [20]. Only a few publications give a broader overview [20,35,42,67,68] but do not cover all related topics in detail.

The first study considering the effects of land-use changes on epiphytes was published in Selbyana by Michael Madison in 1979 [69], addressing the distribution of epiphytes in a rubber plantation in Malaysia (Figure 2). During the following two decades, only 10 additional studies were published. This number increased in the 2000s to 59 and reached 143 in the 2010s, while during the last five years, the annual number of papers ranged between 10 and 20, with a yearly mean of 14.4 published studies.

The major regions/continents and countries where these studies were conducted (Figure 3) show a strong geographic bias towards Central America (107 papers). In Mexico alone 81 studies were conducted, slightly less than those conducted in eight countries of South America (87), of which in turn almost half were situated in Brazil (43). In ten countries of Asia, 33 studies were conducted, led by India (7), and in nine countries of Africa, 21 studies were conducted, led by Ethiopia (7). In three Caribbean countries, six studies were conducted, and eight were conducted in Oceania. Furthermore, there are 23 studies, mainly review papers, with no clear geographic focus.

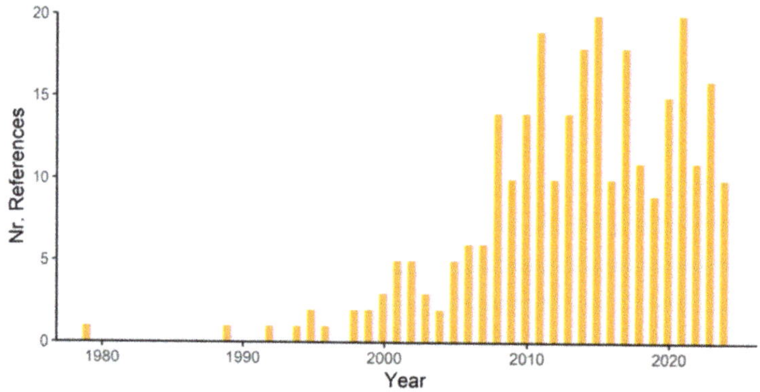

Figure 2. The number of references considering the effect of land-use changes on epiphytes per year.

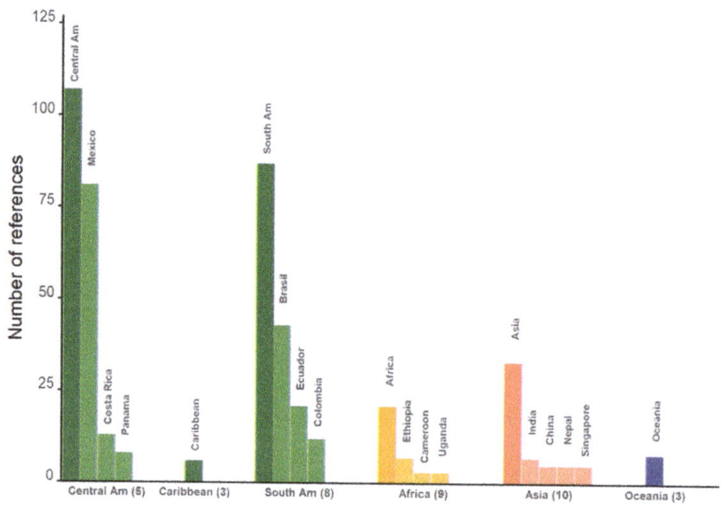

Figure 3. The number of articles/studies per major region/continent (first bars; indicating the number of countries in brackets on the x-axis) and main countries (following bars).

3.2. *Identifying the Main Aspects of Research*

The text-mining analysis based on almost 300 abstracts (summing up 4080 words) resulted in four clusters (Figure 1; Supplementary Material, Figures S1 and S2), which we classified into (A) *trade, use and conservation*, (B) ecological effects of *climate and land-use change*, (C) diversity across *human-modified habitats*, and (D) ecological effects of *responses to disturbance*. This classification was based on the most frequent words grouped per cluster and the majority of abstract titles contained in each cluster in the ordination. The clusters grouped a very uneven number of abstracts, with the largest number indicating that most research focuses on evaluating epiphyte diversity across *human-modified habitats* (68%, 193 abstracts). The other three clusters grouped a third (32%) of the abstracts, with the cluster of *trade, use and conservation* having the lowest number of abstracts (8), followed by the clusters grouping research on the ecological effects of *climate and land-use change* (32) and *responses to disturbance* (52).

In the cluster of diversity across *human-modified habitats*, research focuses on evaluating the different aspects of diversity (species richness, abundance, species composition, biomass and others) across different anthropogenic habitats (e.g., secondary forests, plantations, urban areas, etc.) with a focus on specific taxa (e.g., orchids or bromeliads) but with diverse approaches. The cluster *responses to disturbance* concentrates on comparative research where differences at the community level are evaluated between old-growth forests and habitats that have undergone land-use changes of different types; this is similar to the cluster of ecological effects of *climate and land-use change,* with the difference that in most studies the ecological effects of land-use are evaluated but are mostly used as a proxy for possible effects of climatic changes. Finally, the cluster *trade, use and conservation,* is the most consistent, focusing on research on the use of species and the different aspects of their trade in the context of conservation.

The total number of words per cluster varied somewhat according to the number of abstracts (number of words for *trade, use and conservation,* 91; *climate and land-use change,* 352; *human-modified habitats,* 3575; *responses to disturbance,* 62; Supplementary Material, Figure S3), but this could also be influenced by the length of the abstracts per group, the stemming process whereby several words could be grouped under their stem (e.g., epiphyte, epiphytic, forest, forestry, etc.), or the fact that some abstracts contained a large number of deleted terms (e.g., prepositions, articles, adverbs, and conjunctions). The word abundance structures of clusters *trade, use and conservation* and *responses to disturbance* were more similar than they were to *climate and land-use change* and *human-modified habitats,* as they contained several common words (those with the highest frequency; for *trade, use and conservation* the three most frequent words were repeated between 210 and 353 times, and for *responses to disturbance* the three most frequent words were repeated from 419 to 1298 times, Supplementary Material, Figure S3 and Table S2), and these clusters had very few to no rare words (those with low frequency, e.g., repeated ≤ 20 times); while *climate and land-use change* and *human-modified habitats* had very many rare words (*climate and land-use change* with 43 singletons and 13 doubletons, and *human-modified habitats* with 1853 and 571, respectively, Supplementary Material, Table S2).

This means that clusters on *trade, use and conservation* and *responses to disturbance* are relatively consistent, independently of the number of publications/abstracts contained, or that the abstracts within these clusters somewhat consistently mainly reflect research on the impact of *trade, use and conservation* and *climate and land-use change,* respectively. The clusters on *human-modified habitats* and *climate and land-use change* grouped research abstracts with far fewer thematic or words in common. From these clusters, the one on *human-modified habitats* had the most heterogeneous content, and is almost miscellaneous, perhaps because the research into the anthropogenic effects on diversity can be very broad and is very heterogeneous for vascular epiphytes (e.g., lack of sampling protocols or general recommendations in the specific field). While the cluster on *climate and land-use change,* which mainly uses the ecological effects of land-use as a proxy to evaluate the effects of climatic changes on diversity, contains less explored avenues of research directly evaluating the impacts of climate change effects on vascular epiphyte diversity (one from 2002 and all others from 2009 till 2023), e.g., the modeling of forest dynamics, long-term dynamics, and experimental testing of the climate change effect (Supplementary Material, Figure S2, Table S1).

3.3. *Tropical Forest Conversion and Epiphytes*

Habitat loss caused by the growing human pressure on terrestrial ecosystems is one of the most important threats to biodiversity, especially in the tropics [9,70,71]. The planet is suffering rapid and dramatic changes across most biomes [72]. Considering the current high rates of deforestation in most tropical countries [73], it is projected that the few existing areas with extensive, undisturbed forests will become scarce and fragmented soon [5,9]. Human population growth and the intensification of agriculture are the major factors threatening primary forests in the tropics and their associated biodiversity [74]; due to conversion into cropland, grassland for cattle, and human settlements, many plant taxa have disappeared [75]. In spite of this bleak scenario, botanists should investigate poorly explored areas where species new to science can still be found, and possibly, species listed as extinct may be rediscovered [76].

The conversion of primary or old-growth forests, i.e., forests that have never been clear-felled and have been impacted by little or no known recent human disturbance [5], in the tropics to treeless vegetation will result in an almost complete loss of epiphytes in the affected area [42], although an unknown number of facultative epiphyte species might survive on the ground or on human-made structures like walls, roofs or electricity wires [77–79]. In many human-modified landscapes (Figure 4), potential hosts may still be present, e.g., in the form of fragments of disturbed or secondary forests, isolated remnant or planted trees, or in plantations [46]. Although in the last two decades the number of studies considering the effects of various human activities on epiphyte diversity has increased considerably (Figure 2; Supplementary Material, Table S1), the fate of epiphytes in anthropogenic tropical habitats remains insufficiently understood. Therefore, a central question of tropical plant conservation remains whether human-dominated landscapes can provide suitable refuges for epiphytes, or whether these challenging habitat conditions result in slowly declining communities with strongly diminished regeneration, rendering them 'living dead' [80].

Figure 4. Views of human-modified landscapes in the Neotropics, including old-growth forest fragments, secondary forests, cattle pastures or plantations with remnant trees, and agricultural lands in (**A**) Los Tuxtlas, Mexico; (**B**) Central Veracruz, Mexico; (**C**) Azuero, Panama; (**D**) Alto Beni, Bolivia. Photographs by (**A**,**B**,**D**) Thorsten Krömer; (**C**) Helena J. R. Einzmann.

A few of these studies included three or more relevant types of anthropogenic habitats, comparing them with old-growth forests along gradients of human impact at a single study site, while considering all vascular epiphyte taxa in standardized study units in humid forests in the Neotropics [39,41,81]. We will take these comprehensive investigations as a starting point to explore the impact of land-use changes on epiphyte diversity in human-modified landscapes. It was observed that species richness was consistently the lowest in young secondary forests, followed by plantations or isolated remnant trees in pastures, disturbed forest fragments, and it was the highest in old-growth forest. We here conduct our review of the literature following this gradient in the decline of epiphyte diversity with increasing degrees of disturbance by starting with the most affected habitat type.

3.4. Secondary Forests

The conservation value of tropical secondary forests has been debated for decades [12,82,83]. Here, we define a secondary forest as a habitat that has recovered or regenerated largely through natural processes after significant human or natural disturbance (e.g., clearcut logging or hurricanes) that caused a complete removal of the original forest vegetation [82]. The remaining open area usually has been used as cattle pasture or for the cultivation of cash crops such as corn or sugar cane, but then due to various reasons (e.g., infertility of soils, low crop prices), these lands were not intensively used anymore, and as the result of secondary succession, a young fallow forest has started developing on the abandoned ground, which initially lacks resident epiphytes [82,84].

Considering trees or lianas, secondary forests may attain many aspects of the structure and species richness of old-growth forests within a few decades if there are propagule sources left at a feasible distance, but reaching a similar species composition may take centuries [15,85,86]. Some plant species, including many vascular epiphytes, however, appear to be old-growth specialists that only establish on the large host tree species present in old-growth forests [39,40,87]. This is likely because of the specific abiotic conditions needed (e.g., stable microclimate with high humidity and low insolation), which are only found in structurally complex forests [88,89].

Generally, the destruction of old-growth forests results in a considerable species loss and a major decrease in total species richness, as is found in human-modified habitats [72,90], but this reduction may be more severe for vascular epiphytes than for other plant groups [37]. For example, in the few remaining patches of natural vegetation and mostly secondary rain forest fragments of Singapore, Turner et al. [19] found that epiphyte diversity was reduced by 62% compared to original flora, and was thus affected much more severely by local extinction (i.e., extirpation) than any other life form. This considerable reduction in the species richness of epiphytes following deforestation is corroborated by several studies, which indicate that fragments of young secondary forests (less than 20 years old; Figure 5A,B) usually show lower species richness of epiphytes in comparison with nearby sites of primary forests, while older secondary forests can recover a mayor part of that diversity [41,87,91].

For example, in three 15-year-old fallows, there was a reduction in species richness of between 61% and 73% in humid montane forests of Bolivia [40], and between 46% and 66% in comparable young secondary forests in Ecuador [41] and Mexico [81] and lowland rain forests of Colombia [91] (Figure 6; Supplementary Material, Table S3). In contrast, Barthlott et al. [39] found a 96% reduction in species-richness in a 23-year-old fallow in humid montane forests in Venezuela, whereas species numbers were less reduced in a similar aged secondary forest in lowland rain forests of Mexico [92], as well as in slightly older secondary forests in other humid forests in South America [41,91]. Finally, in the oldest secondary forests studied, epiphyte diversity was still reduced by 64% and 29% after 85 and 115 years of abandonment, respectively, in the humid lowlands of Panama [87], whereas in a 60-year-old subtropical montane forest in Argentina, epiphyte diversity had almost completely recovered [93].

Figure 5. Views of two young secondary forests at (**A**) Central Veracruz, Mexico; (**B**) Veraguas, Panama; and cattle pastures with shade trees at (**C**) Azuero, Panama; (**D**) Central Veracruz, Mexico. Photographs by (**A**,**D**) Thorsten Krömer; (**B**,**C**) Helena J. R. Einzmann.

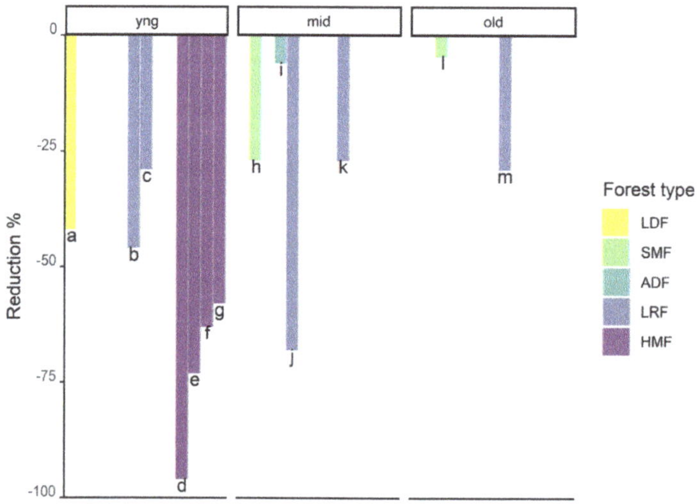

Figure 6. Epiphyte species-richness reduction in secondary forest fragments (young—yng < 25 years; middle—mid 25–50 years; old > 50 years) compared to old-growth forests (ADF—Andean dry forest; HMF—humid montane forest; LDF—lowland dry forest; SMF—subtropical montane forest; LRF—lowland rain forest) in different study regions in tropical countries: (a) Mexico [94]; (b, k) Colombia [91]; (c) Mexico [92]; (d) Venezuela [39]; (e) Bolivia [40]; (f) Ecuador [41]; (g) Mexico [81]; (h, l) Argentina [93]; (i) Ecuador [44]; (j, m) Panama [87].

Generally, the recolonization of secondary forest by epiphytes seems to be a slow process [46,87,95], but available data are hard to compare because of differences in the spatial context (e.g., distance of seed sources) [96–98]. Dispersal limitations may hamper the successful establishment of epiphytic bromeliads in human-modified habitats [99]. However, it is challenging to disentangle dispersal limitations from establishment limitations [100], although the clumped distribution of many epiphytes indicates that dispersal limitation might be involved [101].

The difference in both vascular epiphyte richness and species composition between young and old secondary forests is probably due to the unfavorable characteristics of young trees. On the one hand, they provide a more homogeneous structure with fewer microhabitats, a drier microclimate and lower bryophyte cover, and on the other hand, a shorter period for recolonization [81]. Differences in structural variables mainly lead to a loss of drought-sensitive hygrophilous species, with specific habitat requirements, such as shade- and humidity-adapted understory orchids and ferns, which might not be able to persist under modified microclimatic conditions (higher temperature and less moisture) in the open young fallow forests [102,103]. Similarly, *Socratea* palm trees were estimated to be c. 20 years old before colonization with vascular epiphytes began [104], while it took at least 10 years for epiphyte seedlings to recolonize stripped branches in humid montane forests [105]. In addition, most studied epiphytes show inherently slow growth; it may take more than 10 years for certain species to reach reproductive maturity [106,107].

However, two studies conducted in dry secondary forests indicated that the reduction in epiphyte richness, especially in young fallows, may be less severe (Supplementary Material, Table S3), as richness in 5–8 year-old and 10–20 year-old forests in the lowlands of Mexico was reduced by only 62% and 42%, respectively [94], and the epiphyte diversity of inter-Andean secondary dry forests of 13–28 years was almost comparable to that in a closed-canopy forests [44]. This relatively rapid recovery of epiphyte richness may be attributable to the physiological and morphological traits of the local drought-tolerant species that help them to cope with the harsh conditions in dry forests [108,109], whose canopies are usually low and open and provide little microclimatic buffering [110]. Many of the local species in these forests are widely distributed in Neotropical forests, because they are able to survive in a wide range of environmental conditions [93]. Moreover, dry forests in general have relatively poor epiphyte flora [21,44] that may recover after disturbance much faster than the highly diverse epiphyte flora with many rare species in humid montane forests [39,41,81].

Concerning the specific responses of major taxonomic groups, there are some trends, but substantial variance between studies warns against premature generalizations. For example, orchid richness in secondary forests of Bolivia decreased by ca. 90% compared to old-growth forests and by ca. 45% compared to ferns [40]. This study corroborates the findings from Singapore, where Turner et al. [19] found only 10 of the 110 inland epiphytic orchids (91% decrease), whereas the epiphytic pteridophytes were much less reduced in diversity (37%). In contrast, Barthlott et al. [39] documented a total loss of orchids and a 92% reduction in ferns in secondary forests in Ecuador, which had been completely cleared 23 years prior. Orchids were also the most affected epiphyte groups in two secondary forests in Mexico, showing a loss of >70% compared to adjacent old-growth forests [81,92]. However, the fate of ferns differed markedly; while the loss was about 65% in the humid montane forest, no reduction was found in the lowland rainforest. Also, Carvajal-Hernández et al. [111] found that in 10- and 20-year-old secondary forests in humid montane Mexico, epiphytic fern richness was reduced by 36% and 27%, respectively. The establishment of epiphytic orchids generally might be hampered by their dependence on mycorrhizal fungi for seed germination [112], as well as due to the unfavorable microclimatic and structural conditions in fallow forests [40]. The latter also affects many drought-vulnerable grammitid, vittarioid, and filmy ferns [39,89], while some species of the genera *Pleopeltis* and *Polypodium* have traits associated with drought resistance (e.g., poikilohydry, dense scales; [113–115]).

Similarly, studies on the human impact on bromeliads also showed mixed results. In humid montane areas, species richness was usually more reduced in young fallow forests (<20 years; 75–88% reduction) than in old-growth forests [39,40], while secondary forests of 30–40 years could recover some of the species richness present in old-growth forests (53–30% reduction; [116,117]). In contrast, in moist lowland areas in which the number of bromeliad species is generally lower, the species-richness of secondary forests was only slightly reduced compared to old-growth forests [91,92]. Moreover, secondary forests in dry lowlands can have even more species richness than old-growth forests [94,118].

Some bromeliad species, specifically atmospheric or xeromorphic species of the genus *Tillandsia*, can be more abundant in secondary than in adjacent old-growth forests [102,119]. These are well-adapted to anthropogenic habitats, which are more open and drier, through succulence, foliar trichomes, and CAM [114,120]. Apparently, the increased light levels in the canopies of secondary forests can favor the survival and growth of some less specialized, drought-tolerant *Tillandsia* species [121–123], which can dominate and increase their biomass, causing the epiphyte community to become less rich and less even [117,124,125].

3.5. Tree Plantations

Over the last several decades, commercial agriculture expansion has become a key driver of forest loss in the tropics [126]. Currently, agricultural production in tropical biomes has undergone a dramatic shift that has led to increases in the production of crops and wood products from plantations [127]. Agriculture plantations are defined as areas that are typically monocropped with perennials, producing tropical or subtropical products that commonly require prompt initial processing, and for which there is an export market [128]. Some of the most important tropical plantation crops are bananas, pineapple, sugar cane, and tobacco [129]. However, these agriculture plantations comprise herbaceous plants that can hardly host epiphytes, and typically are monocultures without remnant forest or cultivated shade trees.

Therefore, our discussion is focused on the impacts of different tree plantations on epiphyte species richness. Tree plantations can be defined as mono- or polycultures of agricultural arborescent species established and managed by humans for wood, fruit, luxury food, fiber and other products [130], e.g., oil and coconut palms, or coffee, cacao and tea under shade trees [131]. The expansion of tree plantation areas in the tropics has frequently come at the expense of intact forests, and increased from about 6.7 million ha in 1965 to 109 million ha in 2005 [132].

Generally, these kinds of plantations are not known for their high capacity to maintain or harbor rich epiphyte communities, as studies mainly show reductions > 50% compared to nearby old-growth forest fragments (Figure 7; Supplementary Material, Table S4). First, we discuss structurally impoverished timber plantations with exotic pine or eucalypt species that usually host very few epiphytes [46,133]. Pines have been considered poor epiphyte hosts, not only because of the contents of phenolic and resinous substances [134–136], but also because of the instability and low water-holding capacity of their bark [137]. Additionally, the monopodial growth and lack of large horizontal branches of these conifers might be a constraining factor limiting epiphyte abundance and diversity [109].

Boelter et al. [133] compared epiphyte richness in a natural *Araucaria* forest of the Atlantic Forest in Brazil with that in managed plantations of *Araucaria*, *Pinus*, and *Eucalyptus*.

The plantations of native *A. angustifolia* (Bertol.) Kuntze had a much higher conservation value than the exotic monocultures of *P. taeda* L. and *E. saligna* Sm. with reductions of 52% vs. 85% and 89%, respectively. While host trees of *A. angustifolia* are characterized by well-developed secondary branching that might favor epiphyte establishment and growth, the physiognomic and bark characteristics (e.g., crown morphology, textural, absorptive and physiochemical properties) of *Pinus* and *Eucalyptus* trees are mostly unsuitable [133]. In contrast, in an Andean 32-year-old plantation of the native *Cedrela montana* Moritz ex Turcz., epiphyte richness was reduced by >90% [39]. This was explained by the uniform tree structure, the absence of an understory and the poor development of plantation trees with a crown openness of more than 50%, which reduced the number of suitable micro-habitats for epiphytes.

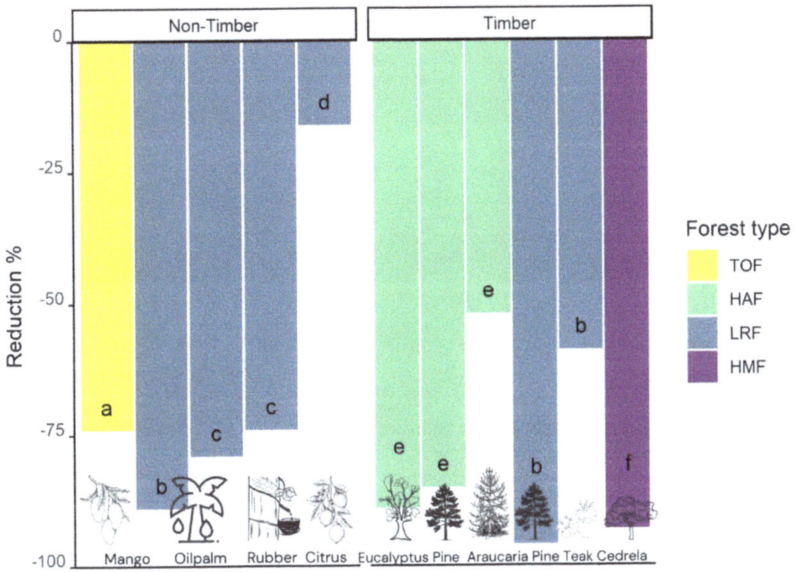

Figure 7. Epiphyte species richness reductions in different tree plantations compared to old-growth forests (HAF—humid Araucaria forest; HMF—humid montane forest; LRF—lowland rain forest; TOF—tropical oak forest) at different study regions in tropical countries: (a) Mexico [113]; (b) Panama [46]; (c) Indonesia [138]; (d) Mexico [92]; (e) Brazil [133]; (f) Venezuela [39].

In a comprehensive study, Einzmann and Zotz [46] found that pine (Figure 8A; *Pinus caribaea* Morelet) and teak (*Tectona grandis* L.f.; (Figure 8B) plantations hosted considerably fewer epiphytes than the tree stands surrounding them. The short rotation times of the timber crop leave little time for the colonization of these plantations. Consequently, the epiphyte individuals observed in the teak (and pine) plantations were mostly juveniles, and there were hardly any adult plants with reproducing structures [46].

The second types of tree plantations discussed are rubber (*Hevea brasiliensis* (Willd.) Müll. Arg.) and oil palm (*Elaeis guineensis* Jacq.; Figure 8C). The former produces latex sap, which is the basis for the natural rubber used mostly in the manufacture of automobile and aircraft tires, while palm oil, obtained from fruits and seeds, is mainly used in making cosmetics, biofuels, pharmaceuticals, and edible products. These are the two main crops threatening biodiversity and natural habitats in Southeast Asia and Latin America [139,140]; however, in the last decade, the decreasing prices and high labor costs in contrast to higher profitability are driving the conversion of rubber to oil palm plantations [141].

The first study considering the effect of land-use changes on epiphytes was published by Madison [69], who studied their distribution in a rubber plantation in Malaysia. When a rubber plantation is established, the forest is cleared and burned, and young trees from a nursery are planted. As the plantation matures, individual trees are colonized by epiphytes whose seeds arrive from populations in nearby primary forests or older plantations [69]. Of the 25 recorded epiphytes, orchids and ferns with minute diaspores dispersed by wind were the most species-rich and showed a random distribution in the plantation, similar to a few fleshy-fruited species dispersed by birds.

Figure 8. Views of different tree plantations of (**A**) pine in Veraguas, Panama; (**B**) teak in Veraguas, Panama; (**C**) oil palm in Chiriquí, Panama; (**D**) citrus in Los Tuxtlas, Mexico; (**E**) shaded coffee in central Veracruz, Mexico; (**F**) shaded cocoa in Tabasco, Mexico. Photographs by (**A–C**) Helena J. R. Einzmann; (**D**,**E**) Thorsten Krömer; (**F**) Jonas Morales-Linares.

Almost four decades later, Böhnert et al. [138] studied the epiphyte richness in rubber and oil palm plantations in central Sumatra, Indonesia and found a reduction in epiphyte richness of 74% and 79%, respectively, compared with adjacent rainforest. They suggested three main factors as responsible, as follows: 1. a less favorable microclimate with hotter and drier conditions and more limited ecological niche space for a greater number of specialized epiphyte species; 2. the age and characteristics of the host trees, as these plantations are generally replanted after about 25–30 years, which abruptly stops the age-related increase in epiphyte species richness; 3. the occasional removal of epiphytes by plantation workers who commonly assume that epiphytes reduce yields by parasitizing the trees. Rubber trees are sparsely colonized by epiphytes anyway, and the removal drastically curtails any successful establishment of a new epiphyte community [138]. In contrast, the complex structure of oil palm stems hampers complete epiphyte removal, and larger individuals might be the most affected by this practice. In conclusion, the values of both monotypic tree plantations for epiphyte conservation are very low [138].

Apart from these studies, there is little and mostly anecdotal information on the suitability of oil palms as a host for epiphytes [142,143], although the accumulation of organic material in leaf bases makes them unique hosts compared to other plantation trees [144]. However, as epiphytes are mentioned as problematic in some management practice guides (e.g., [145]), they are regularly removed from oil palms [142], although Prescott et al. [144] showed that their removal did not improve the productivity of the crop and thus should be avoided, even from an economic point of view. Interestingly, oil palms are invasive in parts of Brazil, at least in part due to fruits that are attractive to many disperser animals, and have high germination success and establishment rates [146]. Notably, about a third of the epiphytes in a sub-spontaneous stand of *E. guineensis* were accidentals, terrestrial species that establish in the organic matter accumulated in their leaf sheaths.

Another case of cultivated palm trees is coconut (*Cocos nucifera* L.), which provides food, fiber, biofuels, cosmetics, medicine, and building material. However, coconut palms were shown to be poor epiphyte hosts, because of their smooth trunks, low structural diversity and usual occurrence in relatively dry, exposed contexts providing harsh microclimatic conditions; although a small number of ferns and at least one orchid species have been found on them [147,148].

The third type of tree plantation is cultivated only for the production of edible fruits, including important crops such as mango (*Magnifera indica* L.) and different species of *Citrus* (e.g., oranges, mandarins, lemons, grapefruits). Information on the suitability of mango trees for epiphytes is contradictory: Schimper [149] called them "poor" hosts in the West Indies, but "good" hosts around Rio de Janeiro. While the first categorization was supported in Johansson's study in Africa [150], the latter was supported by observations in Costa Rica [151]. The study of Nir [151] provided evidence that large/old individuals hold disproportionately large epiphyte loads, while many smaller mango trees were entirely free of epiphytes. Hietz [42] found an epiphyte richness reduction of 74% in mango compared to nearby forest trees. One reason appears to be the very dense foliage concentrated in the outer crown, which is a strong barrier to propagules [149]. However, once colonization has occurred as a chance event, plants can thrive because humid conditions inside the crown are conducive to epiphyte growth, although little light might reach the inner branches [42,152].

There are also ambivalent results regarding the suitability of citrus trees for epiphytes. Hietz-Seifert et al. [43] noted that young (<25 years) *Citrus* spp. and *Cedrela odorata* L. trees cultivated together in a pasture were poor hosts. On a total of 45 individuals of both trees, only 15 epiphytic species could be found, equivalent to a reduction of 57% with respect to a nearby rainforest fragment in Mexico. In contrast, in the same region, Pérez-Peña and Krömer [92] recorded only a 14% reduction by recording 51 epiphyte species in up to 40-year-old *Citrus* plantations (Figure 8D) without epiphyte removal as a management practice, where mostly drought-resistant bromeliads and ferns had a higher richness than in the forest. Interestingly, the number of orchid species was also similar, which is usually the taxonomic group most negatively affected by disturbance and land-use change (e.g., [19,39,40,81]). However, it is known that citrus trees can be particularly good hosts for certain orchid species that can become very abundant on the smaller branches and twigs, such as *Ionopsis utricularioides* (Sw.) Lindl. [153,154].

The epiphyte assemblages in human-modified systems are typically qualitatively and quantitatively very different from those found in old-growth forest [20]; however, epiphyte diversity in plantations is not invariably low. Traditional coffee polycultures with shade trees in Latin America can have low reduction values compared to the other tree plantations described above (Figure 9; Supplementary Material, Table S4). The cultivation of coffee (different species of *Coffea*) can be considered as a specific type of tree plantation, as these consist of small shrubs or bushes that are periodically pruned to improve productivity and facilitate harvesting the ripened coffee cherries [155]. These are cultivated either in intensively managed sun plantations or in traditional agroforestry systems, grown under a more or less dense canopy of various natural humid montane forest or cultivated shade tree species [156,157].

While epiphytes almost never grow on coffee shrubs in sun coffee plantations [158], in shaded polycultures in southern Mexico, Mondragón et al. [155] found 23 species of vascular epiphytes, mostly polypodioid ferns and orchids; in Puerto Rico, Nir [151] found nine species of epiphytic orchids, and in a review on orchids in Mexican coffee plantations, Espejo-Serna et al. [159] reported 18 species that grew on the coffee shrubs themselves. In Ethiopia, the origin of coffee (*Coffea arabica* L.), Hylander and Nemomissa [160] found that large, ancient coffee bushes in home gardens carried a low diversity of vascular epiphytes similar to the surrounding old shade trees. Here, coffee occurs as a natural understory shrub within more or less intensively managed forest fragments [161]. Despite the generally low value of the coffee bushes themselves as hosts for epiphytes, plantations of shade-grown coffee carry great importance for epiphyte conservation because of the old shade trees that might host a similar diversity of epiphytes to trees in surrounding old-growth forests ([158]; Figure 8E). For instance, commercial coffee mono- or polycultures with small or a low density of shade trees showed a reduction of >50% compared to nearby forest fragments, whereas in traditional polycultures with many old shade trees, the reduction was only 12–28% [81,158,162,163]. However, epiphyte species richness can also be notably reduced (68%) on remnant shade trees in coffee agroecosystems as compared with forest trees in Afromontane forests of Ethiopia [164].

In general, epiphyte communities in coffee plantations are more homogeneous than in the forest, possibly because the agroecosystem structure is uniform (same tree species and size), with few long-lived trees and a drier microclimate, which makes them unsuitable for drought-vulnerable epiphytes [158]. Therefore, large shade trees and remnants of the original forest represent an essential refuge for some species of epiphytes that have adapted to the ecological and environmental conditions of the agroecosystem [158,165–167]. For example, shade coffee agroecosystems in Mexico harbored fewer species of hygrophilous orchids and ferns, which are more susceptible to disturbance, but more drought-tolerant bromeliad species, compared to humid montane forest habitats [81,167,168].

Unfortunately, the deliberate removal of epiphytes from shade trees in coffee plantations is a common management practice in Latin America [169]. This activity forms part of the maintenance of the shade trees to increase the availability of light for the coffee plants, and because epiphytes are considered as harmful parasites. Indeed, epiphyte removal from shade trees and coffee bushes might have a positive effect on coffee productivity as plants produce more flowers and fruits [169,170]. However, this practice leads to a depauperate agroecosystem, with negative impacts on the epiphyte community and its associated fauna [168,171,172].

Similarly, Hundera et al. [161] indicated a negative effect of forest fragmentation size and management on the diversity of epiphytic orchids in shade trees of semiforest coffee systems, as in large and small managed fragments, epiphyte richness was reduced by 30% and 55%, respectively, compared to old-growth forest. In these agroecosystems, the canopy layer is manipulated, shrubs are removed, and the herbaceous understory is cleared to reduce interspecific competition and increase coffee yield quality and quantity [161].

Even though some endangered orchid species may persist in small managed fragments, these cannot compete with the conservation benefit generated by extensive unmanaged Afromontane forests. Besides this, De Beenhouwer et al. [173] showed that coffee shrubs and their shade trees in managed forest fragments are a suitable habitat for only a limited set of orchid species. Thus, to conserve orchid diversity, it is necessary to avoid coffee management intensification in the remaining old-growth forest. At the same time, farmers should keep old canopy trees and tolerate epiphytes on their coffee shrubs.

Considering the numerous studies on shade coffee agroecosystems, there is minimal information regarding the diversity of epiphytes in shaded cocoa (*Theobroma cacao* L.; Figure 8F) plantations. These can generally be found where tropical rainforests have been lost or disturbed, and the current landscape is a mixture of secondary vegetation. In this pastureland for cattle grazing and agroforestry plantations, cacao grows under shade trees of native or cultivated exotic species [174]. While a study on non-vascular epiphyte diversity in Ecuador has shown that species richness was usually lower on cocoa trees than on natural tropical rainforest trees [175], there is no similar research for vascular epiphytes. In the same country, Haro-Carrión et al. [176] found that epiphyte richness on relict shade trees in rustic shade cacao plantations was reduced by 30% compared to large trees in adjacent forest fragments. Aroids, *Peperomia*, and ferns were less species-rich in plantations than in forests, while there were no differences in orchids and bromeliads, which shows that cacao agroforests may preserve a portion of epiphyte diversity but do not fully compensate for the loss of forest. Similar results were reported by Morales-Linares et al. [174] in Mexico, where the reduction in epiphytic orchids in cocoa plantations was 31% compared to tropical rainforest fragments. However, this study considered shade and cocoa trees, which both contributed to maintaining orchid diversity. Interestingly, more than 50% of these phorophyte species harbored ant gardens composed of up to 22 epiphytic species [177].

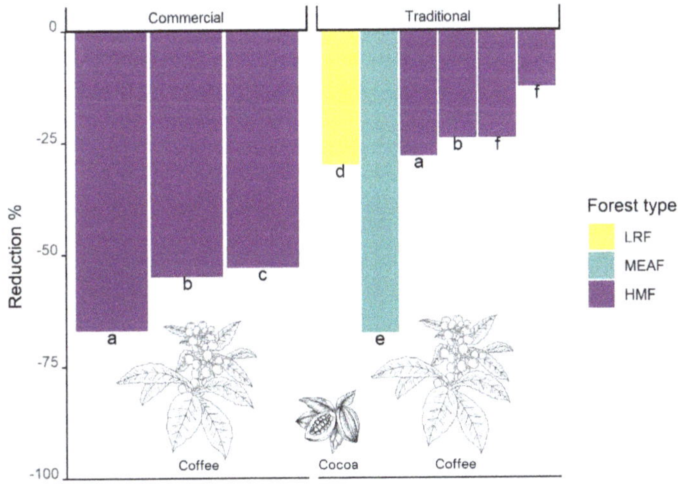

Figure 9. Epiphyte species richness reduction in different shaded coffee and cocoa plantations (Commercial monoculture with low density of shade trees; Traditional coffee polyculture with high density of shade trees) compared to old-growth forests (HMF—humid montane forest; LRF—lowland rain forest; MEAF—moist evergreen afromontane forest) at different study regions in tropical countries (a) Mexico [158]; (b) Mexico [162]; (c) Mexico [81]; (d) Ecuador [176]; (e) Ethiopia [164]; (f) Nicaragua [163].

Similar to traditional coffee and cacao polycultures, ancient tea gardens in the Yunnan Province of Southwest China were established by the partial clearing of the original vegetation, where tea (*Camellia sinensis* (L.) Kuntze) has been planted under shade tree remnants of the natural tropical montane evergreen forests [178]. These agroforestry systems combine biodiversity conservation and natural resource utilization, and are essential habitats for many native plant species, including epiphytic orchids [178,179]. Wang et al. [180] recorded a total of 19 vascular epiphyte species, including 13 orchids, on a total of 343 top-pruned tea trees, while Wu et al. [181] identified 85 epiphytic orchid species (in 33 genera) on pruned tea trees, including several threatened species meeting the IUCN Red List criteria (in two 40×100 m^2 plots each in three ancient tea gardens). The high richness of orchids on tea trees may be due to the preservation of a large number of shade trees, which can serve as a reservoir for propagules and possibly even host a greater variety of orchid species than the shorter tea trees [181]. In each case, ancient tea gardens were shown to maintain a high epiphytic orchid diversity, and to provide pollinator services for reproductive success and a suitable micro-environment for seed germination. Besides this, to date, only one study, conducted in northern Bengal, India, has investigated epiphyte diversity on shade trees of tea plantations [182]. The authors found a total of 6704 individuals belonging to 74 species of 20 families of vascular epiphytes, mostly holoepiphytes (62%), hemiepiphytes (20%), and a few accidentals. The study shows that epiphyte assemblages on shade trees of tea gardens have a high potential to contribute to epiphytic diversity in non-forest ecosystems of this region.

Although some plantations have great potential to host diverse epiphyte communities, this will almost always go along with the loss of species that are very vulnerable to changes in microclimate. As epiphytes are mainly forest species, the majority will cope better in plantations with a structure that resembles an old-growth forest with little intervention by humans. Unfortunately, this stands diametrically against the intention and practices mostly applied in this context. Timber production is often executed at a large scale and with relatively short rotation times, which will drastically reduce the potential for epiphyte assemblages to fully establish. Furthermore, fruit production involves an even more regular human intervention in the plantation, be it for harvest or maintenance, and epiphytes are usually seen as pests in this context, even further precluding their establishment.

3.6. *Pastures with Remnant Trees*

Much land in the tropics worldwide has been converted to pastures with remnant trees and planted isolated trees, interspersed with scattered forest patches (Figure 5C,D). These pasture trees are spared as a source of shade and fodder for cattle and other pasture animals [183], and incidentally act as islands for epiphytes in human-modified environments. Thus, individual trees are often described as the minimum habitat unit of epiphytes (e.g., [184]), which is certainly a valid concept in the case of isolated trees in pastures [20].

Studies of epiphyte communities on such isolated trees usually find a reduction in species richness compared to undisturbed forests (however, see [43]). However, this reduction is very heterogeneous (Figure 10; Supplementary Material, Table S5). By far the greatest loss reported, with almost no epiphytes found growing on isolated trees, comes from an area formerly covered by Atlantic Forest in Brazil [185]. It is likely that these isolated pasture trees were not remnants of natural forests, because the trees had significantly smaller diameters at breast height and smaller statures than trees in the studied forest. The long history of human modification of the Atlantic Forest region [185] might not only have led to the loss of the original forest, but also to the greater turnover of pasture trees, precluding the successful establishment of epiphyte communities in pasture trees.

Species loss at >50% was reported for epiphyte communities growing in pasture trees compared to their forest counterparts in montane forests of Mexico and Ecuador [186–188]. In contrast, Hietz-Seifert et al. [43] found more epiphyte species on pasture trees than on trees close to lowland rainforests in Veracruz, Mexico. The variability in vegetation zones, from dry to humid lowlands [43,189], and regions of lower humid montane forest [116,190], montane dry forest [44,191] and humid montane forest [41,45,192], could potentially explain the observed variation. However, even in the same vegetation zone, i.e., humid montane forest, in the same country, the recorded species loss varied by an order of magnitude between two studies, i.e., Larrea and Werner [45] vs. Werner et al. [186].

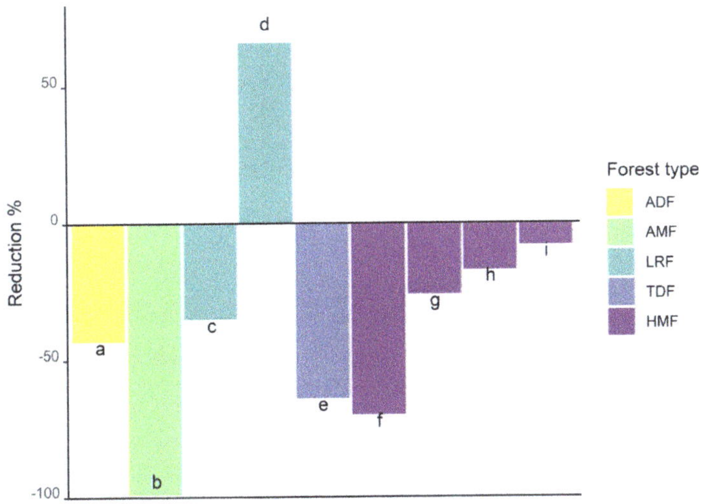

Figure 10. Epiphyte species richness reduction in isolated remnant trees in pastures compared to old-growth forests (ADF, Andean dry forest; AMF, Atlantic montane forest; HMF, humid montane forest; LRF, lowland rain forest; TDF, tropical dry forest) at different study regions in tropical countries: (a) Ecuador [44]; (b) Brazil [185]; (c) Panama [189]; (d) Mexico [43]; (e) Mexico [188]; (f) Ecuador [186]; (g) Mexico (2008) [192]; (h) Ecuador [41]; (i) Ecuador [45].

All studies, however, reported some extent of change in epiphyte composition, showing that assemblages on isolated remnant trees had a reduced floristic heterogeneity compared to old-growth forests, which is arguably an important aspect of land-use change in the context of epiphyte conservation [45]. Most studies concluded that a highly important driver is the change in microclimate towards higher levels of light and desiccation stress between the forest habitat and isolated trees that the epiphyte communities must cope with (e.g., [41,45,186,193]). An isolated tree cannot provide the same microclimate as a tree growing in a matrix of other trees; light air movement will diffuse humidity much more easily in an open pasture, and light will not only mainly penetrate the crown from above but also from all sides, thus reducing the diversity of habitat niches within the crown and changing the microclimate of the trunk considerably.

Hygrophilous species are usually the first species lost, such as filmy and grammitid ferns and other understory taxa (e.g., [40,192]), whereas more xerophytic taxa, adapted to drier climates (e.g., Bromeliaceae, Orchidaceae, Piperaceae, Cactaceae), usually cope well, and even increase in abundance, in such pasture habitats (e.g., [45,189]). Particularly drought-adapted bromeliads that thrive in the drier microclimates of isolated growing host trees are repeatedly observed making up a much higher percentage of the epiphyte communities than in trees from closed forests (e.g., [188,194,195]). This floristic turnover results in epiphyte communities with distinct species compositions compared to those of undisturbed forest habitats [45]. However, some communities appear to be dynamic, growing assemblages [194] that could be stepping-stones in time and space for re-populating nearby or newly forming secondary forests.

The diversity of the community of species on these hosts will depend highly on whether the trees are remnants of cut forests that already hosted epiphytes, or if the communities must establish from scratch. In both cases, the chance to increase diversity will depend on whether there are seed sources from which successful dispersal to the isolated trees is possible [99,105,117]. Also relevant is the constancy of the pasture trees. Much human intervention in the system will likely lead to the impoverishment of epiphyte communities, which could be the reason for the poor epiphyte diversity in pastures of the Atlantic Forest [185]. As short rotation times of tree plantations hamper the development of epiphyte communities, cutting isolated trees on pastures or in other human-modified landscapes will diminish their usefulness as stepping-stones for epiphyte communities, while closer seed sources will most likely increase the probability of re-establishment of epiphytes in any potentially growing secondary forest, as most epiphytes seem to have a high potential for long-distance dispersal [98]. Genetic evidence suggests that enough seeds of wind-dispersed epiphyte species enter the air column, allowing dispersal over several kilometers in fragmented landscapes [196]. On average, epiphytes have larger geographic range sizes than closely related terrestrial species, supporting the hypothesis that epiphytism favors dispersal into larger geographic areas. However, species in families where epiphytism is prevalent tend to have small range sizes regardless of their lifeform [197]. Nevertheless, dispersal in open landscapes may be much less limited than in forests [99]. In line with this notion, although dispersal limitation was repeatedly mentioned as a potential driver of their diminished diversity on isolated trees, it seemed to be a far less important driver than the change in microclimate (e.g., [41,45,186,193,198]).

3.7. *Fragmented, Disturbed, Degraded and Managed Forests*

The widespread deforestation in the tropics does not generally result in a straight frontier between old-growth forest and deforested land, but human activities encroach on the forest from many sides, which leads to a complex mosaic of old-growth forest fragments, negatively affected by edge effects, surrounded by secondary forests, agricultural land and deforested areas [42]. Understanding how the fragmentation and degradation of their natural habitats are affecting plants is critical for the conservation of biodiversity [199]. Thus, pertinent studies are urgently needed, as the conservation value of fragmented, selectively logged or otherwise degraded forest should not be taken for granted [45].

Forest fragmentation creates strong edge effects, abrupt ecological changes in previously continuous vegetation [200–202]. Haddad et al. [203] showed that approximately 20% of the world's remaining forest area is within 100 m of an edge, while more than 70% of the planet's forest areas are within 1 km from the next forest edge. The consequences of the edge effect include lower relative humidity, higher light intensity, and increased temperature and wind exposure [200,202,204]. Although the effect on the species richness of plant and animal communities has been extensively studied [205], only a few studies have focused on epiphytes [206–209], indicating reductions in species richness of up to 50% at the forest edge compared to inner forest sites in Brazil.

In a similar way, forest degradation occurs as a result of human activities; it can be defined as a state of anthropogenically induced arrested succession, where ecological processes that underlie forest dynamics are diminished or severely constrained [210]. The degradation process can be rapid, or may take place over a long period and only become evident gradually [211], but it finally leads to a reduction in biodiversity and changes to the structure and species composition of the forest [212]. It is estimated that more than 2 billion ha of forests are degraded globally [213], while from a remaining area of 1071 million hectares of tropical moist forests, about 10% were degraded in 2019 [214]. The primary causes of forest degradation in the tropics are unsustainable exploitation, such as harvesting for timber, firewood and charcoal (Figure 11A), uncontrolled forest fires, and grazing [215]. Furthermore, the overharvesting of non-timber forest products (NTFPs), i.e., wild plants that are used for food, medicine, construction materials, and fibers, can have negative effects on the plant communities and lead to the disturbance of tropical forests [216]. Using the classification of Guzmán-Jacob et al. [109], disturbed or degraded forests are defined as fragments with clear signs of previous logging, sometimes with ongoing cattle grazing, the removal of understory and/or the harvesting of NTFPs. In contrast to secondary forests, these have not been clear-cut, and thus maintain at least some of the original canopy trees, but the structure, processes, functions and dynamics in degraded forests have been altered [212]. One of the main drivers of the degradation process in a forest is the selective harvesting of economically important trees [212]. Consequently, there is a need to adopt sustainable forest management practices that promote biodiversity conservation together with the production of timber [217,218].

Figure 11. Views of (**A**) forest degradation by logging for timber in Chiapas, Mexico; (**B**) wild epiphyte traffic in the streets of central Veracruz, Mexico; (**C**) ceremonial use of epiphytes in Oaxaca, Mexico; (**D**) fallen epiphytes in Los Tuxtlas, Mexico. Photographs by (**A**) Nayely Martínez-Meléndez; (**B**) Thorsten Krömer; (**C**) Gabriela Cruz Lustre; (**D**) Esteban Francisco-Ventura.

Forestry management consists of programmed harvesting to achieve continual timber yield while maintaining considerable forestry carbon stock. In this way, a significant proportion of the forest's structural attributes are preserved, allowing for the provision of those ecosystem services unmanaged forests provide [218]. Nevertheless, different methods of forest management, such as selective harvesting and rotation with intermediate felling, modify the composition and structure of tree vegetation [218–220].

The impact of forest disturbance in the context of different silvicultural management practices on the diversity of vascular epiphytes has been investigated in several studies with mixed results (Figure 12; Supplementary Material, Table S6). For disturbed humid montane forests, severe species richness reductions of about 50% and higher were reported. In the Atlantic Forest of Brazil, Parra-Sanchez and Banks-Leite [185] found that besides the drastically reduced number of epiphytic species (−84%), human-modified forests had low numbers of seedlings and juveniles, which will exacerbate the differences compared with continuous old-growth forest in the future, because of the high seedling mortality and lower tolerance to drought of early ontogenetic stages.

In the Venezuelan Andes, Barthlott et al. [39] found a sharp difference in species richness between a primary and a disturbed forest, although the last selective logging occurred some 50 years ago. The decrease in species richness of 76% at the disturbed site was driven mainly by the loss of epiphytes that grow in large stands and mats, which build up reservoirs of nutrients and water exploitable by many other epiphytes [39]. Consequently, epiphytes in the disturbed forest mostly grew on the bare bark.

Another severe reduction in epiphytic species (49%), mainly orchids, was found by Krömer et al. [81] in a disturbed humid montane forest in eastern Mexico, located in close vicinity to a large urban area, which had been used for the extraction of firewood, construction timber and NTFPs for decades. Specifically, the mostly illegal harvest of orchids for commercial and cultural purposes further reduced epiphyte abundance.

Due to their horticultural and ceremonial value, wild epiphytes are often sold in markets and streets in Mexico ([221–225]; Figure 11B,C), and in many other tropical countries [226]. In Africa and Asia, many epiphytic orchids are used in traditional medicine or as food supplements [227–230]; in addition, wild-collected orchids are commercially traded globally [231]. However, only a few studies evaluated the ecological effects of harvesting epiphytes [232–235]. Most of them show that sustainable use is difficult [236]. Other studies recommended taking advantage of the naturally fallen epiphytes in forests and agroecosystems ([169,233,237,238]; Figure 11D).

In another study from southern Mexico, Wolf [135] compared epiphyte diversity between sites of least disturbed old-growth pine-oak forests with cyclically clear-cut and selectively logged forests. Epiphyte richness in the ca. 25-year-old oak coppices with regular clear-cutting was 51% lower than in old-growth forests. Thus, cyclical clear-cutting should be avoided, as the restorative period is apparently too short for epiphytes to fully re-colonize coppices. In contrast, selectively logged pine-oak forests, in which large trees were spared, sustain epiphytic vegetation very similar in richness and biomass to old-growth forests [135]. Such remnant trees are essential for epiphytes that require the presence of accumulated suspended soil, and they may also serve as epiphyte seed sources for the re-growing trees in the vicinity. The author concludes that selective logging may help to protect the epiphyte vegetation locally, but if such management is not adopted over a larger region, much of the epiphyte species diversity in commercially exploited pine-oak forests will be lost.

In the same area, Martínez-Meléndez et al. [218] studied the epiphyte diversity in two pine-oak forest stands that had been subjected to thinning and release cutting (methods of silvicultural management) in comparison to an old secondary forest, with no timber harvest since the 1960s. While thinning only slightly reduced species richness by 15%, release cutting led to a 25% reduction, compared to the secondary forest. The latter is related to the lack of mature oak trees, which provide more favorable structural conditions (e.g., a variety of bark types) for successful epiphyte establishment [218]. In contrast, the predominating pine trees (e.g., *Pinus oocarpa* Schiede ex Schltdl., *P. maximinoi* H.E. Moore) are considered less suitable host trees for most epiphytes [134,135,137], although Jiménez-Bautista et al. [239] found some orchids established on *P. ayacahuite* C. Ehrenb. ex Schltdl.

In humid montane forest in northeastern Ecuador, Larrea and Werner [45] found that epiphyte species richness was only slightly reduced by 10% in managed forest with altered under- and midstory due to cattle grazing relative to old-growth forest without major human interference for at least 50 years. However, the managed habitat differed considerably in floristic composition from the old-growth forest, suggesting rapid species turnover in the course of only six years since cattle disturbance started, as shown by the loss of numerous drought-sensitive species, e.g., the replacement of ferns by more xerotolerant taxa [45]. It is likely that drought-sensitive species, as a functional group, hold a disproportionate share of rare and range-restricted species, thus the conservation value of intensively managed habitats should be viewed with caution.

Similarly, Seshadri et al. [240] showed a reduction of 11% in species richness between selectively logged (40 years ago) and unlogged wet-evergreen montane forests in the southern Western Ghats, India. Although there was higher epiphytic diversity and relative abundance in the selectively logged forest, the vascular epiphyte assemblage did not match that of the unlogged forest, e.g., there was a greater abundance of drought-resistant species. This goes in line with the study of Padmawathe et al. [241] in a moist lowland forest of the Eastern Himalaya, India, where they found in selectively logged forests a slight reduction of 15% in epiphyte richness and abundance, except for orchids, whose survival was ensured by large remnant trees. However, logging reduced the forest structural complexity and altered their microclimate, which negatively affected the abundance and species composition of pteridophytes and non-orchid angiosperm epiphytes, whose diversity can be maintained only if patches of forests are left uncut in logged areas.

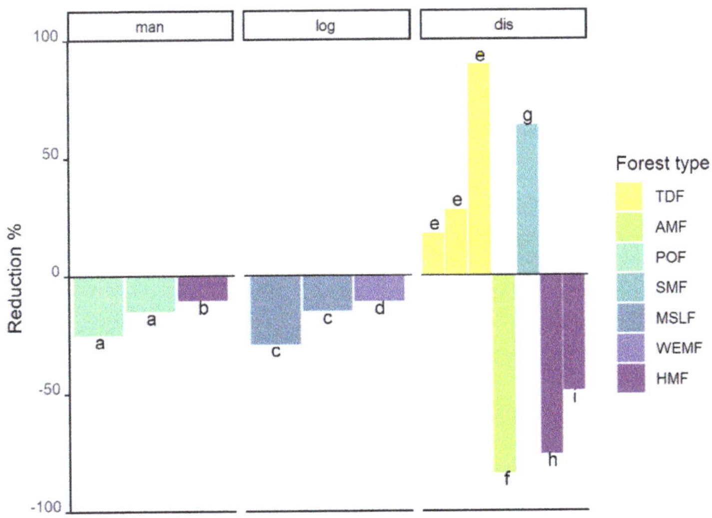

Figure 12. Epiphyte species-richness reduction in managed (man), selectively logged (log) or disturbed (dis) forests compared to old-growth forests (AMF—Atlantic montane forest; HMF—humid montane forest; MSLF—moist, semi-evergreen lowland forest; POF—pine–oak forest; SMF—*Sabal mexicana* forest; TDF—tropical dry forest; WEMF—wet-evergreen montane forests) in different study regions in tropical countries: (a) Mexico [218]; (b) Ecuador [45]; (c) India [241] (d) India [240]; (e) Mexico [119]; (f) Brazil [185]; (g) Mexico [242]; (h) Venezuela [39]; (i) Mexico [81].

In contrast, Flores-Argüelles et al. [119] found that in the three types of tropical deciduous and semideciduous lowland dry forests in western Mexico, surprisingly, selectively logged and grazed sites presented a greater number (18%, 28% and 90% increase) and abundance of epiphyte species than the adjacent old-growth forest sites. This could indicate that disturbances rather diversified the microhabitat types available in these dry forests, providing the possibility for other species types to establish (i.e., other than drought-tolerant species; [108,109]). A similar observation, i.e., slightly higher epiphyte richness in degraded vs. adjacent old-growth forest, was also made by Guzmán-Jacob et al. [109] at a 500 m elevation at the Cofre de Perote in central Veracruz, Mexico. Furthermore, Aguirre et al. [242] recorded a greater richness of epiphytes (64% increase) in disturbed sites compared with conserved sites in a forest dominated by the palm *Sabal mexicana*, on the Gulf Coast in Veracruz State. This pattern was mainly due to the accumulation of hemiepiphytic species of low abundance (e.g., *Ficus* spp.) and the presence of accidental epiphytes, whose growth was favored by the rich and abundant aerial substrate retained in old leaf bases of these palms located in open areas.

The impacts of forest fragmentation due to land-use change and degradation on vascular epiphyte species richness and composition boil down to significant reductions in epiphyte richness when compared to possible old-growth forests nearby, and shifts in species composition. Generally, the degree to which epiphytes are negatively affected depends on the severity of disturbance and the age or size of the remnant trees [125]. In addition to total species numbers, disturbance affects species composition, as anthropogenic disturbances force a shift from hygrophilous or mesic to more drought-tolerant species [45], but also vertical distribution, as epiphyte species from shady canopy strata decline in more open vegetation types compared with old-growth forests [43,158]. Activities such as selective logging, clear-cutting, cattle grazing, and the overharvesting of non-timber forest products (NTFPs) are actions that are largely carried out in already impacted ecosystems, or may be the direct reason for their conversion. Indirectly or actively, these activities create edge effects, and alter the structure and microclimate of forests. Because recovering ecosystems or secondary forests are here to stay, and because they have a high conservation value for terrestrial vegetation, it is still important to evaluate what possible changes in selective logging practices may be put into place to avoid the further degradation of epiphyte assemblages, along with research that can really establish what is the actual conservation value of remnant trees, as they are implied to be critical for preserving epiphyte diversity and maintaining ecosystem functions (although there is no evidence of the latter).

Generally, the degree to which epiphytes are negatively affected depends on the severity of disturbance and the age or size of the remnant trees [125]. In addition to total species numbers, disturbance affects species composition, as anthropogenic disturbances force a shift from hygrophilous or mesic to more drought-tolerant species [45], but also their vertical distribution, as epiphyte species from shady canopy strata decline in more open vegetation types compared with old-growth forests [43,158]. For epiphyte conservation purposes, it is recommended to adopt a management of selective logging instead of periodic clear-cutting, and to spare large trees during logging [135]. These remnant trees may be a "refuge" where a substantial portion of local floristic richness can survive and later serve as propagules, which otherwise would be lost during logging activity if not properly managed [243]. However, it is recommended that fallen epiphytes be rescued from the logged trees or the ground for possible translocation and reintroduction into nearby forest fragments as an enrichment strategy within ecological restoration projects [244–248].

3.8. *Epiphytes in Urban Settings*

Human settlements and infrastructure account for an ever-increasing, albeit still relatively small, portion of the global land surface (c. 0.2×10^9 ha or c. 3% of the total), although the boundary between rural and urban areas is often blurred. Recent decades have seen numerous studies on the plant communities that appear spontaneously in these urban habitats [249]. Vascular epiphytes have not escaped the attention of scientists interested in the relatively new discipline of urban ecology, and over the last two decades, epiphytes in urban settings have been the focus of a considerable number of studies.

These studies cover numerous types of urban habitats (Figure 13), from urban forest fragments [250] to parks [251], college campuses [252,253], botanical gardens [254], private gardens [255], and single roadside trees [256,257], or deal with epiphytes on artificial substrates like electricity wires [79] or on buildings [78]. Although most of these studies were performed in America (e.g., [252,258–260]), there are reports from all over the globe, e.g., from Africa [261], Europe [262], Asia [257], Australia [263] and New Zealand [264]. The studies provide important baseline information, but a common disadvantage of most of these studies is their entirely descriptive nature, along with a lack of context—in most cases, there is no comparison of the urban epiphytic vegetation with that of an old-growth forest in the vicinity standardized for area or number of trees. Comparisons are also hampered by the frequent taxonomic bias, e.g., on orchids [253,265], bromeliads [266,267] or ferns [268,269].

Figure 13. Views of (**A**) *Tillandsia schiedeana* Steud. on a mesh wire fence in central Veracruz, Mexico. (**B**) Epiphytes on power cables in central Veracruz, Mexico. (**C**) Epiphytes on park trees in central Veracruz, Mexico. (**D**) Epiphytes on fence tree in Las Lajas, Panama. Photographs by (**A**,**C**,**D**) Helena J. R. Einzmann; (**B**) Thorsten Krömer.

A few studies, however (e.g., [152,257]), applied a comparative approach. Adhikari et al. [257] compared the species numbers and abundance of epiphytic orchids along a gradient of naturalness from single trees in urban settings to trees in nature reserves along Kathmandu Valley—urban settings' median richness was 5% that of the nature reserves, while median abundance was also much lower (with c. 30%). Einzmann et al. [152] used a similar approach in southwestern rural Panama, and compared epiphyte diversity and abundance on trees in human settlements with those on trees in pastures and forests. Species numbers derived from rarefaction curves suggest a two-fold difference between the species numbers of urban habitat and nearby forest, but neither the average number of species per tree nor their abundance were significantly different. Aoki-Gonçalves et al. [266] analyzed epiphyte diversity and abundance in the city of Xalapa along a woody cover gradient from 10 to 100% in a 100 m buffer. As expected, epiphyte species richness was positively related to woody cover, but surprisingly, community composition was not. The most abundant species was *Tillandsia recurvata* (L.) L., a generalist species that is very common in urban contexts in much of the Americas. A study by Alvim et al. [270], which compared epiphytes in 26 urban green areas in Brazil, is highly instructive in this context. They reported a set of dominant species like *T. recurvata* or *Pleopeltis pleopeltifolia* (Raddi) Alston, with low diversity and a large proportion of accidental epiphytes, suggesting that the urban environment acts as a strong filter for widely distributed species, thus resulting in the homogenization of the potential regional epiphyte flora.

Most studies report the results of single censuses without information on size structure, which does not allow us to predict the longer-term perspectives of epiphytes in urban habitats. Again, there are exceptions; Mondragón and Mora-Flores [271] analyzed the size class structure of two *Tillandsia* species in the city of Oaxaca, Mexico. At least for *T. recurvata*, the large proportion of seedlings and juveniles suggests a dynamic population [272]. Clearly, repeated censuses are needed to obtain information on the dynamics of urban epiphytes. Results will certainly differ among species. A genetic study with *T. recurvata* highlighted the different consequences of the low connectivity of urban populations for selfing and outcrossing species [267].

Among the tree flora of many cities, exotic species are not uncommon. There is no indication that these differ from native trees as regards suitability as potential hosts (e.g., [253,271,273]). Quite a few species even seem to be particularly good hosts. For example, *Samanea saman* (Jacq.) Merr. from South America, which is frequently planted along roads in the city of Singapore, has a very rich epiphyte flora [274]. Similarly, locally exotic palm trees of the genera *Phoenix* or *Elaeis* were the focus of several studies because of their conspicuously rich epiphyte flora [275–277]. Noteworthy, a large proportion of the recorded epiphytes were actually accidentals. This observation is not idiosyncratic for palms—numerous studies of epiphytes in urban settings report a large proportion of accidental epiphytes (Table 1). Further increasing the heterogeneity of the epiphyte communities in urban settings is the occurrence of exotics, e.g., *Dendrobium nobile* Lindl. or *Platycerium* spp. in the Americas, where they have escaped from cultivation, or ornamentals may have intentionally been attached to trees [274,278–280].

In summary, the currently available evidence leads to the unsurprising conclusion that urban areas are quite impoverished in epiphyte diversity and abundance compared to a forest of similar size and location. Moreover, accidentals play a much larger role than in undisturbed vegetation. Nevertheless, there is a substantial number of epiphyte species, which led Neo et al. [281] to call cities "a botanical oasis rather than a biological desert". We partially agree that, far from being diverse or hotspots, urban areas can play a certain role in epiphyte conservation.

Table 1. Epiphyte communities in urban settings. Given are the total numbers (#) of species growing epiphytically, the total and relative numbers of accidental epiphytes and exotics, the study site and the source.

Total # Species	Accidental #	Accidental %	Exotics	Study Site	Source
10	1	10	-	Piratininga, Brazil	[256]
8	2	25	1	Port Harcourt, Nigeria	[260]
16	4	25	-	Bogor, Indonesia	[282]
49	18	36.7	8	Mar de Espanha, Brazil	[283]
43	16	37.2	-	Juiz de Fora, Brazil	[252]
110	46	41.8	30	Juiz de Fora, Brazil	[270]
47	21	44.7	-	Juiz de Fora, Brazil	[258]
15	9	60	-	Buenos Aires, Argentina	[284]
72	48	66.7	-	Douala, Cameroon	[276]
34	23	67.6	12	Santo Domingo, R Dominicana	[285]
71	60	84.5	32	Quito, Ibarra, Riobamba, Mendoza; South America	[275]

4. Conclusions

This review provides a comprehensive overview of the literature on the impact of land-use change on vascular epiphytes, and thus spans individual studies that often focus on a locally and temporally restricted scale. Most studies across forest types agree that the conversion of tropical old-growth forests, typically into a mosaic of secondary forests, cattle pastures with remnant trees, and plantations of native exotic tree species leads to a general decline in epiphyte richness (e.g., [39,41,46,81]). However, the magnitude of the impact depends on the magnitude of disturbance and the type of vegetation that replaces the original forest, and varies with functional types. The associated changes in microclimate are much more severe for hygrophilous species in constantly humid ecosystems, such as lowland rain or montane cloud forests, while more drought-resistant species in less rich dry forests seem to be less affected, or may even benefit from these changes [44]. Three large groups, bromeliads, ferns, and orchids, seem to show distinct patterns regarding human disturbance when analyzed separately [40,45]. This is not very surprising; epiphytes are simply defined by their structural dependence on trees, but are functionally highly diverse [286]. Thus, we cannot expect a consistent response across all species.

The impact of climatic changes on vascular epiphytes, though still a relatively underexplored field, is increasingly being studied experimentally. The existing small body of literature indicates that epiphytes are affected by shifts in climatic conditions, particularly changes in precipitation patterns and the frequency of extreme weather events [287,288]. This will influence epiphyte diversity and community composition, with observations suggesting that canopy-dwelling plants are often, but not always, more vulnerable to such changes than terrestrial counterparts [35]. Although few studies have directly addressed the effects of climate change on epiphytes, evidence from various sources supports the notion that sustained microclimatic changes, regardless of disturbance intensity, will adversely affect many epiphyte populations [21,48,289]. Considering this, disturbances, which have long been used as a proxy for understanding potential climate change effects, are examined as factors contributing to the vulnerability of epiphytes to both natural and human-induced environmental shifts [290]. This review highlights how land-use changes—such as deforestation and conversion to agricultural land—impact epiphyte diversity across a wide range of systems. We assume the mechanisms in many cases are related to the drastic alteration of the microclimate, making the habitat unsuitable for epiphytes while allowing terrestrial vegetation to persist. This is because, unlike terrestrial plants, epiphytes are more dependent on their host trees and specific microclimatic conditions [95]. Hence, the regeneration of vascular epiphytes is slower than that of terrestrial plants, and this difference in recovery rates is essential when considering the ecological consequences of land-use changes for epiphyte populations.

While the value of secondary forests for biodiversity is increasingly acknowledged, none of these ecosystems entirely replicate the biodiversity of undisturbed primary forests. Nevertheless, they provide critical habitats for metapopulation persistence, but also contribute to carbon sequestration. Over time, secondary forests may assist in the recovery of a substantial portion of species richness, taxonomic diversity, and functional roles, including threatened and endemic species. Additionally, they complement primary forest remnants by enhancing landscape connectivity and offering alternative habitats. Even so, land-use change in tropical forests typically leads to substantial biodiversity loss, reduced ecosystem services, and decreased soil health. Primary forests are irreplaceable for biodiversity conservation, and efforts to mitigate land-use impacts should focus on preserving these areas and implementing sustainable land-management practices [291–293]. This irreplaceability paints a bleak scenario for plant biodiversity in tropical forests. The scenario is even worse for vascular epiphytes, which face several challenges regarding reestablishment after disturbance related to low dispersal, drought sensitivity, high mortality, and unsuitable substrates, microsites, or habitats. Also, epiphytes need not only terrestrial vegetation, but also a degree of habitat diversity in the forest so as to find suitable habitats, colonize, and recover once the challenges they face are overcome.

5. Future Directions

The plethora of methods and sampling designs applied when evaluating the effects of land-use on epiphyte diversity prevents a quantitative analysis of the different results [294,295]. To allow such analyses in the future, ecological studies on epiphytes require standardization; also, greater replication within and between trees, sites, and habitats is needed for a more rigorous statistical and quantitative assessment of the impact. Researchers should take measurements of climatic variables at sampling locations both within and outside of tree crowns, along the vertical gradient, with attention paid to the edge effects on epiphyte communities. Much more long-term research is needed to understand the consequences of disturbance on epiphytes at larger spatial scales and longer timescales [68]. The results of the few published long-term studies (e.g., [194,296,297]) suggest that epiphyte communities at a plot scale of, e.g., 1 ha are not at equilibrium. Thus, changes in secondary habitats may be unrelated to disturbance, but simply represent typical dynamics of epiphyte communities. An interpretation of such studies is only possible with similar studies in undisturbed forests.

Degraded habitats are becoming increasingly prevalent in tropical ecosystems. Therefore, it is key to understand the baseline of vascular epiphyte assemblages in natural habitats and across degraded ecosystems, as well as what temporal changes occur in degraded or non-natural ecosystems, so as to best inform restoration efforts. Although it is difficult to draw an overall picture of the impact and, therefore, solutions, some studies on the ecological restoration of degraded and secondary forests with naturally fallen epiphytes or those collected from felled trees reported encouraging results [244–247]. Active interventions or restorations seem to be more effective than a passive approach. For instance, active interventions are mentioned for areas with slow natural regeneration due to long disturbance histories, whereby actively planting vegetation nuclei may produce seed sources, accelerate recovery, and establish vegetation nuclei to mitigate dispersal limitation. Any effort should recognize and acknowledge the value of vascular epiphytes as secondary foundation species, whereby any active restoration would consider the species' ecotypes and requirements.

For successful restoration with or without transplants, it is essential to understand abiotic and biotic influences (e.g., rainfall, drought, substrate conditions, branch and bark quality) on survival, colonization, and extinction. Vegetation restoration may enable the restoration of canopy vegetation. For instance, the transplantation of vascular epiphytes to host trees can be effective, with high survival rates influenced by species traits and environmental conditions. Focusing on particular "ecotypes" according to the forest's stage of recovery could be a strategy, whereby tank bromeliad transplantation offers cost-effective support by creating stable microhabitats. Case studies, such as in the Colombian Andes [247], highlight the importance of host tree selection and considering epiphyte species-specific responses when transplanting.

None of these measures guarantee success without knowing the spatio-temporal dynamics. Temporal dynamics are almost unexplored, whether in old-growth forests or in degraded or secondary forests. We know very little about the long-term temporal changes of vascular epiphyte communities; therefore, long-term monitoring to evaluate the temporal variation of epiphyte assemblages at a fine-grain and broad-grained level is key to understanding how to aid regeneration and restoration best. Epiphyte survival and demographics show substantial variation over time, driven by both abiotic and biotic factors.

As a final thought, although our review focuses on vascular epiphytes, it seems essential that the research on the impact of land-use on vascular epiphytes be expanded to understand the cascading effects that the loss of vascular epiphytes may have on other components of the ecosystem.

Supplementary Materials: The following supporting information can be downloaded at: https://www.mdpi.com/article/10.3390/plants14081188/s1, Figure S1: Visual assessment of the different methods for determining the optimal number of clusters for k-means; we used two types: total within sum of square or "wws" (upper graph) and gap statistics or "gap_stat" (lower graph) as a function of clusters. These methods include a direct method, consisting of optimizing a criterion, such as within-cluster sums of squares, and statistical testing, consisting of comparing evidence against null hypothesis, respectively; Figure S2: Four aspects of research: A, *Trade, use and conservation*; B, *Climate and land-use change*; C, *Human-modified habitats*; D, *Responses to disturbance*. Here we show the 20 most frequent words per cluster derived from the overall word matrix (based on all abstracts), independently of their frequency or the number of papers represented in each cluster. Words varied in their frequency per cluster; these are depicted in the Supplementary Material, Figure S3; Figure S3: Word clouds of the four aspects of research (upper graph): *Trade, use and conservation* (upper left, word input from eight references, see Table S1), *Climate and land-use change* (upper right, word input from 33 references), *Human-modified habitats* (lower left, word input from 195 references; an inset (lower graph) is provided for this category for the sake of visibility), *Responses to disturbance* (lower right, word input from 52 references). Word size reflects word frequency in relation to other words of the respective word cloud; Table S1: List of references resulting in four clusters of the following aspects of research: A, *Trade, use and conservation*; B, *Climate and land-use change*; C, *Human-modified habitats*; D, *Responses to disturbance*; Table S2: Comparison of the frequency of words in the different clusters (A, B, C, D), the frequency indicates the number of times the word is repeated in a given cluster, and it ranges from 1 (singletons) to 1298, whereby some cluster have a large number of common words (A and D) and others a large number of "rare" words or singletons (B and C); the differences in the abundance structure of the words per cluster can be contrasted graphically with Figure S3; Table S3: Comparison of epiphyte species richness reduction in old-growth vs. secondary forest fragments (SFF) at different study regions, indicating vegetation type (VT), elevational range (ER), annual precipitation (AP), and references in chronological order; Table S4: Comparison of epiphyte species richness reduction in old-growth forest vs. different types of plantations at different study regions, indicating vegetation type (VT), elevational range (ER), annual precipitation (AP), and references in chronological order; Table S5: Comparison of epiphyte species-richness reduction in old-growth forest vs. isolated remnant trees in pastures at different study regions, indicating vegetation type (VT), elevational range (ER), annual precipitation (AP), and references in chronological order; Table S6: Comparison of epiphyte species richness reduction in old-growth forest vs. disturbed, managed or selectively logged forests at different study regions, indicating vegetation type (VT), elevational range (ER), annual precipitation (AP), and references in chronological order.

Author Contributions: Conceptualization, T.K., G.M.-L. and G.Z.; methodology, T.K., G.M.-L. and G.Z.; formal analysis, T.K. and G.M.-L.; writing—original draft preparation, T.K., H.J.R.E., G.M.-L. and G.Z.; writing—review and editing, T.K., H.J.R.E., G.M.-L. and G.Z. All authors have read and agreed to the published version of the manuscript.

Funding: G.M.-L. acknowledges funding from the DFG Project CONTINENT (ME 5087/1-1).

Data Availability Statement: In accordance with Open Science communication practices, the authors declare that all data are available within the manuscript and Supplementary Materials.

Acknowledgments: We thank Esteban Francisco-Ventura, Gabriela Cruz-Lustre, Jonas Morales-Linares, and Nayely Martínez-Meléndez for providing photographs, Eire Ramírez-García for curation of references.

Conflicts of Interest: The authors declare no conflicts of interest.

References

1. Newbold, T.; Hudson, L.N.; Hill, S.L.L.; Contu, S.; Lysenko, I.; Senior, R.A.; Börger, L.; Bennett, D.J.; Choimes, A.; Collen, B.; et al. Global effects of land use on local terrestrial biodiversity. *Nature* **2015**, *550*, 45–50. [CrossRef] [PubMed]
2. Bellard, C.; Bertelsmeier, C.; Leadley, P.; Thuiller, W.; Courchamp, F. Impacts of climate change on the future of biodiversity. *Ecol. Lett.* **2012**, *15*, 365–377. [CrossRef] [PubMed]
3. Díaz, S.; Settele, J.; Brondízio, E.S.; Ngo, H.T.; Agard, J.; Arneth, A.; Balvanera, P.; Brauman, K.A.; Butchart, S.H.M.; Chan, K.M.A.; et al. Pervasive human-driven decline of life on Earth points to the need for transformative change. *Science* **2019**, *366*, eaax3100. [CrossRef]
4. Laurance, W.F. Reflections on the tropical deforestation crisis. *Biol. Conserv.* **1999**, *91*, 109–117. [CrossRef]
5. Gibson, L.; Lee, T.M.; Koh, L.P.; Brook, B.W.; Gardner, T.A.; Barlow, J.; Peres, C.A.; Bradshaw, C.J.A.; Laurance, W.F.; Lovejoy, T.E.; et al. Primary forests are irreplaceable for sustaining tropical biodiversity. *Nature* **2011**, *478*, 378–381. [CrossRef]
6. Taubert, F.; Fischer, R.; Groeneveld, J.; Lehmann, S.; Müller, M.S.; Rödig, E.; Wiegand, T.; Huth, A. Global patterns of tropical forest fragmentation. *Nature* **2018**, *554*, 519–522. [CrossRef]
7. Achard, F.; Beuchle, R.; Mayaux, P.; Stibig, H.J.; Bodart, C.; Brink, A.; Carboni, S.; Desclee, B.; Donnay, F.; Eva, H.D.; et al. Determination of tropical deforestation rates and related carbon losses from 1990 to 2010. *Glob. Chang. Biol.* **2014**, *20*, 2540–2554. [CrossRef]
8. Keenan, R.J.; Reams, G.A.; Achard, F.; de Freitas, J.V.; Grainger, A.; Lindquist, E. Dynamics of global forest area: Results from the FAO Global Forest Resources Assessment 2015. *For. Ecol. Manag.* **2015**, *352*, 9–20. [CrossRef]
9. Food and Agriculture Organization of the United Nations. *State of the World's Forests 2014*; Food and Agriculture Organization of the United Nations: Rome, Italy, 2014; p. 133.
10. Oberleitner, F.; Egger, C.; Oberdorfer, S.; Dullinger, S.; Wanek, W.; Hietz, P. Recovery of aboveground biomass, species richness and composition in tropical secondary forests in SW Costa Rica. *For. Ecol. Manag.* **2021**, *479*, 118580. [CrossRef]
11. Chazdon, R.L.; Peres, C.A.; Dent, D.; Sheil, D.; Lugo, A.E.; Lamb, D.; Stork, N.E.; Miller, S.E. The potential for species conservation in tropical secondary forests. *Conserv. Biol.* **2009**, *23*, 1406–1417. [CrossRef]
12. Dent, D.H.; Wright, S.J. The future of tropical species in secondary forests: A quantitative review. *Biol. Conserv.* **2009**, *142*, 2833–2843. [CrossRef]
13. Norden, N.; Chazdon, R.L.; Chao, A.; Jiang, Y.H.; Vilchez-Alvarado, B. Resilience of tropical rain forests: Tree community reassembly in secondary forests. *Ecol. Lett.* **2009**, *12*, 385–394. [CrossRef]
14. Dent, D.H.; DeWalt, S.J.; Denslow, J.S. Secondary forests of central Panama increase in similarity to old-growth forest over time in shade tolerance but not species composition. *J. Veg. Sci.* **2013**, *24*, 530–542. [CrossRef]

15. DeWalt, S.J.; Schnitzer, S.A.; Denslow, J.S. Density and diversity of lianas along a chronosequence in a central Panamanian lowland forest. *J. Trop. Ecol.* **2000**, *16*, 1–19. [CrossRef]
16. Schnitzer, S.A.; Bongers, F.; Burnham, R.J.; Putz, F.E. *Ecology of Lianas*; John Wiley & Sons Ltd.: Chichester, UK, 2015; p. 504.
17. Schnitzer, S.A. Testing ecological theory with lianas. *New Phytol.* **2018**, *220*, 366–380. [CrossRef] [PubMed]
18. Schnitzer, S.A.; DeFilippis, D.M.; Visser, M.; Estrada-Villegas, S.; Rivera-Camaña, R.; Bernal, B.; Peréz, S.; Valdéz, A.; Valdéz, S.; Aguilar, A.; et al. Local canopy disturbance as an explanation for long-term increases in liana abundance. *Ecol. Lett.* **2021**, *24*, 2635–2647. [CrossRef] [PubMed]
19. Turner, I.M.; Tan, H.T.W.; Wee, Y.C.; Ibrahim, A.B.; Chew, P.T.; Corlett, R.T. A study of plant species extinction in Singapore: Lessons for the conservation of tropical biodiversity. *Conserv. Biol.* **1994**, *8*, 705–712. [CrossRef]
20. Zotz, G. *Plants on Plants—The Biology of Vascular Epiphytes*; Springer: Cham, Switzerland, 2016; p. 282.
21. Gentry, A.H.; Dodson, C.H. Diversity and biogeography of neotropical vascular epiphytes. *Ann. Missouri Bot. Gard.* **1987**, *74*, 205–233. [CrossRef]
22. Krömer, T.; Kessler, M.; Gradstein, S.R.; Acebey, A. Diversity patterns of vascular epiphytes along an elevational gradient in the Andes. *J. Biogeogr.* **2005**, *32*, 1799–1809. [CrossRef]
23. Taylor, A.; Zotz, G.; Weigelt, P.; Cai, L.; Karger, D.N.; König, C.; Kreft, H. Vascular epiphytes contribute disproportionately to global centres of plant diversity. *Glob. Ecol. Biogeogr.* **2022**, *31*, 62–74. [CrossRef]
24. Zotz, G.; Weigelt, P.; Kessler, M.; Kreft, H.; Taylor, A. EpiList 1.0: A global checklist of vascular epiphytes. *Ecology* **2021**, *102*, e03326. [CrossRef] [PubMed]
25. Kelly, D.L.; Tanner, E.V.J.; Lughadha, E.N.; Kapos, V. Floristics and biogeography of a rain forest in the Venezuelan Andes. *J. Biogeogr.* **1994**, *21*, 421–440. [CrossRef]
26. Nadkarni, N.M.; Sumera, M.M. Old-growth forest canopy structure and its relationship to throughfall interception. *For. Sci.* **2004**, *50*, 290–298. [CrossRef]
27. Gotsch, S.G.; Nadkarni, N.M.; Amici, A.A. The functional roles of epiphytes and arboreal soils in tropical montane cloud forests. *J. Trop. Ecol.* **2016**, *32*, 455–468. [CrossRef]
28. Mendieta-Leiva, G.; Ramos, F.N.; Elias, J.P.C.; Zotz, G.; Acuña-Tarazona, M.; Alvim, F.S.; Barbosa, D.E.F.; Basílio, G.A.; Batke, S.P.; Benavides, A.M.; et al. EpIG-DB: A database of vascular epiphyte assemblages in the Neotropics. *J. Veg. Sci.* **2020**, *31*, 518–528. [CrossRef]
29. Aguilar-Cruz, Y.; García-Franco, J.G.; Zotz, G. Litter-trapping tank bromeliads in five different forests: Carbon and nutrient pools and fluxes. *Biotropica* **2021**, *54*, 170–180. [CrossRef]
30. Stuntz, S.; Ziegler, C.; Simon, U.; Zotz, G. Diversity and structure of the arthropod fauna within three canopy epiphyte species in central Panama. *J. Trop. Ecol.* **2002**, *18*, 161–176. [CrossRef]
31. Cestari, C. Epiphyte plants use by birds in Brazil. *Oecol. Bras.* **2009**, *13*, 689–712. [CrossRef]
32. Méndez-Castro, F.E.; Bader, M.Y.; Mendieta-Leiva, G.; Rao, D. Islands in the trees: A biogeographic exploration of epiphyte-dwelling spiders. *J. Biogeogr.* **2018**, *45*, 2262–2271. [CrossRef]
33. Aguilar-Rodríguez, P.A.; Krömer, T.; Tschapka, M.; García-Franco, J.G.; Escobedo-Sarti, J.; MacSwiney, G.M.C. Bat pollination in Bromeliaceae. *Plant Ecol. Divers.* **2019**, *12*, 1–19. [CrossRef]
34. Thomsen, M.S.; Altieri, A.H.; Angelini, C.; Bishop, M.J.; Gribben, P.E.; Lear, G.; He, Q.; Schiel, D.R.; Silliman, B.R.; South, P.M.; et al. Secondary foundation species enhance biodiversity. *Nat. Ecol. Evol.* **2018**, *2*, 634–639. [CrossRef]
35. Zotz, G.; Bader, M.Y. Epiphytic plants in a changing world- global change effects on vascular and non-vascular epiphytes. In *Progress in Botany*; Lüttge, U., Beyschlag, W., Büdel, B., Francis, D., Eds.; Springer: Berlin/Heidelberg, Germany, 2009; Volume 70, pp. 147–170.
36. Rozendaal, D.M.A.; Bongers, F.; Aide, T.M.; Alvarez-Dávila, E.; Ascarrunz, N.; Balvanera, P.; Becknell, J.M.; Bentos, T.V.; Brancalion, P.H.S.; Cabral, G.A.L.; et al. Biodiversity recovery of Neotropical secondary forests. *Sci. Adv.* **2019**, *5*, eaau3114. [CrossRef]
37. Sodhi, N.S.; Koh, L.P.; Peh, K.S.-H.; Tan, H.T.W.; Chazdon, R.L.; Corlett, R.T.; Lee, T.M.; Colwell, R.K.; Brook, B.W.; Sekercioglu, C.H.; et al. Correlates of extinction proneness in tropical angiosperms. *Divers. Distrib.* **2008**, *14*, 1–10. [CrossRef]
38. Carmona-Higuita, M.J.; Mendieta-Leiva, G.; Gómez-Díaz, J.A.; Villalobos, F.; Nunes Ramos, F.; Costa Elias, J.P.; Jiménez-López, D.A.; Zuluaga, A.; Holst, B.; Kessler, M.; et al. Conservation status of vascular epiphytes in the neotropics. *Biodivers. Conserv.* **2024**, *33*, 51–71. [CrossRef]
39. Barthlott, W.; Schmit-Neuerburg, V.; Nieder, J.; Engwald, S. Diversity and abundance of vascular epiphytes: A comparison of secondary vegetation and primary montane rain forest in the Venezuelan Andes. *Plant Ecol.* **2001**, *152*, 145–156. [CrossRef]
40. Krömer, T.; Gradstein, S.R. Species richness of vascular epiphytes in two primary forests and fallows in the Bolivian Andes. *Selbyana* **2003**, *24*, 190–195.

41. Köster, N.; Friedrich, K.; Nieder, J.; Barthlott, W. Conservation of epiphyte diversity in an Andean landscape transformed by human land use. *Conserv. Biol.* **2009**, *23*, 911–919. [CrossRef]
42. Hietz, P. Diversity and conservation of epiphytes in a changing environment. *Pure Appl. Chem.* **1999**, *70*, 1–11.
43. Hietz-Seifert, U.; Hietz, P.; Guevara, S. Epiphyte vegetation and diversity on remnant trees after forest clearance in southern Veracruz, Mexico. *Biol. Conserv.* **1996**, *75*, 103–111. [CrossRef]
44. Werner, F.A.; Gradstein, S.R. Diversity of dry forest epiphytes along a gradient of human disturbance in the tropical Andes. *J. Veg. Sci.* **2009**, *20*, 59–68. [CrossRef]
45. Larrea, M.L.; Werner, F.A. Response of vascular epiphyte diversity to different land-use intensities in a neotropical montane wet forest. *For. Ecol. Manag.* **2010**, *260*, 1950–1955. [CrossRef]
46. Einzmann, H.J.R.; Zotz, G. How diverse are epiphyte assemblages in plantations and secondary forests in tropical lowlands? *Trop. Conserv. Sci.* **2016**, *9*, 629–647. [CrossRef]
47. Werner, F.A. Reduced growth and survival of vascular epiphytes on isolated remnant trees in a recent tropical montane forest clear-cut. *Basic Appl. Ecol.* **2011**, *12*, 172–181. [CrossRef]
48. Benzing, D.H. Vulnerabilities of tropical forests to climate change: The significance of resident epiphytes. *Clim. Change* **1998**, *39*, 519–540. [CrossRef]
49. Zotz, G. "Hemiepiphyte": A confusing term and its history. *Ann. Bot.* **2013**, *111*, 1015–1020. [CrossRef]
50. Zotz, G.; Almeda, F.; Bautista-Bello, A.P.; Eskov, A.; Giraldo-Cañas, D.; Hammel, B.; Harrison, R.; Köster, N.; Krömer, T.; Lowry II, P.P.; et al. Hemiepiphytes revisited. *Perspect. Plant Ecol. Evol. Syst.* **2021**, *51*, 125620. [CrossRef]
51. Hoeber, V.; Zotz, G. Accidental epiphytes: Ecological insights and evolutionary implications. *Ecol. Monogr.* **2022**, *92*, e1527. [CrossRef]
52. Muenchow, J.; Dieker, P.; Kluge, J.; Kessler, M.; von Wehrden, H. A review of ecological gradient research in the Tropics: Identifying research gaps, future directions, and conservation priorities. *Biodivers. Conserv.* **2018**, *27*, 273–285. [CrossRef]
53. R Core Team. R: A Language and Environment for Statistical Computing. R Foundation for Statistical Computing. 2024. Available online: https://www.R-project.org/ (accessed on 15 October 2024).
54. Rinker, T. qdap: Bridging the Gap Between Qualitative Data and Quantitative Analysis. R Package Version 2.4.6. 2023. Available online: https://CRAN.R-project.org/package=qdap (accessed on 15 October 2024).
55. Wickham, H. stringr: Simple, Consistent Wrappers for Common String Operations. R Package Version 1.5.1. 2023. Available online: https://CRAN.R-project.org/package=stringr (accessed on 15 October 2024).
56. Mori, K. striprtf: Extract Text from RTF File. R Package Version 0.6.0. 2023. Available online: https://CRAN.R-project.org/package=striprtf (accessed on 15 October 2024).
57. Feinerer, I.; Hornik, K. tm: Text Mining Package. R Package Version 0.7-14. 2024. Available online: https://CRAN.R-project.org/package=tm (accessed on 15 October 2024).
58. Wickham, H.; Averick, M.; Bryan, J.; Chang, W.; McGowan, L.D.; François, R.; Grolemund, G.; Hayes, A.; Henry, L.; Hester, J.; et al. Welcome to the tidyverse. *J. Open Source Softw.* **2019**, *4*, 1686. [CrossRef]
59. Barrett, T.; Dowle, M.; Srinivasan, A.; Gorecki, J.; Chirico, M.; Hocking, T.; Schwendinger, B. data.table: Extension of "data.frame". R Package Version 1.16.2. 2024. Available online: https://CRAN.R-project.org/package=data.table (accessed on 15 October 2024).
60. Hill, M.O.; Gauch, H.G. Detrended correspondence analysis: An improved ordination technique. *Vegetatio* **1980**, *42*, 47–58. [CrossRef]
61. Oksanen, J.; Simpson, G.; Blanchet, F.; Kindt, R.; Legendre, P.; Minchin, P.; O'Hara, R.; Solymos, P.; Stevens, M.; Szoecs, E.; et al. vegan: Community Ecology Package. R Package Version 2.6-8. 2024. Available online: https://CRAN.R-project.org/package=vegan (accessed on 15 October 2024).
62. Charrad, M.; Ghazzali, N.; Boiteau, V.; Niknafs, A. NbClust: An R Package for Determining the Relevant Number of Clusters in a Data Set. *J. Stat. Softw.* **2014**, *61*, 1–36. [CrossRef]
63. James, G.; Witten, D.; Hastie, T.; Tibshirani, R. *An Introduction to Statistical Learning*; Springer: New York, NY, USA, 2013; p. 426.
64. Wickham, H. *g gplot2: Elegant Graphics for Data Analysis*, 2nd ed.; Springer: New York, NY, USA, 2016; p. 213. [CrossRef]
65. Kassambara, A. ggpubr: "ggplot2" Based Publication Ready Plots. R Package Version 0.6.0. 2023. Available online: https://CRAN.R-project.org/package=ggpubr (accessed on 15 October 2024).
66. Slowikowski, K. ggrepel: Automatically Position Non-Overlapping Text Labels with 'ggplot2'. 2024. Available online: https://github.com/slowkow/ggrepel (accessed on 15 October 2024).
67. Gradstein, S.R. Epiphytes of tropical montane forests-impact of deforestation and climate change. In *The Tropical Mountain Forest, Patterns and Processes in Biodiversity Hotspot*; Gradstein, S.R., Homeier, J., Gansert, D., Eds.; Göttingen Centre for Biodiversity and Ecology: Göttingen, Germany, 2008; pp. 51–65.

68. Nadkarni, N.M. Complex consequences of disturbance on canopy plant communities of world forests: A review and synthesis. *New Phytol.* **2023**, *240*, 1366–1380. [CrossRef] [PubMed]
69. Madison, M. Distribution of epiphytes in a rubber plantation in Sarawak. *Selbyana* **1979**, *5*, 207–213.
70. Godfray, H.C.J.; Beddington, J.R.; Crute, I.R.; Haddad, L.; Lawrence, D.; Muir, J.F.; Pretty, J.; Robinson, S.; Thomas, S.M.; Toulmin, C. Food security: The challenge of feeding 9 billion people. *Science* **2010**, *327*, 812–818. [CrossRef]
71. Melo, F.P.L.; Arroyo-Rodríguez, V.; Fahrig, L.; Martínez-Ramos, M.; Tabarelli, M. On the hope for biodiversity-friendly tropical landscapes. *Trends Ecol. Evol.* **2013**, *28*, 462–468. [CrossRef]
72. Foley, J.A.; DeFries, R.; Asner, G.P.; Barford, C.; Bonan, G.; Carpenter, S.R.; Chapin, F.S.; Coe, M.T.; Daily, G.C.; Gibbs, H.K.; et al. Global consequences of land use. *Science* **2005**, *309*, 570–574. [CrossRef]
73. Lindenmayer, D.B.; Franklin, J.F.; Fischer, J. General management principles and a checklist of strategies to guide forest biodiversity conservation. *Biol. Conserv.* **2006**, *131*, 433–445. [CrossRef]
74. Wright, S.J. Tropical forests in a changing environment. *Trends Ecol. Evol.* **2005**, *20*, 553–560. [CrossRef]
75. Corlett, R.T. Plant diversity in a changing world: Status, trends, and conservation needs. *Plant Divers.* **2016**, *38*, 10–16. [CrossRef]
76. Christenhusz, M.J.M.; Govaerts, R. Plant extinction in the Anthropocene. *Bot. J. Linn. Soc.* **2024**, *207*, 183–196. [CrossRef]
77. Jim, C.Y. Ecology and conservation of strangler figs in urban wall habitats. *Urban Ecosyst.* **2014**, *17*, 405–426. [CrossRef]
78. Ruas, R.d.B.; de Godoy, S.M.; Feliciano, D.C.; Ruas, C.d.F.; Bered, F. A bromeliad living in the city: A case of a native species resilient to urbanization in South Brazil. *Bot. J. Linn. Soc.* **2024**, *205*, 161–176. [CrossRef]
79. Zotz, G.; Cascante-Marín, A. Life on the wire—Plant growth on power lines in the Americas. *Diversity* **2024**, *16*, 573. [CrossRef]
80. Janzen, D.H. Latent Extinction—The Living Dead. In *Encyclopedia of Biodiversity*, 2nd ed.; Levin, S.A., Ed.; Academic Press: Cambridge, MA, USA, 2013; pp. 590–598.
81. Krömer, T.; Viccon-Esquivel, J.; Gómez-Díaz, J.A. Efectos antrópicos sobre la diversidad de epífitas vasculares y orquídeas en el centro de Veracruz. In *Las Orquídeas de Veracruz*; Viccon-Esquivel, J., Castañeda Zárate, M., Castro Cortés, R., Cetzal Ix, W., Eds.; Editorial de La Universidad Veracruzana: Xalapa, Mexico, 2021; pp. 235–252.
82. Brown, S.; Lugo, A.E. Tropical secondary forests. *J. Trop. Ecol.* **1990**, *6*, 1–32. [CrossRef]
83. Chazdon, R.L. Tropical forest recovery: Legacies of human impact and natural disturbances. *Perspect. Plant Ecol. Evol. Syst.* **2003**, *6*, 51–71. [CrossRef]
84. Poorter, L.; Craven, D.; Jakovac, C.C.; Van Der Sande, M.T.; Amissah, L.; Bongers, F.; Chazdon, R.L.; Farrior, C.E.; Kambach, S.; Meave, J.A.; et al. Multidimensional tropical forest recovery. *Science* **2021**, *374*, 1370–1376. [CrossRef]
85. Guariguata, M.R.; Ostertag, R. Neotropical secondary forest succession: Changes in structural and functional characteristics. *For. Ecol. Manag.* **2001**, *148*, 185–206. [CrossRef]
86. Chazdon, R.L.; Letcher, S.G.; van Breugel, M.; Martínez-Ramos, M.; Bongers, F.; Finegan, B. Rates of change in tree communities of secondary Neotropical forests following major disturbances. *Philosphical Trans. R. Soc. Lond. B Biol. Sci.* **2007**, *362*, 273–289. [CrossRef]
87. Woods, C.L.; DeWalt, S.J. The conservation value of secondary forests for vascular epiphytes in Central Panama. *Biotropica* **2013**, *45*, 119–127. [CrossRef]
88. DeWalt, S.J.; Maliakal, S.K.; Denslow, J.S. Changes in vegetation structure and composition along a tropical forest chronosequence: Implications for wildlife. *For. Ecol. Manag.* **2003**, *182*, 139–151. [CrossRef]
89. Carvajal-Hernández, C.I.; Krömer, T.; López-Acosta, J.C.; Gómez-Díaz, J.A.; Kessler, M. Conservation value of disturbed and secondary forests for ferns and lycophytes along an elevational gradient in Mexico. *Appl. Veg. Sci.* **2017**, *20*, 662–672. [CrossRef]
90. Van Vuuren, D.P.; Sala, O.E.; Pereira, H.M. The future of vascular plant diversity under four global scenarios. *Ecol. Soc.* **2006**, *11*, 25. [CrossRef]
91. Benavides, A.-M.; Wolf, J.H.D.; Duivenvoorden, J.F. Recovery and succession of epiphytes in upper Amazonian fallows. *J. Trop. Ecol.* **2006**, *22*, 705–717. [CrossRef]
92. Pérez-Peña, A.; Krömer, T. ¿Qué pueden aportar los acahuales y las plantaciones de cítricos a la conservación de las epífitas vasculares en los Tuxtlas, Veracruz? In *Avances y Perspectivas en la Investigación de los Bosques Tropicales y Sus Alrededores: La Región de Los Tuxtlas*; Reynoso, V.H., Coates, R.I., Vázquez-Cruz, M.d.L., Eds.; Instituto de Biología, Universidad Nacional Autónoma de México: Ciudad de México, Mexico, 2017; pp. 569–580.
93. Ceballos, S.J. Vascular epiphyte communities in secondary and mature forests of a subtropical montane area. *Acta Oecol.* **2020**, *105*, 103571. [CrossRef]
94. Cuevas Reyes, P.; Vega Gutiérrez, J.I. Cambios en la estructura, composición y fenología de plantas epífitas bajo diferentes estadios de sucesión vegetal en un bosque tropical seco. *Biol. Rev. Cienc. Biol. Agropecu. Univ. Michoacana San Nicolás Hidalgo* **2012**, *14*, 37–44.

95. Reid, J.L.; Chaves-Fallas, J.M.; Holl, K.D.; Zahawi, R.A. Tropical forest restoration enriches vascular epiphyte recovery. *Appl. Veg. Sci.* **2016**, *19*, 508–517. [CrossRef]
96. Bacles, C.F.; Lowe, A.J.; Ennos, R.A. Effective seed sispersal across a fragmented landscape. *Science* **2006**, *311*, 628. [CrossRef]
97. Thomson, F.J.; Moles, A.T.; Auld, T.D.; Kingsford, R.T. Seed dispersal distance is more strongly correlated with plant height than with seed mass. *J. Ecol.* **2011**, *99*, 1299–1307. [CrossRef]
98. Einzmann, H.J.R.; Zotz, G. Dispersal and establishment of vascular epiphytes in human-modified landscapes. *AoB Plants* **2017**, *9*, plx052. [CrossRef]
99. Cascante-Marín, A.; von Meijenfeldt, N.; de Leeuw, H.M.H.; Wolf, J.H.D.; Oostermeijer, J.G.B.; den Nijs, J.C.M. Dispersal limitation in epiphytic bromeliad communities in a Costa Rican fragmented montane landscape. *J. Trop. Ecol.* **2009**, *25*, 63–73. [CrossRef]
100. Werth, S.; Wagner, H.H.; Gugerli, F.; Holderegger, R.; Csencsics, D.; Kalwij, J.M.; Scheidegger, C. Quantifying dispersal and establishment limitation in a population of an epiphytic lichen. *Ecology* **2006**, *87*, 2037–2046. [CrossRef] [PubMed]
101. Burns, K.C.; Zotz, G. A hierarchical framework for investigating epiphyte assemblages: Networks, meta-communities, and scale. *Ecology* **2010**, *91*, 377–385. [CrossRef] [PubMed]
102. Krömer, T.; García-Franco, J.G.; Toledo-Aceves, T. Epífitas vasculares como bioindicadores de la calidad forestal: Impacto antrópico sobre su diversidad y composición. In *Bioindicadores: Guardianes de Nuestro Futuro Ambiental*; González-Zuarth, C.A., Vallarino, A., Pérez-Jimenez, J.C., Low-Pfeng, A.M., Eds.; Instituto Nacional de Ecología y Cambio Climático (INECC)—El Colegio de la Frontera Sur (ECOSUR): Ciudad de México, México, 2014; pp. 605–623.
103. Costa Elias, J.P.; Borges e Silva, B.A.; Gonçalves de Carvalho, R.; Sampaio, M.B.; Mendieta-Leiva, G.; Nunes Ramos, F. Tree structure instead of microclimatic zones determines differences in vascular epiphyte assemblages between forest and pasture. *For. Ecol. Manag.* **2024**, *552*, 121567. [CrossRef]
104. Zotz, G.; Vollrath, B. The epiphyte vegetation of the palm *Socratea exorrhiza*—Correlations with tree size, tree age and bryophyte cover. *J. Trop. Ecol.* **2003**, *19*, 81–90. [CrossRef]
105. Nadkarni, N.M. Colonization of stripped branch surfaces by epiphytes in a Lower Montane Cloud Forest, Monteverde, Costa Rica. *Biotropica* **2000**, *32*, 358–363. [CrossRef]
106. Hietz, P.; Ausserer, J.; Schindler, G. Growth, maturation and survival of epiphytic bromeliads in a Mexican humid montane forest. *J. Trop. Ecol.* **2002**, *18*, 177–191. [CrossRef]
107. Schmidt, G.; Zotz, G. Inherently slow growth in two Caribbean epiphytic species: A demographic approach. *J. Veg. Sci.* **2002**, *13*, 527–534. [CrossRef]
108. Werner, F.A.; Köster, N.; Kessler, M.; Gradstein, S.R. Is the resilience of epiphyte assemblages to human disturbance a function of local climate. *Ecotropica* **2011**, *17*, 15–20. [CrossRef]
109. Guzmán-Jacob, V.; Zotz, G.; Craven, D.; Taylor, A.; Krömer, T.; Monge-González, M.L.; Kreft, H.; Pfeifer, M. Effects of forest-use intensity on vascular epiphyte diversity along an elevational gradient. *Divers. Distrib.* **2020**, *26*, 4–15. [CrossRef]
110. Graham, E.A.; Andrade, J.L. Drought tolerance associated with vertical stratification of two co-occurring epiphytic bromeliads in a tropical dry forest. *Am. J. Bot.* **2004**, *91*, 699–706. [CrossRef]
111. Carvajal-Hernández, C.I.; Krömer, T.; Vázquez-Torres, M. Riqueza y composición florística de pteridobiontes en bosque mesófilo de montaña y ambientes asociados en el centro de Veracruz, México. *Rev. Mex. Biodivers.* **2014**, *85*, 491–501. [CrossRef]
112. McCormick, M.K.; Jacquemyn, H. What constrains the distribution of orchid populations? *New Phytol.* **2014**, *202*, 392–400. [CrossRef]
113. Hietz, P.; Briones, O. Correlation between water relations and within-canopy distribution of epiphytic ferns in a Mexican cloud forest. *Oecologia* **1998**, *114*, 305–316. [CrossRef]
114. Zotz, G.; Andrade, J.L. La ecología y fisiología de las epífitas y hemiepífitas. In *Ecología y Conservación de Bosques Neotropicales*; Guariguata, M.R., Kattan, G.H., Eds.; Libro Universitario Regional del Instituto Tecnológico de Costa Rica, San José: San José, Costa Rica, 2002; pp. 271–296.
115. Zotz, G.; Hietz, P.; Einzmann, H.J.R. Functional ecology of vascular epiphytes. *Annu. Plant Rev. Online* **2021**, *4*, 869–906. [CrossRef]
116. Dunn, R.R. Bromeliad communities in isolated trees and three successional stages of an Andean cloud forest in Ecuador. *Selbyana* **2000**, *21*, 137–143.
117. Cascante-Marín, A.; Wolf, J.H.D.; Oostermeijer, J.G.B.; den Nijs, J.C.M.; Sanahuja, O.; Durán-Apuy, A. Epiphytic bromeliad communities in secondary and mature forest in a tropical premontane area. *Basic Appl. Ecol.* **2006**, *7*, 520–532. [CrossRef]
118. Alanís-Méndez, J.L.; Muñoz-Arteaga, F.O.; López-Ortega, M.; Cuervo-López, L.; Raya-Cruz, B.E. Aportes al conocimiento de las epífitas (Bromeliaceae, Cactaceae y Orchidaceae) en dos tipos de vegetación del Municipio de Pánuco, Veracruz, México. *Rev. Cient. UDO Agríc.* **2007**, *7*, 160–174.
119. Flores-Argüelles, A.; Espejo-Serna, A.; López-Ferrari, A.R.; Krömer, T. Diversity and vertical distribution of epiphytic angiosperms, in natural and disturbed forest on the Northern Coast of Jalisco, Mexico. *Front. For. Glob. Change* **2022**, *5*, 828851. [CrossRef]

120. Griffiths, H.; Smith, J.A.C. Photosynthetic pathways in the Bromeliaceae of Trinidad: Relations between life-forms, habitat preference and the occurrence of CAM. *Oecologia* **1983**, *60*, 176–184. [CrossRef]
121. Winkler, M.; Huelber, K.; Hietz, P. Effect of canopy position on germination and seedling survival of epiphytic bromeliads in a Mexican humid montane forest. *Ann. Bot.* **2005**, *95*, 1039–1047. [CrossRef]
122. Winkler, M.; Hülber, K.; Hietz, P. Population dynamics of epiphytic bromeliads: Life strategies and the role of host branches. *Basic Appl. Ecol.* **2007**, *8*, 183–196. [CrossRef]
123. Köster, N.; Kreft, H.; Nieder, J.; Barthlott, W.; Jetz, W. Range size and climatic niche correlate with the vulnerability of epiphytes to human land use in the tropics. *J. Biogeogr.* **2013**, *40*, 963–976. [CrossRef]
124. Flores-Palacios, A.; García-Franco, J.G. Effect of isolation on the structure and nutrient content of oak epiphyte communities. *Plant Ecol.* **2004**, *173*, 259–269. [CrossRef]
125. Hietz, P.; Buchberger, G.; Winkler, M. Effect of forest disturbance on abundance and distribution of epiphytic bromeliads and orchids. *Ecotropica* **2006**, *12*, 103–112.
126. Curtis, P.G.; Slay, C.M.; Harris, N.L.; Tyukavina, A.; Hansen, M.C. Classifying drivers of global forest loss. *Science* **2018**, *361*, 1108–1111. [CrossRef] [PubMed]
127. Pendrill, F.; Persson, U.M.; Godar, J.; Kastner, T. Deforestation displaced: Trade in forest-risk commodities and the prospects for a global forest transition. *Environ. Res. Lett.* **2019**, *14*, 055003. [CrossRef]
128. Stephens, W.; Hamilton, A.P.; Carr, M.K.V. Plantation crops. In *Agriculture in the Tropics*, 3rd ed.; Webster, C.C., Williams, C.N., Eds.; Blackwell: Oxford, UK, 1998; pp. 200–221.
129. Hartemink, A.E. Plantation agriculture in the tropics: Environmental issues. *Outlook Agric.* **2005**, *34*, 11–21. [CrossRef]
130. Fagan, M.E.; Kim, D.-H.; Settle, W.; Ferry, L.; Drew, J.; Carlson, H.; Slaughter, J.; Schaferbien, J.; Tyukavina, A.; Harris, N.L.; et al. The expansion of tree plantations across tropical biomes. *Nat. Sustain.* **2022**, *5*, 681–688. [CrossRef]
131. Evans, J. *Plantation Forestry in the Tropics*, 2nd ed.; Cambridge University Press: Cambridge, UK, 1992; p. 403.
132. Onyekwelu, J.C.; Stimm, B.; Evans, J. Review Plantation Forestry. In *Silviculture in the Tropics (Tropical Forestry)*; Günter, S., Weber, M., Stimm, B., Mosandl, R., Eds.; Springer: Berlin, Germany, 2011; Volume 8, pp. 399–454.
133. Boelter, C.R.; Zartman, C.E.; Fonseca, C.R. Exotic tree monocultures play a limited role in the conservation of Atlantic Forest epiphytes. *Biodivers. Conserv.* **2011**, *20*, 1255–1272. [CrossRef]
134. Hietz, P.; Hietz-Seifert, U. Composition and ecology of vascular epiphyte communities along an altitudinal gradient in central Veracruz, Mexico. *J. Veg. Sci.* **1995**, *6*, 487–498. [CrossRef]
135. Wolf, J.H.D. The response of epiphytes to anthropogenic disturbance of pine-oak forests in the highlands of Chiapas, Mexico. *For. Ecol. Manag.* **2005**, *212*, 376–393. [CrossRef]
136. Zotz, G. Vascular epiphytes in the temperate zones—A review. *Plant Ecol.* **2005**, *176*, 173–183. [CrossRef]
137. Callaway, R.M.; Reinhart, K.O.; Moore, G.W.; Moore, D.J.; Pennings, S.C. Epiphyte host preferences and host traits: Mechanisms for species-specific interactions. *Oecologia* **2002**, *132*, 221–230. [CrossRef]
138. Böhnert, T.; Wenzel, A.; Altenhövel, C.; Beeretz, L.; Tjitrosoedirdjo, S.S.; Meijide, A.; Rembold, K.; Kreft, H. Effects of land-use change on vascular epiphyte diversity in Sumatra (Indonesia). *Biol. Conserv.* **2016**, *202*, 20–29. [CrossRef]
139. Carrasco, L.R.; Larrosa, C.; Milner-Gulland, E.J.; Edwards, D.P. A double-edged sword for tropical forests. *Science* **2014**, *346*, 38–40. [CrossRef]
140. Furumo, P.R.; Aide, T.M. Characterizing commercial oil palm expansion in Latin America: Land use change and trade. *Environ. Res. Lett.* **2017**, *12*, 024008. [CrossRef]
141. Jayathilake, H.M.; Jamaludin, J.; De Alban, J.D.T.; Webb, E.L.; Carrasco, L.R. The conversion of rubber to oil palm and other landcover types in Southeast Asia. *Appl. Geogr.* **2023**, *150*, 102838. [CrossRef]
142. Piggott, A.G. The fern flora of oil palm plantations in West Malaysia. *Fern Gaz.* **1980**, *12*, 93–102.
143. Gill, L.S.; Onyibe, H.I. Phytosociological studies of epiphytic flora of oil palm (*Elaeis guineensis* Jacq.) in Benin City, Nigeria. *Feddes Repert.* **1986**, *97*, 691–695. [CrossRef]
144. Prescott, G.W.; Edwards, D.P.; Foster, W.A. Retaining biodiversity in intensive farmland: Epiphyte removal in oil palm plantations does not affect yield. *Ecol. Evol.* **2015**, *5*, 1944–1954. [CrossRef] [PubMed]
145. Turner, P.D.; Gillbanks, R.A. *Oil Palm Cultivation and Management*; The Incorporated Society of Planters: Kuala Lumpur, Malaysia, 2003; p. 672.
146. Nunes-Freitas, A.F.; de Asis Ribeiro, D.C.; Meireles, A.S.; Azevedo, D.B.; Almeida, G.H.; Oliveira, W.F.; Rocha-Pessôa, T.C. Invasive exotic oil palm trees modify the structure of vascular epiphytes community on an Atlantic Forest Island. In *Invasive Species. Ecology, Impacts, and Potential Uses*; Londe, V., Ed.; Nova Science Publishers: New York, NY, USA, 2020; pp. 157–188.
147. Wagner, W.H. Ferns on Pacific island coconut trees. *Am. Fern J.* **1945**, *35*, 74–76. [CrossRef]

148. Porembski, S.; Biedinger, N. Epiphytic ferns for sale: Influence of commercial plant collection on the frequency of *Platycerium stemaria* (Polypodiaceae) in coconut plantations on the southeastern Ivory Coast. *Plant Biol.* **2001**, *3*, 72–76. [CrossRef]
149. Schimper, A.F.W. *Die Epiphytische Vegetation Amerikas*; Fischer: Frankfurt am Main, Germany, 1888; p. 130.
150. Johansson, D. Ecology of Vascular Epiphytes in West African Rain Forest. Ph.D. Thesis, Uppsala University, Uppsala, Sweden, 1974; p. 138.
151. Nir, M.A. The survivor: Orchids on a Puerto Rican coffee finca. *Am. Orchid. Soc. Bull.* **1988**, *57*, 989–995.
152. Einzmann, H.J.R.; Döcke, L.; Zotz, G. Epiphytes in human settlements in rural Panama. *Plant Ecol. Divers.* **2016**, *9*, 277–287. [CrossRef]
153. Cook, M.T. Epiphytic orchids a serious pest on citrus trees. *J. Agric. Univ. Puerto Rico* **1926**, *10*, 5–9. [CrossRef]
154. García-González, A.; Riverón-Giró, F.B. Organización espacial y estructura de una población de *Lonopsis utricularioides* (Orchidaceae) en un área suburbana de Pinar del Río, Cuba. *Lankesteriana* **2013**, *13*, 419–427. [CrossRef]
155. Mondragón, D.; Santos-Moreno, A.; Damon, A. Epiphyte diversity on coffee bushes: A management question? *J. Sustain. Agric.* **2009**, *33*, 703–715. [CrossRef]
156. Moguel, P.; Toledo, V.M. Biodiversity conservation in traditional coffee systems of Mexico. *Conserv. Biol.* **1999**, *13*, 11–21. [CrossRef]
157. Hernández-Martínez, G.; Manson, R.H.; Contreras Hernández, A. Quantitative classification of coffee agroecosystems spanning a range of production intensities in central Veracruz, Mexico. *Agric. Ecosyst. Environ.* **2009**, *134*, 89–98. [CrossRef]
158. Hietz, P. Conservation of vascular epiphyte diversity in Mexican coffee plantations. *Conserv. Biol.* **2005**, *19*, 391–399. [CrossRef]
159. Espejo-Serna, A.; López-Ferrari, A.R.; Jiménez-Machorro, R.; Sánchez-Saldaña, L. Orchids from coffee-plantations in Mexico: An alternative for the sustainable use of tropical ecosystems. *Rev. Biol. Trop.* **2005**, *53*, 73–84.
160. Hylander, K.; Nemomissa, S. Home garden coffee as a repository of epiphyte biodiversity in Ethiopia. *Front. Ecol. Environ.* **2008**, *6*, 524–528. [CrossRef]
161. Hundera, K.; Aerts, R.; De Beenhouwer, M.; Van Overtveld, K.; Helsen, K.; Muys, B.; Honnay, O. Both forest fragmentation and coffee cultivation negatively affect epiphytic orchid diversity in Ethiopian moist evergreen Afromontane forests. *Biol. Conserv.* **2013**, *159*, 285–291. [CrossRef]
162. Moorhead, L.C.; Philpott, S.M.; Bichier, P. Epiphyte biodiversity in the coffee agricultural matrix: Canopy stratification and distance from forest fragments. *Conserv. Biol.* **2010**, *24*, 737–746. [CrossRef]
163. Richards, J.H.; Luna Torrez, I.M.; Waller, D.M. Tree longevity drives conservation value of shade coffee farms for vascular epiphytes. *Agric. Ecosyst. Environ.* **2020**, *301*, 107025. [CrossRef]
164. Osie, M.; Shibru, S.; Dalle, G.; Nemomissa, S. Habitat fragmentation effects on vascular epiphytes diversity in Kafa biosphere reserve and nearby coffee agroecosystem, southwestern Ethiopia. *Trop. Ecol.* **2022**, *63*, 561–571. [CrossRef]
165. Williams-Linera, G.; Sosa, V.; Platas, T. The fate of epiphytic orchids after fragmentation of a Mexican cloud forest. *Selbyana* **1995**, *16*, 36–40.
166. Solís-Montero, L.; Flores-Palacios, A.; Cruz-Angón, A. Shade-coffee plantations as refuges for tropical wild orchids in central Veracruz, Mexico. *Conserv. Biol.* **2005**, *19*, 908–916. [CrossRef]
167. García-Franco, J.G.; Toledo, T. Vascular epiphytes: Bromeliads and orchids. In *Agroecosistemas Cafetaleros de Veracruz: Biodiversidad, Manejo y Conservación*; Manson, R.H., Hernández-Ortiz, V., Gallina, S., Mehltreter, K., Eds.; Instituto de Ecología A.C. (INECOL) e Instituto Nacional de Ecología (INE-SEMARNAT): Xalapa, México, 2008; pp. 69–82.
168. Toledo-Aceves, T.; García-Franco, J.G.; Hernández-Rojas, A.; MacMillan, K. Recolonization of vascular epiphytes in a shaded coffee agroecosystem. *Appl. Veg. Sci.* **2012**, *15*, 99–107. [CrossRef]
169. Toledo-Aceves, T.; Mehltreter, K.; Garcia-Franco, J.G.; Hernandez-Rojas, A.; Sosa, V.J. Benefits and costs of epiphyte management in shade coffee plantations. *Agric. Ecosyst. Environ.* **2013**, *181*, 149–156. [CrossRef]
170. Solís-Montero, L.; Quintana-Palacios, V.; Damon, A. Impact of moss and epiphyte removal on coffee production and implications for epiphyte conservation in shade coffee plantations in southeast Mexico. *Agroecol. Sustain. Food Syst.* **2019**, *43*, 1124–1144. [CrossRef]
171. Cruz-Angón, A.; Greenberg, R. Are epiphytes important for birds in coffee plantations? An experimental assessment. *J. Appl. Ecol.* **2005**, *42*, 150–159. [CrossRef]
172. Cruz-Angón, A.; Baena, M.L.; Greenberg, R. The contribution of epiphytes to the abundance and species richness of canopy insects in a Mexican coffee plantation. *J. Trop. Ecol.* **2009**, *25*, 453–463. [CrossRef]
173. De Beenhouwer, M.; Aerts, R.; Hundera, K.; Van Overtveld, K.; Honnay, O. Management intensification in Ethiopian coffee forests is associated with crown habitat contraction and loss of specialized epiphytic orchid species. *Basic Appl. Ecol.* **2015**, *16*, 592–600. [CrossRef]

174. Morales-Linares, J.; Garcia-Franco, J.G.; Flores-Palacios, A.; Krömer, T.; Toledo-Aceves, T. The role of shaded cocoa plantations in the maintenance of epiphytic orchids and their interactions with phorophytes. *J. Plant Ecol.* **2020**, *13*, 27–35. [CrossRef]
175. Andersson, M.S.; Gradstein, S.R. Impact of management intensity on non-vascular epiphyte diversity in cacao plantations in western Ecuador. *Biodivers. Conserv.* **2005**, *14*, 1101–1120. [CrossRef]
176. Haro-Carrión, X.; Lozada, T.; Navarrete, H.; de Koning, G.H.J. Conservation of vascular epiphyte diversity in shade cacao plantations in the Chocó region of Ecuador. *Biotropica* **2009**, *41*, 520–529. [CrossRef]
177. Morales-Linares, J.; García-Franco, J.G.; Flores-Palacios, A.; Valenzuela-González, J.E.; Mata-Rosas, M.; Díaz-Castelazo, C. Vascular epiphytes and host trees of ant-gardens in an anthropic landscape in southeastern Mexico. *Sci. Nat.* **2016**, *103*, 96. [CrossRef]
178. Qi, D.-H.; Guo, H.-J.; Sheng, C.-Y. Assessment of plant species diversity of ancient tea garden communities in Yunnan, Southwest of China. *Agrofor. Syst.* **2013**, *87*, 465–474. [CrossRef]
179. Qi, D.-H.; Guo, H.-J.; Cui, J.-Y.; Sheng, C.-I. Plant biodiversity assessment of the ancient tea garden ecosystem in Jingmai of Lancang, Yunnan. *Biodivers. Sci.* **2005**, *13*, 221–231. [CrossRef]
180. Wang, Q.; Guan, W.-B.; Gi Wong, M.H.; Ranjitkar, S.; Sun, W.-N.; Pan, Y.; El-Kassaby, Y.A.; Shen, L.-X. Tree size predicts vascular epiphytic richness of traditional cultivated tea plantations in Southwestern China. *Glob. Ecol. Conserv.* **2017**, *10*, 147–153. [CrossRef]
181. Wu, S.-M.; Li, T.-Q.; Yang, W.-K.; Liu, Q.; Gao, J.-Y. Ancient tea gardens play important role on in situ conservation of epiphytic orchids in southwest Yunnan, China. *Glob. Ecol. Conserv.* **2024**, *49*, e02778. [CrossRef]
182. Chowdhury, R.; Chowdhury, M. Diversity of vascular epiphytes on preferred shade trees in tea gardens of sub-Himalayan tracts in West Bengal, India. *J. Threat. Taxa* **2024**, *16*, 25720–25729. [CrossRef]
183. Guevara, S.; Meave, J.; Moreno-Casasola, P.; Laborde, J. Floristic composition and structure of vegetation under isolated trees in neotropical pastures. *J. Veg. Sci.* **1992**, *3*, 655–664. [CrossRef]
184. Flores-Palacios, A.; García-Franco, J.G. The relationship between tree size and epiphyte species richness: Testing four different hypotheses. *J. Biogeogr.* **2006**, *33*, 323–330. [CrossRef]
185. Parra-Sanchez, E.; Banks-Leite, C. Value of human-modified forests for the conservation of canopy epiphytes. *Biotropica* **2022**, *54*, 958–968. [CrossRef]
186. Werner, F.A.; Homeier, J.; Gradstein, S.R. Diversity of vascular epiphytes on isolated remnant trees in the montane forest belt of southern Ecuador. *Ecotropica* **2005**, *11*, 21–40.
187. Werner, F.A.; Gradstein, S.R. Seedling establishment of vascular epiphytes on isolated and enclosed forest trees in an Andean landscape, Ecuador. *Biodivers. Conserv.* **2008**, *17*, 3195–3207. [CrossRef]
188. Trejo-Cruz, I.A.; Martínez-Camilo, R.; Martínez-Meléndez, N.; Jiménez-López, D.A. Diversidad de epífitas vasculares en árboles remanentes del género *Ficus* (Moraceae) en sistemas silvopastoriles del sureste de México. *Acta Bot. Mex.* **2021**, *128*, e1827. [CrossRef]
189. Poltz, K.; Zotz, G. Vascular epiphytes on isolated pasture trees along a rainfall gradient in the lowlands of Panama. *Biotropica* **2011**, *43*, 165–172. [CrossRef]
190. Nadkarni, N.M.; Haber, W.A. Canopy seed banks as time capsules of biodiversity in pasture-remnant tree crowns. *Conserv. Biol.* **2009**, *23*, 1117–1126. [CrossRef]
191. Werner, F.A.; Gradstein, S.R. Spatial distribution and abundance of epiphytes along a gradient of human disturbance in an Interandean dry valley, Ecuador. *Selbyana* **2010**, *30*, 208–215.
192. Flores-Palacios, A.; García-Franco, J.G. Habitat isolation changes the beta diversity of the vascular epiphyte community in lower montane forest, Veracruz, Mexico. *Biodivers. Conserv.* **2008**, *17*, 191–207. [CrossRef]
193. Amici, A.A.; Nadkarni, N.M.; Williams, C.B.; Gotsch, S.G. Differences in epiphyte biomass and community composition along landscape and within-crown spatial scales. *Biotropica* **2020**, *52*, 46–58. [CrossRef]
194. Einzmann, H.J.R.; Zotz, G. "No signs of saturation": Long-term dynamics of vascular epiphyte communities in a human-modified landscape. *Biodivers. Conserv.* **2017**, *26*, 1393–1410. [CrossRef]
195. Elias, J.P.C.; Mortara, S.R.; Nunes-Freitas, A.F.; van den Berg, E.; Ramos, F.N. Host tree traits in pasture areas affect forest and pasture specialist epiphyte species differently. *Am. J. Bot.* **2021**, *108*, 598–606. [CrossRef]
196. Trapnell, D.W.; Hamrick, J.L. Partitioning nuclear and chloroplast variation at multiple spatial scales in the neotropical epiphytic orchid, *Laelia rubescens*. *Mol. Ecol.* **2004**, *13*, 2655–2666. [CrossRef]
197. Svahnström, V.J.; Lughadha, E.N.; Forest, F.; Leão, T.C. A global study of the geographic range size of epiphytes. *bioRxiv* **2023**. [CrossRef]
198. Winkler, M.; Koch, M.; Hietz, P. High gene flow in epiphytic ferns despite habitat loss and fragmentation. *Conserv. Genet.* **2011**, *12*, 1411–1420. [CrossRef]

199. Laurance, W.F. Have we overstated the tropical biodiversity crisis? *Trends Ecol. Evol.* **2006**, *22*, 65–70. [CrossRef]
200. Murcia, C. Edge effects in fragmented forests: Implications for conservation. *Trends Ecol. Evol.* **1995**, *10*, 58–62. [CrossRef] [PubMed]
201. Fahrig, L. Effects of habitat fragmentation on biodiversity. *Annu. Rev. Ecol. Evol. Syst.* **2003**, *34*, 487–515. [CrossRef]
202. Silva, V.L.; Mehltreter, K.; Schmitt, J.L. Ferns as potential ecological indicators of edge effects in two types of Mexican forests. *Ecol. Indic.* **2018**, *93*, 669–676. [CrossRef]
203. Haddad, N.M.; Brudvig, L.A.; Clobert, J.; Davies, K.F.; Gonzalez, A.; Holt, R.D.; Lovejoy, T.E.; Sexton, J.O.; Austin, M.P.; Collins, C.D.; et al. Habitat fragmentation and its lasting impact on Earth's ecosystems. *Sci. Adv.* **2015**, *1*, e1500052. [CrossRef]
204. Laurance, W.F. Edge effects in tropical forest fragments: Applications of a model for the design of nature reserves. *Biol. Conserv.* **1991**, *57*, 205–219. [CrossRef]
205. Willmer, J.N.G.; Puettker, T.; Prevedello, J.A. Global impacts of edge effects on species richness. *Biol. Conserv.* **2022**, *272*, 109654. [CrossRef]
206. Bataghin, F.A.; Rodrigues Pires, J.S.; de Barros, F. Epifitismo vascular em sítios de borda e interior em Floresta Estacional Semidecidual no Sudeste do Brasil. *Hoehnea* **2012**, *39*, 235–245. [CrossRef]
207. Bianchi, J.S.; Kersten, R.d.A. Edge effect on vascular epiphytes in a subtropical Atlantic Forest. *Acta Bot. Bras.* **2014**, *28*, 120–126. [CrossRef]
208. Parra-Sanchez, E.; Banks-Leite, C. The magnitude and extent of edge effects on vascular epiphytes across the Brazilian Atlantic Forest. *Sci. Rep.* **2020**, *10*, 18847. [CrossRef]
209. Lippert, A.P.U.; Silva, V.L.; Mallmann, I.T.; Müller, A.; Droste, A.; Schmitt, J.L. Edge effect on vascular epiphytes in a subtropical Atlantic Forest fragment. *J. Environ. Anal. Prog.* **2022**, *7*, 135–149. [CrossRef]
210. Ghazoul, J.; Burivalova, Z.; Garcia-Ulloa, J.; King, L.A. Conceptualizing forest degradation. *Trends Ecol. Evol.* **2015**, *30*, 622–632. [CrossRef] [PubMed]
211. Sasaki, N.; Putz, F.E. Critical need for new definitions of "forest" and "forest degradation" in global climate change agreements. *Conserv. Lett.* **2009**, *2*, 226–232. [CrossRef]
212. Vásquez-Grandón, A.; Donoso, P.J.; Gerding, V. Forest degradation: When is a forest degraded? *Forest* **2018**, *9*, 726. [CrossRef]
213. Stanturf, J.A.; Palik, B.J.; Williams, M.I.; Dumroese, R.K.; Madsen, P. Forest restoration paradigms. *J. Sustain. For.* **2014**, *33*, S161–S194. [CrossRef]
214. Vancutsem, C.; Achard, F.; Pekel, J.-F.; Vieilledent, G.; Carboni, S.; Simonetti, D.; Gallego, J.; Aragão, L.E.O.C.; Nasi, R. Long-term (1990–2019) monitoring of forest cover changes in the humid tropics. *Sci. Adv.* **2021**, *7*, eabe1603. [CrossRef]
215. Hosonuma, N.; Herold, M.; De Sy, V.; De Fries, R.S.; Brockhaus, M.; Verchot, L.; Angelsen, A.; Romijn, E. An assessment of deforestation and forest degradation drivers in developing countries. *Environ. Res. Lett.* **2012**, *7*, 44009. [CrossRef]
216. Ticktin, T. The ecological implications of harvesting non-timber forest products. *J. Appl. Ecol.* **2004**, *41*, 11–21. [CrossRef]
217. Torres-Rojo, J.M.; Moreno-Sánchez, R.; Mendoza-Briseño, M.A. Sustainable forest management in Mexico. *Curr. For. Rep.* **2016**, *2*, 93–105. [CrossRef]
218. Martínez-Meléndez, N.; Ramírez-Marcial, N.; García-Franco, J.G.; Cach-Pérez, M.J.; Martínez-Zurimendi, P. Importance of *Quercus* spp. for diversity and biomass of vascular epiphytes in a managed pine-oak forest in Southern Mexico. *For. Ecosyst.* **2022**, *9*, 100034. [CrossRef]
219. Rutten, G.; Ensslin, A.; Hemp, A.; Fischer, M. Forest structure and composition of previously selectively logged and non-logged montane forests at Mt. Kilimanjaro. *For. Ecol. Manag.* **2015**, *337*, 61–66. [CrossRef]
220. Shima, K.; Yamada, T.; Okuda, T.; Fletcher, C.; Kassim, A.R. Dynamics of tree species diversity in unlogged and selectively logged Malaysian forests. *Sci. Rep.* **2018**, *8*, 1024. [CrossRef]
221. Flores-Palacios, A.; Valencia-Díaz, S. Local illegal trade reveals unknown diversity and involves a high species richness of wild vascular epiphytes. *Biol. Conserv.* **2007**, *136*, 372–387. [CrossRef]
222. Cruz-Garcia, G.; Lagunez-Rivera, L.; Chavez-Angeles, M.G.; Solano-Gomez, R. The wild orchid trade in a Mexican local market: Diversity and economics. *Econ. Bot.* **2015**, *69*, 291–305. [CrossRef]
223. Mondragón, D.; del Carmen Méndez-García, E.M.; Morillo, I.R. Prioritizing the conservation of epiphytic bromeliads using ethnobotanical information from a traditional Mexican market. *Econ. Bot.* **2016**, *70*, 29–36. [CrossRef]
224. Krömer, T.; Acebey, A.; Toledo-Aceves, T. Aprovechamiento de plantas epífitas: Implicaciones para su conservación y manejo sustentable. In *De la Recolección a los Agroecosistemas: Soberanía Alimentaria y Conservación de la Biodiversidad*; Silva-Rivera, E., Martínez-Valdéz, V., Lascuráin, M., Rodríguez-Luna, E., Eds.; Editorial de la Universidad Veracruzana: Xalapa, Veracruz, 2018; pp. 175–196.
225. Jiménez-López, D.A.; Pérez-García, E.A.; Martínez-Meléndez, N.; Solano, R. Orquídeas silvestres comercializadas en un mercado tradicional de Chiapas, México. *Bot. Sci.* **2019**, *97*, 691–700. [CrossRef]

226. Elliott, D.D.; Ticktin, T. Epiphytic plants as NTFPs from the forest canopies: Priorities for management and conservation. In *Treetops at Risk*; Springer: Berlin, Germany, 2013; pp. 435–444.
227. Chinsamy, M.; Finnie, J.F.; Van Staden, J. The ethnobotany of South African medicinal orchids. *S. Afr. J. Bot.* **2011**, *77*, 2–9. [CrossRef]
228. Hossain, M.M. Therapeutic orchids: Traditional uses and recent advances—An overview. *Fitoterapia* **2011**, *82*, 102–140. [CrossRef]
229. Subedi, A.; Kunwar, B.; Choi, Y.; Dai, Y.; van Andel, T.; Chaudhary, R.P.; de Boer, H.J.; Gravendee, B. Collection and trade of wild-harvested orchids in Nepal. *J. Ethnobiol. Ethnomed.* **2013**, *9*, 64. [CrossRef]
230. Liu, H.; Luo, Y.-B.; Heinen, J.; Bhat, M.; Liu, Z.-J. Eat your orchid and have it too: A potentially new conservation formula for Chinese epiphytic medicinal orchids. *Biodivers. Conserv.* **2014**, *23*, 1215–1228. [CrossRef]
231. Hinsley, A.; de Boer, H.J.; Fay, M.F.; Gale, S.W.; Gardiner, L.M.; Gunasekara, R.S.; Kumar, P.; Masters, S.; Metusala, D.; Roberts, D.L.; et al. A review of the trade in orchids and its implications for conservation. *Bot. J. Linn. Soc.* **2018**, *186*, 435–455. [CrossRef]
232. Mondragón, D. Population viability analysis for *Guarianthe aurantiaca*, an ornamental epiphytic orchid harvested in Southeast Mexico. *Plant Species Biol.* **2009**, *24*, 35–41. [CrossRef]
233. Mondragón, D.; Ticktin, T. Demographic effects of harvesting epiphytic bromeliads and an alternative approach to collection. *Conserv. Biol.* **2011**, *25*, 797–807. [CrossRef]
234. Emeterio-Lara, A.; García-Franco, J.G.; Hernández-Apolinar, M.; Toledo-Hernández, V.H.; Valencia-Díaz, S.; Flores-Palacios, A. Does extraction of orchids affect their population structure? Evidence from populations of *Laelia autumnalis* (Orchidaceae). *For. Ecol. Manag.* **2021**, *480*, 118667. [CrossRef]
235. Orozco-Ibarrola, O.; Solano, R.; Valverde, T. Sustainable harvesting and conservation of *Laelia furfuracea*, a rare epiphytic orchid from Oaxaca, Mexico. *Biotropica* **2021**, *53*, 142–151. [CrossRef]
236. Ticktin, T.; Charitonidou, M.; Douglas, J.; Halley, J.M.; Hernández-Apolinar, M.; Liu, H.; Mondragón, D.; Pérez-García, E.A.; Tremblay, R.L.; Phelps, J. Wild orchids: A framework for identifying and improving sustainable harvest. *Biol. Conserv.* **2023**, *277*, 109816. [CrossRef]
237. Toledo-Aceves, T.; García-Franco, J.G.; López-Barrera, F. Bromeliad rain: An opportunity for cloud forest management. *For. Ecol. Manag.* **2014**, *329*, 129–136. [CrossRef]
238. Francisco-Ventura, E.; Menchaca-García, R.A.; Toledo-Aceves, T.; Krömer, T. Potencial de aprovechamiento de epífitas vasculares caídas en un bosque mesófilo de montaña de Los Tuxtlas, Veracruz, México. *Rev. Mex. Biodivers.* **2018**, *89*, 1263–1279. [CrossRef]
239. Jiménez-Bautista, L.; Damon, A.; Ochoa-Gaona, S.; Tapia, R.C. Impact of silvicultural methods on vascular epiphytes (ferns, bromeliads and orchids) in a temperate forest in Oaxaca, Mexico. *For. Ecol. Manag.* **2014**, *329*, 10–20. [CrossRef]
240. Seshadri, K.S.; Ganesan, R.; Devy, S.M. Persistent effects of historical selective logging on a vascular epiphyte assemblage in the forest canopy of the Western Ghats, India. *Front. For. Glob. Change* **2021**, *4*, 727422. [CrossRef]
241. Padmawathe, R.; Qureshi, Q.; Rawat, G.S. Effects of selective logging on vascular epiphyte diversity in a moist lowland forest of Eastern Himalaya, India. *Biol. Conserv.* **2004**, *119*, 81–92. [CrossRef]
242. Aguirre, A.; Guevara, R.; García, M.; López, J.C. Fate of epiphytes on phorophytes with different architectural characteristics along the perturbation gradient of *Sabal mexicana* forests in Veracruz, Mexico. *J. Veg. Sci.* **2010**, *21*, 6–15. [CrossRef]
243. Obermüller, F.A.; Silveira, M.; Salimon, C.I.; Daly, D.C. Epiphytic (including hemiepiphytes) diversity in three timber species in the southwestern Amazon, Brazil. *Biodivers. Conserv.* **2012**, *21*, 565–575. [CrossRef]
244. Duarte, M.M.; Gandolfi, S. Diversifying growth forms in tropical forest restoration: Enrichment with vascular epiphytes. *For. Ecol. Manag.* **2017**, *401*, 89–98. [CrossRef]
245. Fernandez-Barrancos, E.P.; Reid, J.L.; Aronson, J. Tank bromeliad transplants as an enrichment strategy in southern Costa Rica. *Restor. Ecol.* **2017**, *25*, 569–576. [CrossRef]
246. Izuddin, M.; Yam, T.W.; Webb, E.L. Specific niche requirements drive long-term survival and growth of translocated epiphytic orchids in an urbanised tropical landscape. *Urban Ecosyst.* **2018**, *21*, 531–540. [CrossRef]
247. Benavides, A.-M.; Calderón-Caro, J.; Canal, D. Surviving in a new host: Eight years of monitoring translocated aroids, bromeliads, and orchids in the Andean forests in Colombia. *For. Glob. Change* **2023**, *6*, 83466. [CrossRef]
248. Orozco Ávila, J.; Valencia Marín, A.; Betancur Pérez, J.F. Estimation of the transfer of vascular epiphytes, as a conservation strategy in the municipality of Aguazul, Casanare, Colombia. *Rev. Investig. Agrar. Ambiental* **2017**, *8*, 27–37. [CrossRef]
249. Piana, M.R.; Aronson, M.F.J.; Pickett, S.T.A.; Handel, S.N. Plants in the city: Understanding recruitment dynamics in urban landscapes. *Front. Ecol. Environ.* **2019**, *17*, 455–463. [CrossRef]
250. Bryan, C.L. Ecology of Vascular Epiphytes in Urban Forests with Special Reference to the Shrub Epiphyte *Griselinia lucida*. Master's Thesis, The University of Waikato, Hamilton, New Zealand, 2011.
251. Chang, C.-R.; Chen, M.-C.; Su, M.-H. Natural versus human drivers of plant diversity in urban parks and the anthropogenic species-area hypotheses. *Landsc. Urban Plan.* **2021**, *208*, 104023. [CrossRef]

252. Furtado Gomes, S.; Menini Neto, L. Diversity of vascular epiphytes in urban environment: A case study in a biodiversity hotspot, the Brazilian Atlantic forest. *CES Rev.* **2015**, *29*, 82–101.
253. Baltazar-Bernal, O.; Zavala-Ruiz, J.; Hernández-García, A. Orchid diversity (Orchidaceae) in two urban sites in the state of Veracruz, Mexico. *Agrociencia* **2024**, *58*, 571–583. [CrossRef]
254. Asharo, R.K.; Novitasari, A.; Azizah, S.D.N.; Saraswati, R.A.; Setyaningsih, F.; Apriliani, P.; Priambodo, R.; Pasaribu, P.O.; Rizkawati, V.; Usman. Araceae floristic and potential study in Bogor Botanical Gardens, West Java, Indonesia. *J. Biol. Res. Appl.* **2022**, *4*, 8–18. [CrossRef]
255. Yam, T.W.; Tay, F.; Ang, P.; Soh, W. Conservation and reintroduction of native orchids of Singapore—The next phase. *Eur. J. Environ. Sci.* **2011**, *1*, 143–147. [CrossRef]
256. Fabricante, J.R.; de Andrade, L.A.; Marques, F.J. Componente epifítico vascular ocorrente em árvores urbanas. *CERNE* **2006**, *12*, 399–405.
257. Adhikari, Y.P.; Fischer, A.; Fischer, H.S. Micro-site conditions of epiphytic orchids in a human impact gradient in Kathmandu valley, Nepal. *J. Mt. Sci.* **2012**, *9*, 331–342. [CrossRef]
258. Santana, L.D.; Gomes Furtado, S.; Nardy, C.; Silveira Leite, F.; Neto Menini, L. Diversity, vertical structure and floristic relationships of vascular epiphytes in an urban remnant of the Brazilian Atlantic Forest. *Hoehnea* **2017**, *44*, 123–138. [CrossRef]
259. Riefner, R.E., Jr.; Smith, A.R. New and noteworthy epiphytic ferns from the urban forests of Coastal Southern California, U.S.A. *Phytologia* **2019**, *101*, 81–112.
260. Alex, A.; Chima, U.D.; Uzoamaka, D.U. Diversity and phorophyte preference of vascular epiphytic flora on avenues within the University of Port Harcourt, Nigeria. *J. For. Environ. Sci.* **2021**, *37*, 217–225. [CrossRef]
261. Kimpouni, V.; Lenga-Sacadura, M.Y.; Kalath, R.S.; Kiangana-Ngoyi, L. Diversité floristique des épiphytes et hémiparasites vasculaires de l'écosystème forestier urbain de Brazzaville, Congo. *J. Appl. Biosci.* **2017**, *117*, 11704–11719. [CrossRef]
262. Brandes, D. Some Observations on the Urban Flora in Albania. 2023. Available online: http://www.ruderal-vegetation.de/epub/Urban_flora_in_Albania_2.pdf (accessed on 15 October 2024).
263. Spennemann, D.H.R. Growth of ornamental palms, *Phoenix* and *Washingtonia*, as epiphytes on suburban street trees in Albury, NSW, Australia. *Cunninghamia* **2019**, *19*, 113–119. Available online: https://www.botanicgardens.org.au/sites/default/files/2023-06/BGD0562_CunninghamiaSPENNEMANN-Epiphyt-palms.pdf (accessed on 8 April 2025).
264. Rogers, H.C.; Clarkson, B.D. Epiphyte-host relationships of remnant and recombinant urban ecosystems in Hamilton, New Zealand: The importance of *Dicksonia squarrosa* (G.Forst.) Sw., wheki̅. *N. Z. J. Bot.* **2023**, *63*, 150–159. [CrossRef]
265. Hall, C.F.; Gomes Klein, V.L.; de Barros, F. Orchidaceae no município de Caldas Novas, Goiás, Brasil. *Rodriguésia* **2013**, *64*, 685–704. [CrossRef]
266. Aoki-Gonçalves, F.; Pena, J.C.; Toledo-Aceves, T.; MacGregor-Fors, I. Urban epiphytes: Bromeliad diversity in a green cover gradient across a Neotropical streetscape. *Urban For. Urban Green.* **2023**, *83*, 127901. [CrossRef]
267. Quail, M.R.; Ramos, F.N.; Dallimore, T.; Ashton, P.; Clayton-Brown, J.; Provan, J.; Batke, S.P. Surrounded by concrete: Genetic isolation of *Tillandsia recurvata* L. in an urban landscape in southeastern Brazil. *Bot. J. Linn. Soc.* **2023**, *203*, 390–400. [CrossRef]
268. Morajkar, S.; Sajeev, S.; Hegde, S. Ferns: A thriving group of urban dwellers. *Bionature* **2015**, *35*, 13–21.
269. Yañez, A.; Marquez, G.J.; Berrueta, P.C.; García, R.A. An urban fern refugium: Municipal Ecological Reserve of Avellaneda (Eco Área) (Buenos Aires, Argentina). *Blumea* **2021**, *66*, 227–235. [CrossRef]
270. Alvim, F.S.; Gomes Furtado, S.; Menini Neto, L. Diversity of vascular epiphytes in urban green areas of Juiz de Fora, Minas Gerais, Brazil. *Floresta Ambiente* **2020**, *27*, e20190116. [CrossRef]
271. Mondragón, D.; Mora-Flores, M.P. First steps to study the demography of vascular epiphytes in cities. *Braz. J. Bot.* **2024**, *84*, e270998. [CrossRef]
272. Landi, M.; Angiolini, C. Population structure of *Osmunda regalis* in relation to environment and vegetation: An example in the Mediterranean area. *Folia Geobot.* **2011**, *46*, 49–68. [CrossRef]
273. Martins, P.L.S.S.; Furtado, S.G.; Menini Neto, L. Could epiphytes be xenophobic? Evaluating the use of native versus exotic phorophytes by the vascular epiphytic community in an urban environment. *Community Ecol.* **2020**, *21*, 91–101. [CrossRef]
274. Izuddin, M.; Webb, E.L. The influence of tree architecture, forest remnants, and dispersal syndrome on roadside epiphyte diversity in a highly urbanized tropical environment. *Biodivers. Conserv.* **2015**, *24*, 2063–2077. [CrossRef]
275. Cerón Martínez, C.E.; Reyes Tello, C.I. Epifitas de *Phoenix canariensis* Chabaud (Arecaceae) en cinco localidades Sudamericanas. *Cinchonia* **2021**, *16*, 197–216.
276. Ngotta, B.J.B.; Bell, D.M.; Mvogo, O.P.B.; Nguimfack, D.J.; Doumbe, M.L.C.; Wafo, T.Y.D.; Betti, J.L.; Priso, R.J. Diversity and composition of the epiphytic flora in an urban agglomeration: The case of city of Douala, Cameroon. *J. Ecol. Nat. Environ.* **2023**, *15*, 9–17. [CrossRef]

277. Moreno-Barreto, E.; Medina-Sánchez, A.M. Caracterización florística y funcional de las epífitas vasculares asociadas a palmas del género *Phoenix* L. en Bogotá, Colombia. *Biota Colomb.* **2024**, *25*, 1–14. [CrossRef]
278. Ritter, C.M.; Rocha Santos, F.; Pezenti Crespão, L.M.; Ardengui, T.C.; Galeazzi Caxambu, M. Levantamento de epífitas presentes na arborização urbana no Município de Farol—Paraná. *Rev. Soc. Bras. Arboriz. Urbana* **2015**, *9*, 18–28. [CrossRef]
279. Devens, K.U.; Geraldini, A.P.B.; Amadeo, R.M.; Caxambu, M.G.; Magnoni, P.H.J. Levantamento de epífitas na arborização urbana do município de Luiziana-Paraná. *Soc. Bras. Arboriz. Urbana* **2016**, *10*, 1–11. [CrossRef]
280. Wu, J.; Brock, J. The invasion of non-native epiphyte *Platycerium bifurcatum* in Auckland's urban forest canopy. *N. Z. J. Ecol.* **2023**, *47*, 3542. [CrossRef]
281. Neo, L.; Chong, K.Y.; Lindsay, S.; Middleton, D.J.; Tan, P.Y.; Er, K.B.H. A botanical oasis rather than a biological desert: Rediscoveries, new species and new records in a tropical city. *Plants People Planet* **2024**, *6*, 697–709. [CrossRef]
282. Praptosuwiryo, T.N.; Sumanto, S.; Cahyaningsih, R. Diversity and host preferences of ferns and lycopods epiphytes on palm trees. *Biodiversitas J. Biol. Divers.* **2019**, *20*, 3731–3740. [CrossRef]
283. dos Santos Kaeser, S.; Chiavegatto, B.; Ulguim Bordoni, P.S.; Furtado Gomes, S.; Menini Neto, L. Composição florística e ecologia de epífitas vasculares na praça central do município de Mar de Espanha, Minas Gerais, Brasil. *Rev. Bras. Arboriz. Urbana* **2020**, *15*, 26–38.
284. Romero Zapiola, A.; Devoto, M. Factores que afectan la riqueza de especies epífitas vasculares del arbolado de alineación en la Ciudad Autónoma de Buenos Aires. *Ecol. Austral* **2024**, *34*, 579–592. [CrossRef]
285. Jiménez-Orozco, C.; Lebrón-Liriano, B.V.; Fernández-Gutiérrez, R.; Urbáez, R.; Guerrero, Á. Caracterización de la flora epífita vascular del Parque Iberoamérica, Santo Domingo, República Dominicana. *Cienc. Ambiente Clima* **2019**, *2*, 23–33. [CrossRef]
286. Hietz, P.; Wagner, K.; Nunes Ramos, F.; Sarmento Cabral, J.; Agudelo, C.; Benavides, A.M.; Cach-Pérez, M.J.; Cardelús, C.L.; Chilpa Galván, N.; Nascimento da Costa, L.E.; et al. Putting vascular epiphytes on the traits map. *J. Ecol.* **2022**, *110*, 340–358. [CrossRef]
287. Martinson, G.O.; Werner, F.A.; Scherber, C.; Conrad, R.; Corre, M.D.; Flessa, H.; Wolf, K.; Klose, M.; Gradstein, S.R.; Veldkamp, E. Methane emissions from tank bromeliads in neotropical forests. *Nat. Geosci.* **2010**, *3*, 766–769. [CrossRef]
288. Rapp, J.M.; Silman, M.R. Epiphyte response to drought and experimental warming in an Andean cloud forest. *F1000Research* **2014**, *3*, 1–29. [CrossRef] [PubMed]
289. Lugo, A.E.; Scatena, F.N. Epiphytes and climate change research in the Caribbean: A proposal. *Selbyana* **1992**, *13*, 123–130.
290. Foster, P. The potential negative impacts of global climate change on tropical montane cloud forests. *Earth-Sci. Rev.* **2001**, *55*, 73–106. [CrossRef]
291. Barlow, J.; Gardner, T.A.; Araujo, I.S.; Ávila-Pires, T.C.; Bonaldo, A.B.; Costa, J.E.; Esposito, M.C.; Ferreira, L.V.; Hawes, J.; Hernandez, M.I.M.; et al. Quantifying the biodiversity value of tropical primary, secondary, and plantation forests. *Proc. Natl. Acad. Sci. USA* **2007**, *104*, 18555–18560. [CrossRef]
292. Crouzeilles, R.; Curran, M.; Ferreira, M.S.; Lindenmayer, D.B.; Grelle, C.E.; Rey Benayas, J.M. A global meta-analysis on the ecological drivers of forest restoration success. *Nat. Commun.* **2016**, *7*, 11666. [CrossRef]
293. Matos, F.A.R.; Magnago, L.F.S.; Chan Miranda, C.A.; de Menezes, L.F.T.; Gastauer, M.; Safar, N.V.H.; Schaefer, C.E.G.R.; da Silva, M.P.; Simonelli, M.; Edwards, F.A.; et al. Secondary forest fragments offer important carbon and biodiversity cobenefits. *Glob. Change Biol.* **2020**, *26*, 509–522. [CrossRef]
294. Foo, Y.Z.; O'Dea, R.E.; Koricheva, J.; Nakagawa, S.; Lagisz, M. A practical guide to question formation, systematic searching and study screening for literature reviews in ecology and evolution. *Methods Ecol. Evol.* **2021**, *12*, 1705–1720. [CrossRef]
295. Fox, J. Why Do Ecologists Publish so Many More Meta-Analyses Than Evolutionary Biologists? 2021. Available online: https://dynamicecology.wordpress.com/2021/07/08/why-do-ecologists-publish-so-many-more-meta-analyses-than-evolutionary-biologists/ (accessed on 15 October 2024).
296. Laube, S.; Zotz, G. Long-term changes of the vascular epiphyte assemblage on the palm *Socratea exorrhiza* in a lowland forest in Panama. *J. Veg Sci.* **2006**, *17*, 307–314. [CrossRef]
297. Einzmann, H.J.R.; Weichgrebe, L.; Zotz, G. Long-term community dynamics in vascular epiphytes on *Annona glabra* along the shoreline of Barro Colorado Island, Panama. *J. Ecol.* **2021**, *109*, 1931–1946. [CrossRef]

plants

MDPI

Article

Influence of Microclimatic Variations on Morphological Traits of Ferns in Urban Forests of Central Veracruz, Mexico

Jessica G. Landeros-López [1], Thorsten Krömer [1], Jorge A. Gómez-Díaz [2], Noé Velázquez-Rosas [1] and César I. Carvajal-Hernández [2,*]

[1] Centro de Investigaciones Tropicales, Universidad Veracruzana, Xalapa 91000, Mexico; jessgll@hotmail.com (J.G.L.-L.); tkromer@uv.mx (T.K.); novelazquez@uv.mx (N.V.-R.)
[2] Instituto de Investigaciones Biológicas, Universidad Veracruzana, Xalapa 91000, Mexico; jorggomez@uv.mx
* Correspondence: ccarvajal@uv.mx

Abstract: Urban forests are remnants of forest habitats within urban areas. Their structural alterations create stressful microclimatic conditions that can influence the morphology of sensitive plants, such as ferns. This study analyzed variations in the morphological traits of ferns in four urban forest sites in central Veracruz, Mexico, considering the microclimatic differences arising from vegetation structure. Temperature, humidity, canopy openness, and radiation were measured, along with eight foliar traits, while assessing the impact of site and habit (terrestrial or epiphytic) on the response. Sites with greater alterations in vegetation structure exhibited increased canopy openness, solar radiation, temperature, and a higher number of days with lower relative humidity. In these sites, leaves showed an increase in dry matter content and vein density, indicating a greater investment in resource storage and structural resistance. In the less-disturbed sites, terrestrial ferns demonstrated larger leaf area and specific leaf area, suggesting greater growth potential. Conversely, epiphytes generally had smaller leaves, which could represent an adaptive advantage for these species. The results also suggest a process of biotic homogenization within this plant group, reflecting a similar morphological response, except for indicator species restricted to less disturbed sites. Thus, this study reveals that microclimatic variations induced by urbanization significantly affect plant morphology and, ultimately, species diversity.

Keywords: plant communities; vegetation structure; pteridophytes; anthropized ecosystems; functional traits; environmental stress

1. Introduction

Urban forests are plant communities with regenerative capacity characterized by a well-defined tree canopy and lower strata. They are formed from remnants of natural habitats or secondary vegetation that arise from natural or induced regeneration, currently surrounded by urban infrastructure [1,2]. These forests exhibit variation in structure and plant species composition, ranging from structurally complex native forest fragments to simplified systems dominated by a few exotic species [3–5]. Their ecological importance lies in their contribution to regulating temperature, as well as water, soil, and air flows in urban ecosystems, while also providing habitat for biodiversity [6,7].

Like other ecosystems, these forests are influenced by a wide range of factors, such as climate, substrate, resident organisms, relief, and the history of the system [8]. However, urban forests are also impacted by anthropogenic effects [9]. Proximity to urban environments results in several negative consequences for these areas due to changes in land use, the influence of pollutants, continuous recreational pressures (which impact the environment through soil compaction, species extraction, alteration of fauna, etc.), separation from propagule sources, reduced regeneration capacity, and the introduction of exotic species [2,9]. These factors affect the structure and density of existing vegetation, which, in turn, impacts the physical environment of these forest ecosystems, leading to increased solar radiation, higher temperatures, greater evaporation, and lower relative humidity, along with the heat island effect and heat retention by urban surfaces [10–12].

The set of these alterations represents stressful environmental conditions, which induce changes in ecosystem processes and the ecology of the organisms present in these forests [13,14]. In plant species, the selection of more tolerant organisms to the adverse conditions of the urban environment described above has been observed, evidenced by their phenotypic plasticity and traits that enhance resilience [9,15]. At the same time, native and sensitive species are eliminated, which ultimately leads to a reduction in diversity, resulting in a process of biotic homogenization [9,16–18].

Ferns are vascular plants that do not depend on any pollination mechanism, since their reproduction is carried out through the dispersion of spores by wind or water [19,20]. Thus, the distribution and establishment of this group allow them to reflect the abiotic conditions of their habitat [21]. Additionally, this group of plants has a close relationship with environmental humidity due to their limited hydraulic capacity and passive stomatal control [22,23]. Many ferns also have a certain dependence on trees, which provide shaded environments and act as hosts for epiphytic species [24,25]. Therefore, it is understood that alterations in light and humidity conditions can significantly affect their development and establishment, making them particularly sensitive to disturbances in their environment [26–29].

The above mentioned has been demonstrated through a remarkable variation in their functional responses to different environmental conditions [30–32]. Specifically, changes at the microenvironmental level can influence variations in their morphological traits [33,34]. For example, conditions of high light intensity and water limitation tend to produce smaller leaves [30,35,36]. Likewise, sites with extreme temperatures, as well as low relative humidity and higher solar radiation, indicate stressful conditions for these organisms, resulting in leaves with greater laminar thickness, moisture content, and dry mass per unit area [31,37,38]. This reflects a protective response to abiotic stressors in this group. However, to our knowledge, there are no studies related to functional or morphological variation in ferns exclusively in urban forests.

The Xalapa–Banderilla conurbation area is situated in central Veracruz, Mexico, originally characterized by humid montane forest (HMF) vegetation and known as the region with the highest diversity of ferns within the state [27,39,40]. However, this area has experienced significant demographic growth in recent decades, leading to a direct transition from woody vegetation to urban usage, resulting in a noticeably fragmented landscape [41]. Currently, the urban vegetation comprises both native and introduced species distributed throughout its network of green areas, among which forest fragments in various stages of anthropization stand out [42–44]. These forests provide essential ecosystem services to the capital city of Xalapa, including climate regulation, carbon sequestration, and serving as a refuge for biodiversity [45–47]. However, recent studies in the area indicate variations in the structure and composition of the urban forests, reflecting their environmental and anthropogenic usage history [42,47,48]. These differences may influence microclimatic conditions, which directly affect the functional responses and distribution of the present fern communities.

Therefore, the objective of this study was to evaluate the variation in morphological traits of terrestrial and epiphytic ferns most representative of four urban forest sites in central Veracruz, Mexico, in response to microclimatic variation associated with the vegetation structure of these forests. This study adopts a functional ecology approach, focusing on how morphological traits reflect the functional responses of ferns to environmental conditions. We hypothesize that the microclimatic conditions of each urban forest are determined by the vegetation structure of the corresponding forest. These conditions, in turn, will influence the functional response of the ferns, which will be manifested in morphological traits adapted to the microclimatic characteristics of each site. Thus, this evaluation will allow us to better understand some of the adaptive mechanisms of this group of plants in urbanized ecosystems.

2. Results

2.1. Microclimate

The evaluated microclimatic variables revealed differences between sites. For temperature (Kruskal–Wallis: H = 244.25, df = 3, $p < 0.0001$), the highest daily mean was recorded at NAT (17.84 °C), while the lowest was found at MAR (15.83 °C; Figure 1a). Sites at approximately the same altitude (CLA, NAT, and KAN) showed a variation of less than 0.5 °C among them, but a difference greater than 1.5 °C in comparison to MAR, the site at the highest elevation (Figure 1a). Regarding relative humidity, variations of less than 2% were observed among the four sites, but with statistical significance (Kruskal–Wallis: H = 41.37, df = 3, $p < 0.0001$; Figure 1b). In this case, MAR recorded the highest mean (95.03%), while CLA had the lowest (93.83%; Figure 1b).

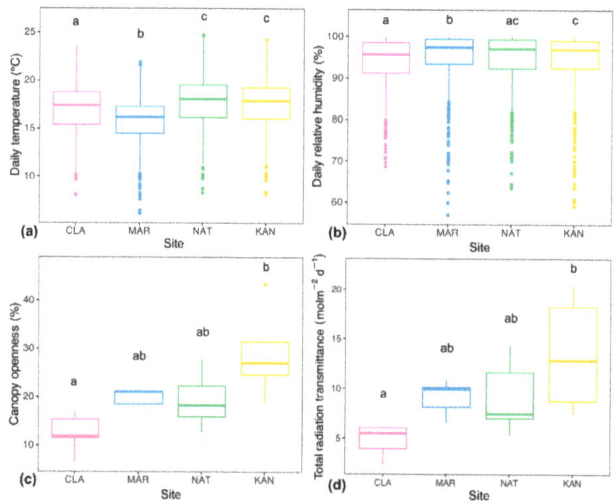

Figure 1. Microclimatic variables measured at the four study sites from August 2021 to August 2023. These include (**a**) daily temperature; (**b**) daily relative humidity; (**c**) canopy openness; and (**d**) total photosynthetic radiation transmittance. CLA: ANP Francisco Javier Clavijero, MAR: ANP La Martinica, NAT: Parque Natura, KAN: Kaná Agroforestry. Different letters above the boxes indicate significant differences between sites ($p < 0.05$).

The canopy openness and radiation transmittance variables responded similarly to one another. In this case, KAN exhibited the highest values (29.0% and 13.5 mol/m^{2}d, respectively; Figure 1c,d), followed by MAR and NAT, while CLA recorded the lowest values (12.4% and 4.77 mol/m^{2}d). This finding was reflected in the statistically significant differences in both variables, according to the ANOVA results (F (3, 16) = 6.81, $p < 0.01$; F (3, 16) = 4.93, $p < 0.05$, respectively), particularly between the CLA and KAN sites (Figure 1c,d).

The time series analysis confirmed that NAT and KAN were the warmest sites, as they recorded the highest number of days with temperatures exceeding 23 °C and the fewest days below 16 °C (Figure 2a; Supplementary Table S1). In contrast, MAR exhibited the opposite pattern, registering the lowest daily temperatures, with values close to 6 °C (Figure 2a). Moreover, MAR had the highest number of days with temperatures below 16 °C and none above 23 °C (Supplementary Table S1). CLA followed MAR in the number of days below 16 °C and had only a minimal proportion of days exceeding 23 °C (Supplementary Table S1).

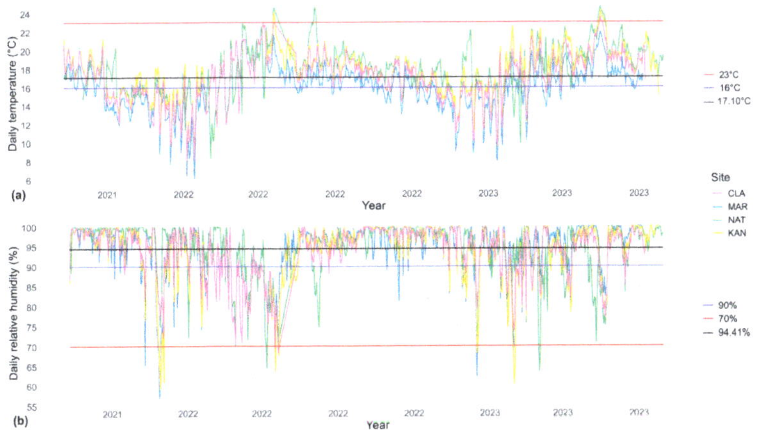

Figure 2. Time series of daily temperature and relative humidity for the four study sites from August 2021 to August 2023. (**a**) Time series of daily temperature means; (**b**) time series of daily relative humidity means. Black lines indicate the mean values across the four sites (temperature: 17.10 °C; relative humidity: 94.41%). Red and blue lines represent environmental thresholds observed in HMF of the region in previous studies [27,49]: red = values typically found in disturbed sites (temperature: 23 °C; relative humidity: 70%), blue = values characteristic of well-conserved sites (temperature: 16 °C; relative humidity: 90%). CLA: Francisco Javier Clavijero, MAR: La Martinica, NAT: Natura Park, KAN: Kaná Agroforest.

Regarding relative humidity, MAR and KAN recorded the highest number of days below 70% (Figure 2b; Supplementary Table S2). Nonetheless, MAR also showed the highest proportion of daily values above the overall mean of the four sites (94.41%; Supplementary Table S2). In the case of NAT, a progressive decrease in relative humidity was observed starting in 2022 (Figure 2b). Meanwhile, CLA showed the least variation over the two-year period, with neither the lowest nor the highest values for temperature or humidity (Figure 2).

2.2. *Principal Component Analysis*

Principal component analysis (PCA) revealed that the first component (Dim. 1) explained most of the variability with a variance of 42.83%. This was primarily influenced by the variables VD and SLA, contributing 46.41% and 30.94%, respectively, and exhibiting opposite orientations according to their correlations (positive for VD: 0.86, negative for

SLA: −0.70; Figure 3). The second component (Dim. 2) accounts for 24.95% of the variance, with the variables LDMC and LA being the most significant, contributing 43.17% and 31.45%, respectively, and positive correlations (0.72 and 0.61, respectively) indicating their influence on vertical separation (Figure 3).

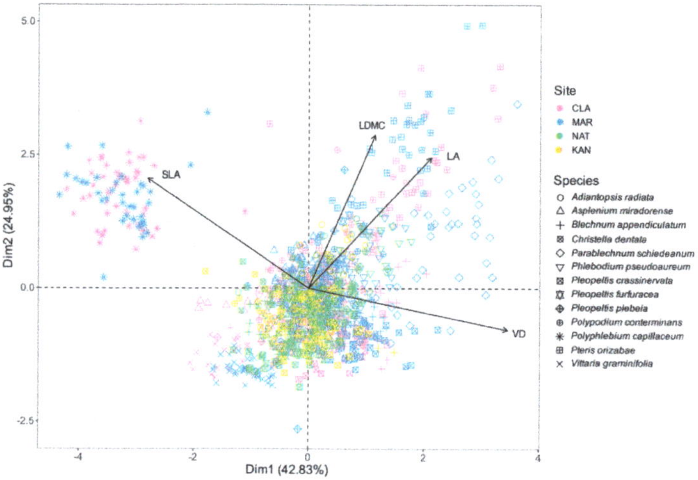

Figure 3. Principal component analysis of the morphological traits of ferns present at the four study sites. CLA: Francisco Javier Clavijero; MAR: La Martinica; NAT: Natura Park; KAN: Kaná Agroforest; LA: leaf area; SLA: specific leaf area; LDMC: leaf dry matter content; VD: vein density.

This analysis also indicated that the ordination reflects patterns associated with both species and sites (Figure 3). The most notable is that of the epiphytic species *Polyphlebium capillaceum*, which demonstrated a strong relationship with the SLA trait, distinguishing itself from the other species. Similarly, the terrestrial species *Parablechnum schiedeanum* and *Pteris orizabae* exhibited strong relationships with LDMC and LA, also setting themselves apart from the rest. In contrast, the epiphytic species *Vittaria graminifolia* was organized in the opposite direction of the LA and LDMC traits. These four species were exclusively found at the CLA and MAR sites, suggesting an association with these locations, while the remaining species, including those from NAT and KAN, did not exhibit clear ordination patterns.

2.3. Generalized Linear Mixed Model

The model indicated that random effects associated with differences among individuals and species carried a significantly greater weight in explaining variability (LA: $R^2c = 0.678$; SLA: $R^2c = 0.162$; LDMC: $R^2c = 0.036$; VD: $R^2c = 0.298$) compared to fixed effects ($R^2m = 0.000$). Nevertheless, it remained evident that morphological traits are substantially influenced by site, habit, and their interactions, with CLA serving as the reference site. For the trait LA, the terrestrial habit exhibited a significant positive effect ($p = 0.006$), indicating that leaves of terrestrial ferns tend to be larger in size, particularly in CLA and MAR (Table 1; Supplementary Figure S1a). Meanwhile, this effect was significantly reduced in NAT ($\beta = -0.75$, $p < 0.001$) and KAN ($\beta = -0.45$, $p = 0.005$), as shown by the negative interaction terms. Regarding SLA, the site displayed a significant positive effect ($p < 0.001$), being highest in KAN ($\beta = 0.17$, $p < 0.001$) and significantly decreasing in the leaves of terrestrial species in MAR ($\beta = -0.10$, $p = 0.017$; Table 1, Supplementary Figure S1b).

Table 1. Generalized linear mixed models for the evaluation of morphological traits of terrestrial and epiphytic ferns present at the four study sites. Estimates, 95% confidence intervals (CIs), t-values, and *p*-values are shown for fixed effects. Marginal and conditional R^2 values (R^2m/R^2c) indicate the proportion of variance explained by fixed effects alone and by both fixed and random effects, respectively. LA: leaf area, SLA: specific leaf area, LDMC: leaf dry matter content, VD: vein density; CLA: Francisco Javier Clavijero; MAR: La Martinica; NAT: Natura Park; KAN: Kaná Agroforestry.

Trait	Predictor	Estimate (β)	CI (95%)	t	*p*	R^2 m/R^2 c
LA	(Intercept)	3.59	2.55–4.63	6.756	<0.001	0.000/0.678
	Site KAN	0.03	−0.16–0.21	0.275	0.783	
	Site MAR	−0.05	−0.17–0.07	−0.781	0.435	
	Site NAT	0.17	0.01–0.33	2.142	0.032	
	Terrestrial Habit	2.15	0.62–3.69	2.748	0.006	
	SiteKAN × TerrestrialHabit	−0.49	−0.83–−0.15	−2.798	0.005	
	SiteMAR × TerrestrialHabit	−0.12	0.34–0.09	−1.158	0.247	
	SiteNAT × TerrestrialHabit	−0.75	−1.05–−0.45	−4.875	<0.001	
SLA	(Intercept)	5.13	4.80–5.46	30.661	<0.001	0.000/0.162
	Site KAN	0.17	0.09–0.25	4.027	<0.001	
	Site MAR	−0.03	−0.07–0.00	−1.823	0.069	
	Site NAT	0.02	−0.07–0.11	0.417	0.677	
	TerrestrialHabit	0.18	−0.31–0.67	0.727	0.467	
	SiteKAN × TerrestrialHabit	0.07	−0.08–0.23	−0.950	0.342	
	SiteMAR × TerrestrialHabit	−0.10	−0.19–−0.02	2.401	0.017	
	SiteNAT × TerrestrialHabit	0.08	−0.07–0.23	1.030	0.303	
LDMC	(Intercept)	5.59	5.44–5.73	75.152	<0.001	0.000/0.036
	Site KAN	−0.16	−0.23–−0.10	−4.953	<0.001	
	Site MAR	0.07	0.03–0.11	3.818	<0.001	
	Site NAT	−0.08	−0.13–−0.03	−3.006	0.003	
	TerrestrialHabit	0.07	−0.15–0.29	0.634	0.526	
	SiteKAN × TerrestrialHabit	0.16	0.04–0.29	2.515	0.012	
	SiteMAR × TerrestrialHabit	0.11	0.05–0.18	3.422	0.001	
	SiteNAT × TerrestrialHabit	−0.02	−0.13–0.09	−0.281	0.779	
VD	(Intercept)	1.68	1.25–2.11	7.649	<0.001	0.050/0.298
	Site KAN	0.03	−0.04–0.10	0.810	0.418	
	Site MAR	0.07	0.02–0.12	2.746	0.006	
	Site NAT	0.07	0.00–0.13	2.048	0.041	
	TerrestrialHabit	0.72	0.08–1.35	2.222	0.027	
	SiteKAN × TerrestrialHabit	−0.20	−0.33–−0.06	−2.884	0.004	
	SiteMAR × TerrestrialHabit	−0.14	−0.22–−0.05	−3.137	0.002	
	SiteNAT × TerrestrialHabit	−0.43	−0.55–−0.31	−7.125	<0.001	

For LDMC, site ($p < 0.001$) and its interaction with habit ($p = 0.012$) exhibited a significant positive effect. In this context, KAN and NAT recorded significantly lower values ($\beta = -0.16$, $p < 0.001$; $\beta = -0.08$, $p = 0.003$), whereas MAR showed higher values ($\beta = 0.07$, $p < 0.001$; Table 1, Supplementary Figure S1c). Additionally, interactions demonstrated a significant increase in KAN ($\beta = 0.16$, $p = 0.012$) and MAR ($\beta = 0.11$, $p = 0.001$) terrestrial leaves. Ultimately, terrestrial leaves displayed the highest VD ($\beta = 0.72$, $p = 0.027$), particularly at the MAR site ($\beta = 0.07$, $p = 0.006$; Table 1, Supplementary Figure S1d). Conversely, a significant decrease was noted in the terrestrial leaves of NAT and KAN ($\beta = -0.43$, $p < 0.001$; $\beta = -0.20$, $p = 0.004$).

3. Discussion

3.1. Microclimatic Variation

The higher values of canopy openness, radiation transmittance, and temperature in KAN and NAT result from their vegetation structure and conservation status [48] (Table 2). Human modifications to the sites have decreased vegetation cover and tree density, leading to more clearings and allowing for increased sunlight penetration [50,51]. This change has increased the total radiation transmittance and ultimately raised the understory temperature [52–54]. The reduced vegetation cover in these sites also leads to lower thermal buffering and evapotranspiration capacity [11,55], which explains why they exhibited low daily relative humidity values and a gradual decline over the two-year period (Figure 2). In addition, NAT and KAN showed the highest frequency of days exceeding 23 °C and below 70% (Supplementary Tables S1 and S2). These specific variations, although moderate individually, accumulate over time and significantly shape the site's microclimate, generating thermal and hydric stress in the understory. Furthermore, days with temperatures exceeding average maximum temperatures [55] often coincide with particularly hot periods during which the average temperature for the region is exceeded (18 °C). This suggests that, during an already warm and stressful season for certain groups, such as ferns, the level of thermal stress intensifies in these sites. It has been documented that in ferns of the humid montane forests of this region, an average increase of 1 °C associated with changes in forest conditions can cause a decrease of up to 37–63% in their richness [27]. Moreover, the urban heat island effect caused by the proximity of these sites to urban areas may also contribute to the temperature rise, which is consistent with previous findings in forests in the area, showing that urban environments significantly raise air temperature due to reduced vegetation and high impermeable surface cover [46,49].

The CLA site, in contrast, exhibits a denser vegetation cover, featuring a forest structure indicative of a higher conservation status [48]. This condition resulted in reduced values of canopy openness and consequently lower levels of radiation transmittance and temperature (Figures 1 and 2a). In this scenario, the site's vegetation functions as a thermal buffer and a direct source of water vapor, helping to mitigate extreme temperature fluctuations and fostering more stable conditions within the forest [10,56,57]. This is further reflected in the fact that it exhibited the lowest daily fluctuations in temperature and relative humidity (Figure 2).

On the other hand, MAR had the lowest daily temperature and the highest daily relative humidity (Figure 1). In this case, altitude was the main factor determining microclimatic variation, surpassing the influence of vegetation structure and cover, as it is the only site located approximately 300 m higher than the others (Table 2). The lower temperature aligns with previous studies indicating a decrease in temperature with increasing altitude, with an average decrease of 0.5 °C for every 100 m of elevation [58,59]. Similarly, MAR's higher altitude favored relative humidity retention, as the persistence of condensation increases in HMF as it approaches the cloud belt, typically found between 2000 and 3000 masl [60–62]. Furthermore, the drop in temperature with altitude leads to cooler air, which promotes moisture retention and condensation, thereby increasing water vapor presence in the environment [63,64].

Table 2. Description of the vegetation structure and composition of the four urban forests comprising the study sites. DBH: diameter at breast height; PNA: protected natural area [48].

Urban Forest	Area (ha)	Altitude (masl)	Basal Area (m^2/ha)	DBH (cm)	Height (m)	Density (tree/ha)	State of Maturity of the Vegetation
PNA Francisco Javier Clavijero (CLA)	22.06	1362	3.3 ± 1.14	19.08 ± 21.89	13.48 ± 7.19	32 ± 46	Mature forest with fragments of secondary forest
PNA La Martinica (MAR)	52.30	1599	1.6 ± 0.72	13.57 ± 14.66	14.20 ± 6.80	49 ± 78	Mature forest with fragments of secondary forest
PNA Natura Park (NAT)	80	1320	1.6 ± 0.40	14.92 ± 13.38	11.20 ± 5.50	46 ± 75	Secondary forest, with few remaining trees of mature forest
Kaná Agroforestry (KAN)	5	1366	1.9 ± 0.56	14.86 ± 16.27	9.88 ± 5.73	37 ± 35	Secondary forest with floristic elements of mature forest

3.2. Microclimatic Influence on Morphological Variation

The statistical model used indicates that species and individuals account for a considerable proportion of the total variability. However, significant differences related to site and habitat confirm their substantial influence on the morphological traits of the ferns in this study. This suggests that these are relevant variables capable of modulating the functional responses of this plant group, especially in urban environments.

Leaf area (LA) is a trait associated with ecological strategies regarding nutrient availability and resource utilization, as well as allometric factors such as plant size [65]. In this study, the leaves of terrestrial species were notably larger than those of epiphytes. This difference may arise from the more stressful conditions that epiphytic species typically face while growing in the tree canopy [66,67]. The increased solar exposure and reduced water availability tend to favor smaller leaves, which help conserve resources, lower water demand, and optimize water use efficiency [66,68,69], possibly explaining these results. A particularly strong positive effect of the terrestrial habit was observed in CLA and MAR, likely associated with the presence of *Pteris orizabae* and *Parablechnum schiedeanum*, two large terrestrial species found exclusively in these sites [69]. In contrast, the interaction effect between terrestrial habit and site was significantly lower in NAT and KAN, indicating that the difference in leaf area between terrestrial and epiphytic species was less pronounced. This reduction may be explained, conversely, by the absence of these large species [70], which considerably raise the mean leaf area in CLA and MAR. These findings suggest that species composition modulates the expression of habit effects at each site and further support the idea that species identity contributes substantially to the variability explained by the model.

Specific leaf area (SLA) indicates how plants allocate resources for leaf growth, reflecting their ability to capture light and photosynthesize [65,71]. Leaves from terrestrial and epiphytic species exhibited the highest values of SLA in CLA and KAN (Table 1). In the case of CLA, this suggests that individuals benefit from the dense canopy and use efficient photosynthetic mechanisms to maximize absorption in low-light conditions, which may indicate a significant adaptation of the associated ferns to thrive in umbrophilous environments. (e.g., *Asplenium miradorense* Liebm.) [25,69,70]. This is consistent with findings that showed that many epiphytic ferns adopt acquisitive strategies characterized by high SLA and low tissue density, particularly in humid, shaded microhabitats where maximizing light capture is advantageous [72–74].

Conversely, in KAN this result may relate to the presence of terrestrial disturbance indicator species, such as *Adiantopsis radiata* (L.) Fée, *Pteris pulchra* Schltdl. and Cham., and *Christella dentata* (Forssk.) Brownsey & Jermy [70]. These species have diverse functional response capacities, enabling them to capitalize on greater light availability, enhancing their photosynthetic capacity, and optimizing growth and development in disturbed environments [71,75]. This is particularly common in species of the Polypodiaceae family, which have been shown to have the ability to occupy more variable or disturbed environments, ranging from humid conditions to more exposed microhabitats [74].

Leaf dry matter content (LDMC) determines the physical resistance of leaves and their resource storage capacity [65,76]. In MAR, the highest LDMC values were found in the leaves of terrestrial and epiphytic species (Table 1, Supplementary Figure S1). This response is typically linked to stressful conditions and, in this case, can be attributed to low temperatures and abrupt decreases in relative humidity at the site (Figures 1 and 2). Evidence suggests that as temperatures drop, plants tend to produce leaves with higher dry matter content as a mechanical support mechanism to endure cold conditions [77–79]. This occurs due to a greater number of cell layers, resulting in structurally denser leaves that are better equipped for heat retention, thanks to the high thermal capacitance provided by the water in these cells [80,81]. Research on ferns has linked increased lamellar thickness to higher LDMC in extreme conditions of greater altitude, which correspond to lower temperatures and increased aridity [31,37], so the situation observed in MAR aligns with this response.

Regarding the significant increase in dry matter in the leaves of terrestrial KAN species, this can also be attributed to stressful site conditions related, conversely, to higher solar radiation, elevated temperatures, and reduced water availability. This can be explained by the increased thickness of cell walls resulting from the compaction of mesophyll cells, which aids in water conservation under water-stress conditions [68,78,79]. This is linked to the high SLA values at this site, suggesting that greater light availability encourages disturbance indicator species. These species also exhibit a relatively robust structure that helps them endure challenging site conditions [68,80]. In contrast, epiphytic species in both CLA and KAN exhibited lower LDMC values, suggesting a different strategy. Epiphytic ferns may prioritize flexibility in water uptake because they are exposed to variable canopy microenvironments, with intermittent moisture sources [66,69,74]. So rather than investing in denser tissues, they may rely on reduced leaf area or increased water storage capacity.

Finally, the density of venation (VD) is often associated with facilitating water transport and thermal regulation in hot environments or those with limited water availability [62,81,82]. In ferns, it has been documented that vein density has a significant relationship with ambient humidity, decreasing as humidity increases [31]. In this study, the highest values of this trait were observed in the leaves of terrestrial species at the site with the highest relative humidity, MAR (Table 1, Supplementary Figure S1d). However, this trait is highly variable among species and is influenced significantly by phylogeny [65]. Thus, this result may be attributed to the influence of the large terrestrial species, *Pteris orizabae* and *Parablechnum schiedeanum*, which are specifically found at the sites with the highest relative humidity, CLA and MAR.

3.3. Morphological Differentiation and Tendency Towards Biotic Homogenization in Urban Forests

Two species exhibiting differential arrangement are *Polyphlebium capillaceum* and *Vittaria graminifolia*, both epiphytic ferns known as indicator species of natural forests, as they typically thrive in humid and shaded environments [27,29,82,83]. In particular, filmy ferns like *P. capillaceum* are sensitive to environmental changes due to their thin leaf structure and lack of cuticle, making them dependent on high relative humidity and shaded conditions [35,84,85]. These characteristics were evident through their corresponding morphological responses in the PCA (Figure 3), where *P. capillaceum* was observed to be strongly associated with the SLA trait, reflecting its thin and delicate leaves, while *V. graminifolia* appeared in the opposite direction, also demonstrating its small leaf size but a lower dry matter content.

Similarly, the terrestrial species *Parablechnum schiedeanum* and *Pteris orizabae* showed evident morphological responses, with a strong association with LDMC and LA traits (Figure 3), indicating a robust leaf structure and higher biomass investment. These species are commonly found in low-disturbance environments, where light, water, and nutrient availability are not limiting for their establishment [27,66,69]. Their ordination patterns in the PCA therefore suggest a strategy oriented towards structural persistence and competitive performance.

These four species (*P. capillaceum*, *V. graminifolia*, *P. schiedeanum*, and *P. orizabae*) were found only in CLA and MAR, the sites with the most mature forests [48,70]. By presenting a morphological response distinct from that of the other species (Figure 3), they indirectly reflect the microclimatic and structural conditions of these sites. This indicates that species sensitive to disturbance and changes in the microclimate are linked to forests with a higher degree of conservation [27,29]. These findings align with previous reports of other conservation indicator fern species at the CLA and MAR sites, such as the epiphytic fern species *Asplenium sphaerosporum* A.R.Sm. and *Didymoglossum reptans* (Sw.) C. Presl, as well as the terrestrial *Asplenium miradorense* Liebm. [70]. Furthermore, these distinct morphological patterns and their contribution to overall trait variability (Figure 3, Table 1) reflect the greater diversity and heterogeneity present in the best-preserved forests. This supports the idea that structurally complex environments promote a broader spectrum of functional strategies, both acquisitive and conservative, especially in ferns [74].

In contrast, the leaves of the NAT and KAN sites showed more clustered positions in the PCA space, indicating less variation in morphological traits across species (Figure 3). This pattern likely reflects the environmental filtering imposed by more open and disturbed forest structures, which tend to select for species with similar ecological strategies, such as tolerance to increased light and temperature, and reduced humidity, and displacing more sensitive ones, such as the four aforementioned species [15,18]. The frequent presence of species identified as disturbance indicators, such as the terrestrials *Adiantopsis radiata* (L.) Fée, *Pteris pulchra* Schltdl. & Cham., and *Christella dentata* (Forssk.) Brownsey & Jermy, common in NAT and KAN but absent in CLA and MAR, reinforces this interpretation [70].

Although variation among species is still present, the reduced morphological diversity and limited ordination patterns in NAT and KAN suggest an early stage of biotic homogenization, where environmental stress filters out species with narrower ecological requirements and favors the persistence of generalist and tolerant taxa. This aligns with observations in other urban green areas in tropical regions worldwide with similar environmental conditions, where the flora underwent homogenization due to environmental filters imposed by urban expansion, resulting in a decline in species richness and diversity [16,86–88].

4. Materials and Methods

4.1. Study Sites

The study was conducted in the Xalapa–Banderilla conurbation area, located in Veracruz State, Mexico, between the coordinates 19°29′ and 19°36′ N latitude and 96°48′ and 96°58′ W longitude, with an altitude ranging from 1120 to 1700 masl [89] (Figure 4). The region's climate is semi-warm–humid, with an average temperature of 18 °C and an average annual rainfall of 1500 mm [90]. The most prevalent vegetation type in the area is the humid montane forest (HMF), although other communities, such as oak, pine, and low deciduous forests, are also present [91,92]. Currently, this area has a combined population of over 500,000 inhabitants, making it one of the fastest-growing urban areas in the state [93,94].

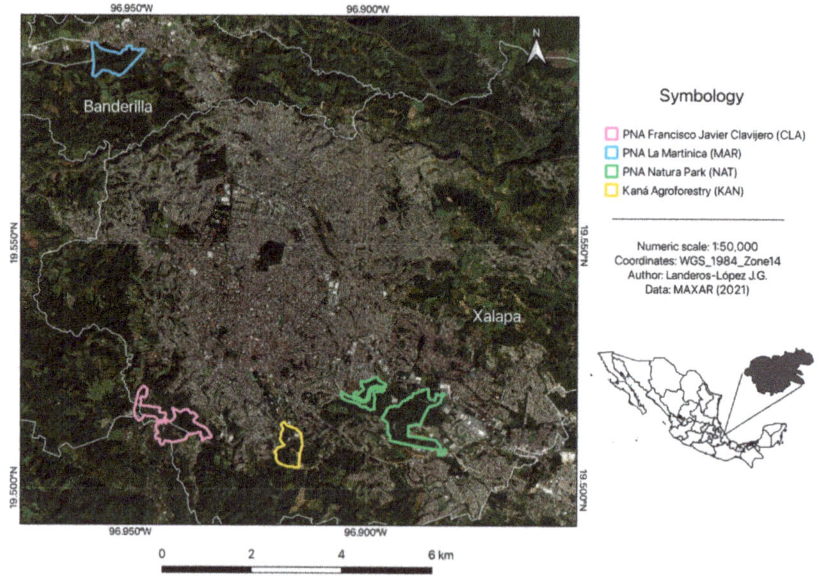

Figure 4. Spatial locations of the four selected urban forest sites in the Xalapa–Banderilla conurbation area, Veracruz State, Mexico.

The four urban forest sites selected for this study include Natura Park (NAT), part of the Protected Natural Area (PNA) "El Tejar Garnica"; Kaná Agroforest (KAN), affiliated with the Campus for Culture, Arts, and Sports of the University of Veracruz; the Francisco Javier Clavijero PNA (CLA), situated in a polygon managed by the Ministry of Environment of Veracruz; and the PNA "La Martinica" (MAR; Figure 4). These forests exhibit remnants of HMF, yet they differ in structure and plant composition due to varying processes of anthropic modification and the influence of the urban environment [48] (Table 2). For instance, NAT and KAN are fully embedded in the urban sprawl, leading to total or partial isolation caused by avenues, shopping centers, and other urban developments. In these areas, secondary tree species dominate (e.g., *Dendropanax arboreus* (L.), *Citharexylum caudatum* L., *Piper amalago* L., *Vachellia pennatula* (Schltdl. & Cham.) Seigler & Ebinger), while primary HMF species are observed to a lesser extent. Conversely, CLA and MAR are located on the periphery, in less developed regions, exhibiting a tree species composition more similar to mature forests (*Liquidambar styraciflua* L., *Quercus* L. spp.) [48] (Figure 4).

4.2. Data Collection

For the evaluation of microclimatic conditions, three digital sensors (Track-It™ RFID Dataloggers) were placed on tree trunks and branches at a height of 2 m per site to record temperature and relative humidity every hour from August 2021 to August 2023. In the case of MAR and KAN, data loss occurred during March and April 2022 due to sensor malfunction. As a result, the total number of days recorded per site was as follows: CLA—674, MAR—621, NAT—699, and KAN—642. During the same period, one hemispherical photograph was taken at each of the five sample plots established in a previous study [48] to describe the light environment. The photographs were captured at a height of 1 to 1.5 m in the central area of each plot to minimize the influence of the edges. In total, five photographs were obtained per site, which were subsequently analyzed using the Gap Light Analyzer application [95] to determine canopy openness and solar radiation transmittance.

For the evaluation of morphological traits in ferns, individuals were collected in the same established plots where the structure and composition of the woody vegetation, as well as the diversity of ferns, have been characterized [48,70]. Since the sites exhibit heterogeneous communities in terms of species count [70], this study considered the most frequent terrestrial and epiphytic species that represent at least 30% of the total fern richness at each site [70]. Therefore, a total of 13 species were selected, distributed as follows: 10 at the CLA site, 11 at MAR, and 7 both at NAT and KAN (Table 3).

Table 3. This study selected species of terrestrial and epiphytic ferns due to their abundance. T: terrestrial, E: epiphytic. CLA: Francisco Javier Clavijero, MAR: La Martinica, NAT: Natura Park, KAN: Kaná Agroforest. The letter X indicates presence at the site.

Family	Species	Habit	CLA	MAR	NAT	KAN
Pteridaceae	*Adiantopsis radiata* (L.) Fée	T			X	
Aspleniaceae	*Asplenium miradorense* Liebm.	T	X	X		
Blechnaceae	*Blechnum appendiculatum* Willd.	T	X	X	X	X
Thelypteridaceae	*Christella dentata* (Forssk.) Brownsey & Jermy	T			X	X
Blechnaceae	*Parablechnum schiedeanum* (Schltdl. ex C. Presl) Gasper & Salino	T		X		
Polypodiaceae	*Phlebodium pseudoaureum* (Cav.) Lellinger	E	X	X	X	X
Polypodiaceae	*Pleopeltis crassinervata* (Fée) T. Moore	E	X	X	X	X
Polypodiaceae	*Pleopeltis furfuracea* (Schltdl. & Cham.) A. R. Sm. & Tejero	E	X	X	X	X
Polypodiaceae	*Pleopeltis plebeia* (Schltdl. & Cham.) A. R. Sm. & Tejero	E	X	X	X	X
Hymenophyllaceae	*Polyphlebium capillaceum* (L.) Ebihara & Dubuisson	E	X	X		
Polypodiaceae	*Polypodium conterminans* Liebm.	E	X	X		
Pteridaceae	*Pteris orizabae Mart. & Galeotti*	T	X	X		
Pteridaceae	*Vittaria graminifolia* Kaulf.	E	X	X		

Thirty fronds per species were obtained by sampling a variable number of individuals according to their availability, with each individual contributing a different number of fronds. Individuals considered healthy and mature were collected with their entire rhizome and allowed to rehydrate for 24 h in resealable plastic bags with the rhizome submerged in water. Subsequently, the fronds were transported to the laboratory, where eight morphological traits were measured following Pérez-Harguindeguy et al. (2013) [65]: total length (L), lamina length (LL), leaf area (LA), specific leaf area (SLA), leaf thickness (LT), leaf dry matter content (LDMC), water content (WC), and vein density (VD) (Figure 5). For compound-leaved fern species, we included the petiole area as part of the total leaf area (LA), as it represents a significant portion of the leaf structure and contributes to its functional surface. This approach is consistent with established protocols, which indicate that including the petiole is valid depending on the study context [65].

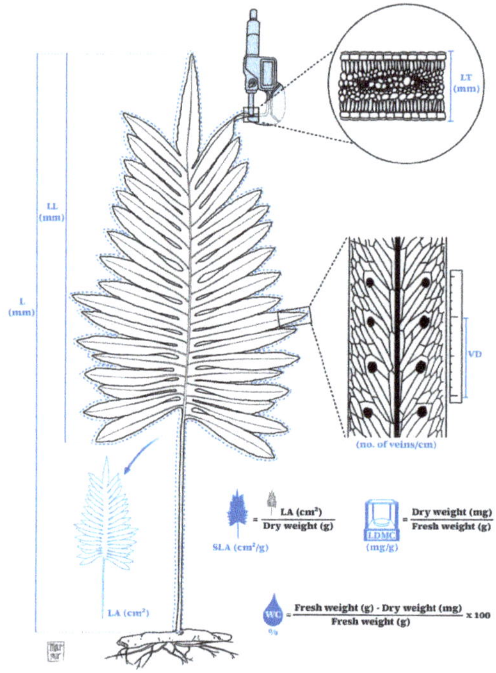

Figure 5. Illustrative scheme for measuring the eight morphological traits of a fern leaf. L: total length, LL: lamina length, LA: leaf area, SLA: specific leaf area, LT: leaf thickness, LDMC: leaf dry matter content, WC: water content, VD: vein density. Concept by Landeros-López, J.G. Illustrated by Muñoz-Velázquez, M.

4.3. Data Analysis

Daily means of temperature and relative humidity, and overall means of canopy openness and radiation transmittance, were evaluated using the Shapiro–Wilk normality test. Since temperature and relative humidity data did not follow a normal distribution (Shapiro–Wilk: $p < 0.0001$), their means were compared using the non-parametric Kruskal–Wallis test. In contrast, canopy openness and radiation transmittance data were normally distributed (Shapiro–Wilk: $p > 0.05$), so their means were analyzed using one-way ANOVA. Post hoc comparisons were performed to identify significant differences between groups. For ANOVA, Tukey's Honest Significant Difference (HSD) test was applied using the TukeyHSD() function, which includes built-in correction for multiple comparisons. For Kruskal–Wallis tests, pairwise comparisons were run with pairwise.wilcox.test() using Holm-adjusted *p*-values (p.adjust.method = "holm"). Letters indicating group differences were added to the graphs based on these post hoc results.

Additionally, time series were generated to visualize daily variation in temperature and relative humidity. These series were constructed by using daily means for each variable across the four study sites and organizing them chronologically. To facilitate interpretation, we included reference lines representing extreme thresholds of temperature and humidity reported for HMF in the region [27,49]. Summary tables indicating the number of days in which values exceeded these thresholds are provided to support the visual interpretation of the time series (see Supplementary Tables S1 and S2).

The morphological variables were also evaluated using the Shapiro–Wilk normality test. Since the data did not follow a normal distribution (Shapiro–Wilk: $p < 0.0001$), a

Spearman correlation analysis was performed for non-parametric data. This allowed us to eliminate highly correlated variables ($\rho > 0.80$) that provided redundant information. Spearman correlation analysis revealed strong correlations among several morphological traits, indicating potential redundancy (See Supplementary Figure S2). In these cases, highly correlated variables may provide overlapping information, which can distort multivariate analyses or inflate the importance of certain trait dimensions. Therefore, to reduce redundancy and improve interpretability, total length (L) and lamina length (LL) were excluded because of their high correlation with leaf area (LA; $r > 0.90$). Leaf thickness (LT) was also removed because of its high negative correlation with specific leaf area (SLA; $r = -0.81$). Similarly, leaf dry matter content (LDMC) and water content (WC) exhibited a strong negative correlation ($r = -0.98$); LDMC was retained due to its greater functional importance. Therefore, the traits retained for further analysis included LA, SLA, LDMC and vein density (VD).

Using the selected variables, a principal component analysis (PCA) was performed to observe the ordination of traits according to species and sites. Prior to the analysis, all variables were standardized using z-score standardization with the scale() function from base R, which centers and scales each variable to have a mean of zero and a standard deviation of one. Finally, a generalized linear mixed model (GLMM) was fitted to evaluate the effects of site, habit (terrestrial or epiphytic), and their interaction on morphological traits, considering a random effect of the leaves of each individual nested within the species to account for the individual effect of species richness at each site. The model employed a Gaussian distribution with a logarithmic link. This choice was made because the morphological trait data were strictly positive and continuous, and the residuals deviated from normality and homoscedasticity. The log link is recommended in such cases to stabilize variance and improve model performance [96,97]. In addition, following the recommendations of previous studies, the transformation was applied within the model framework rather than to the raw data [98]. Therefore, its final form was as follows: glmer((X) ~ Site + Habit + Site:Habit + (1 | Species/Individual), data = Data, family = gaussian(link = "log")), where X corresponds to each trait evaluated. The fit of the variables in the model was evaluated using pseudo-conditional and marginal R^2 values [99]. The corresponding analyses and graphs were executed using the statistical programs RStudio version 2024.09.1 (with the packages nortest, psych, Hmisc, FactoMineR, factoextra, stats, ggplot2, tidyverse, dplyr, grid, lme4, and lubridate) and JAMOVI version 2.3.28 [100,101]. Normality tests (Shapiro–Wilk), as well as ANOVA and Kruskal–Wallis analyses, were performed using the functions from the base R package.

5. Conclusions

In the studied urban forests of central Veracruz, ferns exhibited morphological variation in response to microclimatic and structural changes. Despite the variation caused by species and individuals, both site and habitat were determining factors for the evaluated morphological traits. Generally, responses were linked to microclimatic variation at the four sites, where traits related to resource acquisition influenced morphological variation. This was evident through the optimization of light capture and photosynthetic efficiency in response to changes in light availability, as well as water and biomass conservation due to water limitations. Overall, this provides evidence of the group's ability to respond even to minimal environmental variations, demonstrating that, like other plant groups, they possess mechanisms that allow them to adapt to different environments. This contributes to their persistence in diverse habitats and makes them ideal subjects for studying adaptation and ecological transformation processes.

When analyzing growth habits, terrestrial ferns tended to have larger leaves, a greater specific leaf area, and a higher dry matter content in wetter and shaded sites—traits consistent with better resource acquisition under favorable conditions. In contrast, in drier and more open environments, these species exhibited smaller leaves and lower SLA, likely adapting to water stress. Epiphytic ferns, on the other hand, generally displayed smaller, thicker leaves, with lower SLA and LDMC across all sites, where slight variations suggested a more conservative strategy adapted to limited water access and canopy exposure. Meanwhile, vein density was more closely associated with a taxonomic response. Overall, these patterns indicate that terrestrial species exhibit greater morphological plasticity across environmental gradients, while epiphytes maintain more stable trait expressions, possibly due to greater physiological constraints associated with their habit.

Microclimatic variations in central Veracruz's urban forests primarily result from structural modifications, which appear to favor generalist species while displacing native specialists. This leads to a predominance of similar morphological traits and reduced diversity, potentially limiting the ecological functionality of these urban ecosystems. The principal component analysis (PCA) performed provides initial evidence that most of the studied fern species share similar morphological traits, regardless of the site's degree of urbanization, highlighting a clear trend toward biotic homogenization. Hence, this study also serves as a reminder of the ongoing fundamental role of old-growth forests in biodiversity conservation and ecosystem functioning [102]. While urban forests are often regarded as biodiversity refuges and vital sites for microclimate maintenance, their proximity to urban environments and constant anthropogenic modifications are key factors that make them potential refuges for generalist and tolerant species [4,103].

Finally, this study highlights relevant trends in a group that has been explored little in urban and functional contexts, emphasizing the impact of human modifications on their biodiversity and functionality within urban ecosystems. We acknowledge the limitations of working with this group due to the challenges of measuring morphological traits in ferns; however, these limitations do not diminish the value of the observed patterns, which contribute to a better understanding of their ecological responses. For future related research, we suggest considering the inclusion of species-specific abundance data, as this would allow for a more robust assessment of functional diversity metrics and the processes associated with homogenization.

Supplementary Materials: The following supporting information can be downloaded at https://www.mdpi.com/article/10.3390/plants14111732/s1.

Author Contributions: Conceptualization, C.I.C.-H.; methodology, C.I.C.-H. and J.G.L.-L.; software, J.G.L.-L. and J.A.G.-D.; validation, J.A.G.-D. and N.V.-R.; formal analysis, J.G.L.-L. and J.A.G.-D.; investigation, J.G.L.-L. and C.I.C.-H.; resources, C.I.C.-H.; data curation, J.G.L.-L., C.I.C.-H., and J.A.G.-D.; writing—original draft preparation, J.G.L.-L., T.K., and C.I.C.-H.; writing—review and editing, C.I.C.-H., T.K., J.A.G.-D., and N.V.-R.; visualization, J.G.L.-L., J.A.G.-D., and N.V.-R.; supervision, C.I.C.-H. and T.K.; project administration, C.I.C.-H.; funding acquisition, C.I.C.-H. All authors have read and agreed to the published version of the manuscript.

Funding: J.G.L.-L. acknowledges funding from the Consejo Nacional de Humanidades, Ciencias y Tecnologías for the scholarship awarded to pursue a master's degree (CONAHCYT; No. 1185460). C.I.C.-H. and J.G.L.-L. thank CONAHCYT for funding the project "Socioecological functionality of neotropical urban green areas (No. 64358)".

Data Availability Statement: The original contributions presented in this study are included in the article/supplementary material. Further inquiries can be directed to the corresponding author.

Acknowledgments: We thank Jaén Carvajal-Armenta for his support in the laboratory work. We also thank Israel Huesca Dominguez for his help in the design of the generalized linear mixed model.

Conflicts of Interest: The authors declare no conflicts of interest.

References

1. Alvey, A.A. Promoting and Preserving Biodiversity in the Urban Forest. *Urban For. Urban Green.* **2006**, *5*, 195–201. [CrossRef]
2. Jim, C.Y. Conservation and Creation of Urban Woodlands. In *Greening Cities: Forms and Functions*; Tan, P.Y., Jim, C.Y., Eds.; Springer: Singapore, 2017; pp. 307–330, ISBN 978-981-10-4113-6.
3. Chacón-Castillo, D. Áreas Potenciales Para la Conservación en Xalapa, Veracruz y Conurbados: Cobertura y Almacenes de Carbono. Master's Thesis, Universidad Veracruzana, Xalapa, Mexico, 2020.
4. McKinney, M.L.; Ingo, K.; Kendal, D. The Contribution of Wild Urban Ecosystems to Liveable Cities. *Urban For. Urban Green.* **2018**, *29*, 334–335. [CrossRef]
5. Threlfall, C.G.; Ossola, A.; Hahs, A.K.; Williams, N.S.G.; Wilson, L.; Livesley, S.J. Variation in Vegetation Structure and Composition across Urban Green Space Types. *Front. Ecol. Evol.* **2016**, *4*, 66. [CrossRef]
6. Roeland, S.; Moretti, M.; Amorim, J.H.; Branquinho, C.; Fares, S.; Morelli, F.; Niinemets, Ü.; Paoletti, E.; Pinho, P.; Sgrigna, G.; et al. Towards an Integrative Approach to Evaluate the Environmental Ecosystem Services Provided by Urban Forest. *J. For. Res.* **2019**, *30*, 1981–1996. [CrossRef]
7. Zefferman, E.P.; McKinney, M.L.; Cianciolo, T.; Fritz, B.I. Knoxville's Urban Wilderness: Moving Toward Sustainable Multifunctional Management. *Urban For. Urban Green.* **2018**, *29*, 357–366. [CrossRef]
8. Chapin, F.S., III; Zavaleta, E.S.; Eviner, V.T.; Naylor, R.L.; Vitousek, P.M.; Reynolds, H.L.; Hooper, D.U.; Lavorel, S.; Sala, O.E.; Hobbie, S.E.; et al. Consequences of Changing Biodiversity. *Nature* **2000**, *405*, 234–242. [CrossRef]
9. de Barros Ruas, R.; Costa, L.M.S.; Bered, F. Urbanization Driving Changes in Plant Species and Communities—A Global View. *Glob. Ecol. Conserv.* **2022**, *38*, e02243. [CrossRef]
10. Chen, J.; Saunders, S.C.; Crow, T.R.; Naiman, R.J.; Brosofske, K.D.; Mroz, G.D.; Brookshire, B.L.; Franklin, J.F. Microclimate in Forest Ecosystem and Landscape Ecology: Variations in Local Climate Can Be Used to Monitor and Compare the Effects of Different Management Regimes. *BioScience* **1999**, *49*, 288–297. [CrossRef]
11. Gómez-Sans, V. Forest covers and microclimate response. *For. Syst.* **2004**, *13*, S84–S100. [CrossRef]
12. Laurance, W.F.; Delamônica, P.; Laurance, S.G.; Vasconcelos, H.L.; Lovejoy, T.E. Rainforest Fragmentation Kills Big Trees. *Nature* **2000**, *404*, 836. [CrossRef]
13. Grimm, N.B.; Faeth, S.H.; Golubiewski, N.E.; Redman, C.L.; Wu, J.; Bai, X.; Briggs, J.M. Global Change and the Ecology of Cities. *Science* **2008**, *319*, 756–760. [CrossRef]
14. McDonnell, M.J.; MacGregor-Fors, I. The Ecological Future of Cities. *Science* **2016**, *352*, 936–938. [CrossRef] [PubMed]
15. McKinney, M.L. Effects of Urbanization on Species Richness: A Review of Plants and Animals. *Urban Ecosyst.* **2008**, *11*, 161–176. [CrossRef]
16. Alvim, F.S.; Furtado, S.G.; Menini Neto, L. Are Vascular Epiphytes in Urban Green Areas Subject to the Homogenization of Biodiversity? A Case Study in the Brazilian Atlantic Forest. *Urban Ecosyst.* **2021**, *24*, 701–713. [CrossRef]
17. Knapp, S.; Winter, M.; Klotz, S. Increasing Species Richness but Decreasing Phylogenetic Richness and Divergence over a 320-Year Period of Urbanization. *J. Appl. Ecol.* **2017**, *54*, 1152–1160. [CrossRef]
18. McKinney, M.L. Urbanization as a Major Cause of Biotic Homogenization. *Biol. Conserv.* **2006**, *127*, 247–260. [CrossRef]
19. Mehltreter, K.; Walker, L.R.; Sharpe, J.M. *Fern Ecology*; Cambridge University Press: Cambridge, UK, 2010; ISBN 978-1-139-48768-9.
20. Kato, M. Biogeography of Ferns: Dispersal and Vicariance. *J. Biogeogr.* **1993**, *20*, 265–274. [CrossRef]
21. Barrington, D.S. Ecological and Historical Factors in Fern Biogeography. *J. Biogeogr.* **1993**, *20*, 275–279. [CrossRef]
22. Brodribb, T.J.; Holbrook, N.M.; Zwieniecki, M.A.; Palma, B. Leaf Hydraulic Capacity in Ferns, Conifers and Angiosperms: Impacts on Photosynthetic Maxima. *New Phytol.* **2005**, *165*, 839–846. [CrossRef]
23. Brodribb, T.J.; McAdam, S.A.M. Passive Origins of Stomatal Control in Vascular Plants. *Science* **2011**, *331*, 582–585. [CrossRef]
24. Kawai, H.; Kanegae, T.; Christensen, S.; Kiyosue, T.; Sato, Y.; Imaizumi, T.; Kadota, A.; Wada, M. Responses of Ferns to Red Light are Mediated by an Unconventional Photoreceptor. *Nature* **2003**, *421*, 287–290. [CrossRef] [PubMed]
25. Schneider, H.; Schuettpelz, E.; Pryer, K.M.; Cranfill, R.; Magallón, S.; Lupia, R. Ferns Diversified in the Shadow of Angiosperms. *Nature* **2004**, *428*, 553–557. [CrossRef]
26. Bergeron, A.; Pellerin, S. Pteridophytes as Indicators of Urban Forest Integrity. *Ecol. Indic.* **2014**, *38*, 40–49. [CrossRef]
27. Carvajal-Hernández, C.I.; Krömer, T.; López-Acosta, J.C.; Gómez-Díaz, J.A.; Kessler, M. Conservation Value of Disturbed and Secondary Forests for Ferns and Lycophytes along an Elevational Gradient in Mexico. *Appl. Veg. Sci.* **2017**, *20*, 662–672. [CrossRef]
28. Krömer, T.; García-Franco, J.G.; Toledo-Aceves, T. Epífitas vasculares como bioindicadores de la calidad forestal: Impacto antrópico sobre su diversidad y composición. In *Bioindicadores: Guardianes de Nuestro Futuro Ambiental*; Instituto Nacional de Ecología y Cambio Climático/El Colegio de la Frontera Sur.: Versailles, KY, USA, 2014; pp. 606–623.

29. Silva, V.L.; Mehltreter, K.; Schmitt, J.L. Ferns as Potential Ecological Indicators of Edge Effects in Two Types of Mexican Forests. *Ecol. Indic.* **2018**, *93*, 669–676. [CrossRef]
30. Karst, A.L.; Lechowicz, M.J. Are Correlations among Foliar Traits in Ferns Consistent with Those in the Seed Plants? *New Phytol.* **2007**, *173*, 306–312. [CrossRef]
31. Kessler, M.; Siorak, Y.; Wunderlich, M.; Wegner, C. Patterns of Morphological Leaf Traits among Pteridophytes along Humidity and Temperature Gradients in the Bolivian Andes. *Funct. Plant Biol.* **2007**, *34*, 963–971. [CrossRef] [PubMed]
32. Schellenberger Costa, D.; Zotz, G.; Hemp, A.; Kleyer, M. Trait Patterns of Epiphytes Compared to Other Plant Life-Forms along a Tropical Elevation Gradient. *Funct. Ecol.* **2018**, *32*, 2073–2084. [CrossRef]
33. Guzmán-Jacob, V.; Guerrero-Ramírez, N.R.; Craven, D.; Brant Paterno, G.; Taylor, A.; Krömer, T.; Wanek, W.; Zotz, G.; Kreft, H. Broad- and Small-Scale Environmental Gradients Drive Variation in Chemical, but not Morphological, Leaf Traits of Vascular Epiphytes. *Funct. Ecol.* **2022**, *36*, 1858–1872. [CrossRef]
34. Petter, G.; Wagner, K.; Wanek, W.; Sánchez Delgado, E.J.; Zotz, G.; Cabral, J.S.; Kreft, H. Functional Leaf Traits of Vascular Epiphytes: Vertical Trends Within the Forest, Intra- and Interspecific Trait Variability, and Taxonomic signals. *Funct. Ecol.* **2016**, *30*, 188–198. [CrossRef]
35. Hietz, P.; Briones, O. Correlation Between Water Relations and within-Canopy Distribution of Epiphytic Ferns in a Mexican Cloud Forest. *Oecologia* **1998**, *114*, 305–316. [CrossRef] [PubMed]
36. Hietz, P.; Briones, O. Photosynthesis, Chlorophyll Fluorescence and Within-Canopy Distribution of Epiphytic Ferns in a Mexican Cloud Forest. *Plant Biol.* **2001**, *3*, 279–287. [CrossRef]
37. Kluge, J.; Kessler, M. Morphological Characteristics of Fern Assemblages along an Elevational Gradient: Patterns and Causes. *Ecotropica* **2007**, *13*, 27–43.
38. Susan-Tepetlan, T.M.; Velázquez-Rosas, N.; Krömer, T. Changes in the functional characteristics of vascular epiphytes from montane cloud forest and secondary vegetation in the central region of Veracruz, Mexico. *Bot. Sci.* **2015**, *93*, 153–163. [CrossRef]
39. Castillo-Campos, G. *Vegetación y Flora Del Municipio de Xalapa, Veracruz*; Instituto de Ecología, A.C., MAB UNESCO, H., Ayuntamiento de Xalapa, Veracruz: Xalapa, Mexico, 1991.
40. Tejero-Díez, J.D.; Torres-Díaz, A.; Mickel, J.T.; Mehltreter, K.; Krömer, T. Helechos y Licopodios. In *La biodiversidad en Veracruz: Estudio de estado*; Comisión Nacional para el Conocimiento y Uso de la Biodiversidad, Gobierno del Estado de Veracruz, Universidad Veracruzana, Instituto de Ecología, A.C.: Xalapa, Mexico, 2011; Volume II, pp. 91–115.
41. Lemoine-Rodríguez, R.; MacGregor-Fors, I.; Muñoz-Robles, C. Six Decades of Urban Green Change in a Neotropical City: A Case Study of Xalapa, Veracruz, Mexico. *Urban Ecosyst.* **2019**, *22*, 609–618. [CrossRef]
42. Falfán, I.; MacGregor-Fors, I. Woody Neotropical Streetscapes: A Case Study of Tree and Shrub Species Richness and Composition in Xalapa. *Madera Bosques* **2016**, *22*, 95–110.
43. García-Campos, H.M. Las áreas verdes públicas de Xalapa. In *Ecología Urbana Aplicada a la Ciudad de Xalapa. Xalapa, México*; Instituto de Ecología, A.C., MAB UNESCO, H. Ayuntamiento de Xalapa, Veracruz: Xalapa, Mexico, 1993; pp. 99–132.
44. Hernández-Rivera, M.G.; Torres-Hernández, L. Analysis of Two Protected Natural Areas in Relation to the Growth of the Metropolitan Area of Xalapa, Veracruz. *Investig. Geográficas* **2015**, *87*, 51–61. [CrossRef]
45. Benítez, G.; Ruelas-Monjardín, L.C.; Von Thaden, J.; Acosta-Rosado, I.; Alvarado-Castillo, G.; Equihua, M. Carbon Storage in a Peri-Urban Neotropical Forest: Assessing Its Potential and Patterns of Change over Half a Century. *Urban For. Urban Green.* **2023**, *86*, 128009. [CrossRef]
46. Lemoine-Rodríguez, R.; Inostroza, L.; Falfán, I.; MacGregor-Fors, I. Too Hot to Handle? On the Cooling Capacity of Urban Green Spaces in a Neotropical Mexican City. *Urban For. Urban Green.* **2022**, *74*, 127633. [CrossRef]
47. MacGregor-Fors, I.; Avendaño-Reyes, S.; Bandala, V.M.; Chacón-Zapata, S.; Díaz-Toribio, M.H.; González-García, F.; Lorea-Hernández, F.; Martínez-Gómez, J.; Montes de Oca, E.; Montoya, L.; et al. Multi-Taxonomic Diversity Patterns in a Neotropical Green City: A Rapid Biological Assessment. *Urban Ecosyst.* **2015**, *18*, 633–647. [CrossRef]
48. Jara-Toto, E.; Armenta-Montero, S.; Aquino-Zapata, A.M.; Hernández, C.C. Diversidad y estructura de la vegetación leñosa en cuatro bosques urbanos de la zona conurbada Xalapa-Banderilla, Veracruz, Mexico. *Acta Bot. Mex.* **2023**, *130*, e2214. [CrossRef]
49. Vargas-Huipe, N.D.; Rodríguez-Van Gort, M.F. Clima cambiante en la Zona Metropolitana de Xalapa: Factores naturales y antrópicos. *Investig. Geográficas* **2024**, *114*, e60855. [CrossRef]
50. Chazdon, R.L.; Pearcy, R.W. The Importance of Sunflecks for Forest Understory Plants. *BioScience* **1991**, *41*, 760–766. [CrossRef]
51. Jennings, S.B.; Brown, N.D.; Sheil, D. Assessing Forest Canopies and Understorey Illumination: Canopy Closure, Canopy Cover and Other Measures. *Forestry* **1999**, *72*, 59–74. [CrossRef]
52. Lambers, H.; Chapin, S.; Pons, T.L. *Plant Physiological Ecology*; Springer: New York, NY, USA, 2008.
53. Matlack, G.R. Microenvironment Variation within and among Forest Edge Sites in the Eastern United States. *Biol. Conserv.* **1993**, *66*, 185–194. [CrossRef]

54. Zellweger, F.; Coomes, D.; Lenoir, J.; Depauw, L.; Maes, S.L.; Wulf, M.; Kirby, K.J.; Brunet, J.; Kopecký, M.; Máliš, F.; et al. Seasonal Drivers of Understorey Temperature Buffering in Temperate Deciduous Forests Across Europe. *Glob. Ecol. Biogeogr.* **2019**, *28*, 1774–1786. [CrossRef]
55. Santiago-Pérez, A.L.; Jardel-Peláez, E.J.; Cuevas-Guzmán, R.; Huerta-Martínez, F.M. Vegetación de bordes en un Bosque Mesófilo de Montaña del Occidente de México. *Bol. Soc. Botánica México* **2009**, *85*, 31–49. [CrossRef]
56. Ewers, R.M.; Banks-Leite, C. Fragmentation Impairs the Microclimate Buffering Effect of Tropical Forests. *PLoS ONE* **2013**, *8*, e58093. [CrossRef]
57. Geiger, R.; Aron, R.H.; Todhunter, P. *The Climate Near the Ground*; Rowman & Littlefield: Lanham, MD, USA, 2009.
58. Bach, K.; Schawe, M.; Beck, S.; Gerold, G.; Gradstein, S.R.; Moraes, R.M. Vegetación, suelos y clima en los diferentes pisos altitudinales de un bosque montano de Yungas, Bolivia: Primeros resultados. *Ecol. En Boliv. Rev. Inst. Ecol.* **2003**, *38*, 3–14.
59. Pérez-Rendón, E.P.; Ramírez-Builes, V.H.; Peña-Quiñones, A.J. Variabilidad espacial y temporal de la temperatura del aire en la zona cafetera colombiana. *Investig. Geográficas* **2016**, *89*, 23–40. [CrossRef]
60. Bruijnzeel, L.A.; Proctor, J. Hydrology and Biogeochemistry of Tropical Montane Cloud Forests: What Do We Really Know? In *Tropical Montane Cloud Forests*; Hamilton, L.S., Juvik, J.O., Scatena, F.N., Eds.; Springer: New York, NY, USA, 1995; pp. 38–78.
61. Kitayama, K. Biophysical Conditions of the Montane Cloud Forests of Mount Kinabalu, Sabah, Malaysia. In *Tropical Montane Cloud Forests*; Hamilton, L.S., Juvik, J.O., Scatena, F.N., Eds.; Springer: New York, NY, USA, 1995; pp. 183–197.
62. Schawe, M.; Gerold, G.; Bach, K.; Gradstein, S.R. Hydrometeorological Patterns in Relation to Montane Forest Types along an Elevational Gradient in the Yungas of Bolivia. In *Tropical Montane Cloud Forests: Science for Conservation and Management*; Scatena, F.N., Bruijnzeel, L.A., Hamilton, L.S., Eds.; International Hydrology Series; Cambridge University Press: Cambridge, UK, 2011; pp. 199–207. ISBN 978-0-521-76035-5.
63. Bruijnzeel, L.A.; Mulligan, M.; Scatena, F.N. Hydrometeorology of tropical montane cloud forests: Emerging patterns. *Hydrol. Process.* **2011**, *25*, 465–498. [CrossRef]
64. Foster, P. Changes in mist immersion. In *Tropical Montane Cloud Forests: Science for Conservation and Management*; Scatena, F.N., Bruijnzeel, L.A., Hamilton, L.S., Eds.; International Hydrology Series; Cambridge University Press: Cambridge, UK, 2010; pp. 57–66, ISBN 978-0-521-76035-5.
65. Pérez-Harguindeguy, N.; Diaz, S.; Garnier, E.; Lavorel, S.; Poorter, H.; Jaureguiberry, P.; Bret-Harte, M.; Cornwell, W.; Craine, J.; Gurvich, D.; et al. New Handbook for Standardised Measurement of Plant Functional Traits Worldwide. *Aust. J. Bot.* **2013**, *61*, 167–234. [CrossRef]
66. Benzing, D.H. *Vascular Epiphytes: General Biology and Related Biota*; Cambridge University Press: Cambridge, UK, 1990.
67. Krömer, T.; Kessler, M.; Gradstein, S.R. Vertical Stratification of Vascular Epiphytes in Submontane and Montane Forest of the Bolivian Andes: The Importance of the Understory. *Plant Ecol.* **2007**, *189*, 261–278. [CrossRef]
68. Reich, P.B. The World-Wide 'Fast–Slow' Plant Economics Spectrum: A Traits Manifesto. *J. Ecol.* **2014**, *102*, 275–301. [CrossRef]
69. Zotz, G.; Hietz, P. The Physiological Ecology of Vascular Epiphytes: Current Knowledge, Open Questions. *J. Exp. Bot.* **2001**, *52*, 2067–2078. [CrossRef]
70. Hernández-Zamora, D. Riqueza de epífitas vasculares en las Áreas Verdes Urbanas y Periurbanas de Xalapa, Veracruz. Bachelor's Thesis, Universidad Veracruzana, Xalapa, Mexico, 2022.
71. Poorter, H.; Niinemets, Ü.; Poorter, L.; Wright, I.J.; Villar, R. Causes and Consequences of Variation in Leaf Mass Per Area (LMA): A Meta-Analysis. *New Phytol.* **2009**, *182*, 565–588. [CrossRef]
72. Givnish, T.J. Adaptation to Sun and Shade: A Whole-Plant Perspective. *Funct. Plant Biol.* **1988**, *15*, 63–92. [CrossRef]
73. Valladares, F.; Niinemets, Ü. Shade Tolerance, A Key Plant Feature of Complex Nature and Consequences. *Annu. Rev. Ecol. Evol. Syst.* **2008**, *39*, 237–257. [CrossRef]
74. Hietz, P.; Wagner, K.; Nunes Ramos, F.; Cabral, J.S.; Agudelo, C.; Benavides, A.M.; Cach-Pérez, M.J.; Cardelús, C.L.; Chilpa Galván, N.; Erickson Nascimento da Costa, L.; et al. Putting Vascular Epiphytes on the Traits Map. *J. Ecol.* **2022**, *110*, 340–358. [CrossRef]
75. Wright, I.J.; Reich, P.B.; Westoby, M.; Ackerly, D.D.; Baruch, Z.; Bongers, F.; Cavender-Bares, J.; Chapin, T.; Cornelissen, J.H.C.; Diemer, M.; et al. The Worldwide Leaf Economics Spectrum. *Nature* **2004**, *428*, 821–827. [CrossRef]
76. Vile, D.; Garnier, E.; Shipley, B.; Laurent, G.; Navas, M.-L.; Roumet, C.; Lavorel, S.; Díaz, S.; Hodgson, J.G.; Lloret, F.; et al. Specific Leaf Area and Dry Matter Content Estimate Thickness in Laminar Leaves. *Ann. Bot.* **2005**, *96*, 1129–1136. [CrossRef]
77. He, N.; Liu, C.; Tian, M.; Li, M.; Yang, H.; Yu, G.; Guo, D.; Smith, M.D.; Yu, Q.; Hou, J. Variation in Leaf Anatomical Traits from Tropical to Cold-Temperate Forests and Linkage to Ecosystem Functions. *Funct. Biol.* **2018**, *32*, 10–19. [CrossRef]
78. Mediavilla, S.; Gallardo-López, V.; González-Zurdo, P.; Escudero, A. Patterns of Leaf Morphology and Leaf N Content in Relation to Winter Temperatures in Three Evergreen Tree Species. *Int. J. Biometeorol.* **2012**, *56*, 915–926. [CrossRef]
79. Xu, L.; Zhang, N.; Wei, T.; Liu, B.; Shen, L.; Liu, Y.; Liu, D. Adaptation Strategies of Leaf Traits and Leaf Economic Spectrum of Two Urban Garden Plants in China. *BMC Plant Biol.* **2023**, *23*, 274. [CrossRef]

80. Körner, C.; Neumayer, M.; Menendez-Riedl, S.P.; Smeets-Scheel, A. Functional Morphology of Mountain Plants. *Flora* **1989**, *182*, 353–383. [CrossRef]
81. Niinemets, Ü. Does the Touch of Cold Make Evergreen Leaves Tougher? *Tree Physiol.* **2016**, *36*, 267–272. [CrossRef]
82. Gehrig-Downie, C.; Marquardt, J.; Obregón, A.; Bendix, J.; Gradstein, S.R. Diversity and Vertical Distribution of Filmy Ferns as a Tool for Identifying the Novel Forest Type "Tropical Lowland Cloud Forest". *Ecotropica* **2012**, *18*, 35–44.
83. Krömer, T.; Acebey, A.; Kluge, J.; Kessler, M. Effects of Altitude and Climate in Determining Elevational Plant Species Richness Patterns: A Case Study from Los Tuxtlas, Mexico. *Flora Morphol. Distrib. Funct. Ecol. Plants* **2013**, *208*, 197–210. [CrossRef]
84. Dubuisson, J.-Y.; Hennequin, S.; Rakotondrainibe, F.; Schneider, H. Ecological Diversity and Adaptive Tendencies in the Tropical Fern *Trichomanes* L. (Hymenophyllaceae) with Special Reference to Climbing and Epiphytic Habits. *Bot. J. Linn. Soc.* **2003**, *142*, 41–63. [CrossRef]
85. Saldaña, A.; Parra, M.J.; Flores-Bavestrello, A.; Corcuera, L.J.; Bravo, L.A. Effects of Forest Successional Status on Microenvironmental Conditions, Diversity, and Distribution of Filmy Fern Species in a Temperate Rainforest. *Plant Species Biol.* **2014**, *29*, 253–262. [CrossRef]
86. Larrea, M.L.; Werner, F.A. Response of Vascular Epiphyte Diversity to Different Land-Use Intensities in a Neotropical Montane Wet Forest. *For. Ecol. Manag.* **2010**, *260*, 1950–1955. [CrossRef]
87. Wolf, J.H.D. The Response of Epiphytes to Anthropogenic Disturbance of Pine-Oak Forests in the Highlands of Chiapas, Mexico. *For. Ecol. Manag.* **2005**, *212*, 376–393. [CrossRef]
88. Zhang, H.; Zhu, S.; John, R.; Li, R.; Liu, H.; Ye, Q. Habitat Filtering and Exclusion of Weak Competitors Jointly Explain Fern Species Assemblage Along a Light and Water Gradient. *Sci. Rep.* **2017**, *7*, 298. [CrossRef]
89. Instituto Nacional de Estadística y Geografía (INEGI). Compendio de Información Geográfica Municipal de Los Estados Unidos Mexicanos 2010. *Xalapa, Veracruz de Ignacio de la Llave.* 2010. Available online: https://www.inegi.org.mx (accessed on 16 March 2025).
90. Sistema de Información Estadística y Geográfica del Estado de Veracruz de Ignacio de la Llave (SIEGVER). Cuadernillos Municipales 2020: Xalapa. *Gobierno del Estado de Veracruz: Veracruz, Mexico.* 2020. Available online: https://ceieg.veracruz.gob.mx (accessed on 22 March 2025).
91. Miranda, F.; Hernández-X, E. Los tipos de vegetación de México y su clasificación. *Bot. Sci.* **1963**, *28*, 29–179. [CrossRef]
92. Rzedowski, J. *La Vegetación de México*; Limusa S.A.: Mexico City, Mexico, 1978.
93. Instituto Nacional de Estadística y Geografía (INEGI). ¿Cuántos Habitantes Tiene Xalapa?—Censo de Población y Vivienda 2020. Available online: https://www.inegi.org.mx/app/cpv/2020/resultadosrapidos/default.html?texto=Xalapa (accessed on 10 February 2025).
94. Instituto Nacional de Estadística y Geografía (INEGI). Censo de Población y Vivienda 2020: Población Total de Banderilla. Available online: https://censo2020.mx (accessed on 10 February 2025).
95. Frazer, G.W.; Canham, C.D.; Lertzman, K.P. *Gap Light Analizer (GLA)*, Version 2.0; Simon Fraser University: Burnaby, BC, Canada, 1999.
96. Zuur, A.F.; Ieno, E.N.; Walker, N.; Saveliev, A.A.; Smith, G.M. *Mixed Effects Models and Extensions in Ecology with R*, 1st ed.; Statistics for Biology and Health; Springer: New York, NY, USA, 2009; ISBN 978-0-387-87457-9.
97. Fox, J. *Applied Regression Analysis and Generalized Linear Models*; SAGE Publications: Thousand Oaks, CA, USA, 2015; ISBN 978-1-4833-2131-8.
98. O'Hara, R.; Kotze, J. Do not Log-Transform Count Data. *Nat. Preced.* **2010**, *1*, 118–122. [CrossRef]
99. Nakagawa, S.; Schielzeth, H. A General and Simple Method for Obtaining R2 from Generalized Linear Mixed-Effects Models. *Methods Ecol. Evol.* **2013**, *4*, 133–142. [CrossRef]
100. RStudio Team. *RStudio: Integrated Development Environment for R, Version 2024.09.1*; RStudio Team: Boston, MA, USA, 2024.
101. The Jamovi Project. *Jamovi*, Versión 2.3.28.0; Jamovi: Sydney, Australia, 2023.
102. McKinney, M.L. Urbanization, Biodiversity, and Conservation: The Impacts of Urbanization on Native Species Are Poorly Studied, but Educating a Highly Urbanized Human Population about These Impacts Can Greatly Improve Species Conservation in All Ecosystems. *BioScience* **2002**, *52*, 883–890. [CrossRef]
103. MacGregor-Fors, I.; Falfán, I.; García-Arroyo, M.; Lemoine-Rodríguez, R.; Gómez-Martínez, M.A.; Marín-Gómez, O.H.; Pérez-Maqueo, O.; Equihua, M. A Novel Approach for the Assessment of Cities through Ecosystem Integrity. *Land* **2022**, *11*, 3. [CrossRef]

plants

MDPI

Article

Species Richness, Abundance, and Vertical Distribution of Epiphytic Bromeliads in Primary Forest and Disturbed Forest

Sugeidi S. Siaz Torres [1], Edilia de la Rosa-Manzano [1,2,*], Leonardo U. Arellano-Méndez [1,2], Karla M. Aguilar-Dorantes [3], José Guadalupe Martínez Ávalos [1] and María Cruz Juárez Aragón [1]

1 Instituto de Ecología Aplicada, Universidad Autónoma de Tamaulipas, Avenida División del Golfo Núm, 356, Colonia Libertad, Ciudad Victoria 87019, Mexico; sugey_9876@hotmail.com (S.S.S.T.); luarellano@uat.edu.mx (L.U.A.-M.); jmartin@uat.edu.mx (J.G.M.Á.); mcjuarez@uat.edu.mx (M.C.J.A.)
2 Facultad de Ingeniería y Ciencias, Universidad Autónoma de Tamaulipas, Ciudad Victoria 87149, Mexico
3 Centro de Investigación en Biodiversidad y Conservación, Universidad Autónoma del Estado de Morelos, Cuernavaca 62209, Mexico; karla.aguilar@uaem.mx
* Correspondence: ermanzano@uat.edu.mx

Abstract: Epiphytes represent a key component in tropical forests. They are affected by anthropogenic and natural disturbances suffered by forests, since they depend on their hosts and the microclimatic conditions they generate. We analyzed differences in abundance, species richness, and vertical distributions of epiphytic bromeliads in primary and disturbed forests. We found a higher abundance (5316 individuals) and species richness (8 species) of bromeliads in disturbed forest than in primary forest (1360 individuals and 4 species, respectively). Most bromeliads (97%) were found on *Taxodium mucronatum*, a dominant tree with rough bark in the disturbed forest (gallery forest). Bromeliads were more abundant in the middle of the tree and diminished towards the trunk base and the upper crown. *Tillandsia baileyi* was the most abundant bromeliad, and the size categories of this species differentially colonize trees in gallery forest according to Johansson zones; seedlings of *T. baileyi* abundantly colonize the upper canopy, and juveniles colonize the middle canopy or secondary branches. Gallery forest represents an important reservoir for epiphytic bromeliads. Hence, it is important to extend this kind of study to wetland sites to understand the role they play as a habitat for epiphytes, as well as the dynamics and ecological processes that occur in such habitats.

Keywords: gallery forest; Johansson zone; plant size; submontane scrub; *Tillandsia*

1. Introduction

Tropical forests face serious problems due to land-use changes affecting biodiversity. Several studies have demonstrated that the number of epiphyte species is greater in conserved sites or tropical primary forest due to microclimatic conditions such as temperature, light, and humidity being more favorable for the establishment and growth of epiphytes, thereby helping to maintain their relationships with pollinators, dispersers, and mycorrhizas, among others [1–3]. Other studies have reported higher species richness and epiphyte abundance in isolated trees or secondary forests due to drought-tolerant species dominating modified ecosystems [4], while a few works found no differences [5,6]. Although vascular epiphytes are a major element of tropical forest ecosystems, our understanding of how disturbance affects epiphyte diversity is scarce. Epiphytes are considered an indicator of environmental quality, since they are sensible to microclimate shifts caused by anthropic disturbance [7]. In fact, the level of forest disturbance determines the plant composition and species richness.

Vascular epiphytes are relevant components of tropical forest biodiversity; they comprise over 31,000 species in 79 families and 918 genera, accounting for about 10% of total plants [8]. Epiphytes establish relationships with pollinators and dispersers in the canopy and form part of the water and nutrient cycle in tropical forests [9–11]. Among the best-represented families of vascular epiphytes are Orchidaceae, Araceae, Piperaceae, and Bromeliaceae [12,13]. The latter group is a Neotropical family integrated by a high diversity of terrestrial, saxicolous, and epiphytic species distributed from humid to arid environments [14]. In addition to their ecological contribution, bromeliads provide substantial ecosystem services in terms of capturing water, in addition to providing shelter and breeding sites for a variety of organisms, such as amphibians, bats, and mutualistic ants and spiders [10,15]; they are also a biological indicator of environmental disturbance and climate change [16].

A gallery forest or riparian forest is an ecosystem dominated by *Taxodium mucronatum* Ten. trees in the riparian corridors of major Mexican rivers [17,18]. This ecosystem provides landscape diversity and plays an important role in the ecotone dynamics of swamps and other wetlands [19,20]. Vascular epiphytes are rarely studied in gallery forests, maybe because such forests have lower abundance and species richness [21], in spite of individual trees such as *T. mucronatum* being abundant and dominant [22,23], which may increase habitat diversity for epiphytes, combined with the permanent presence of water. For example, members of the *Peperomia* genus have a marked preference for warm or temperate sites but are found in high-humidity environments such as gallery forest [24]. This ecosystem has been affected by agriculture and paddocks that have been established in surrounding areas, altering microclimatic conditions, including water flux [25], forming patches of vegetation on the borders of rivers or water channels, which affect the epiphyte community.

Other ecosystems, such as submontane scrub, are characterized by the prevalence of drought conditions for epiphytes due to low precipitation and high temperatures [26]. Species belonging to the *Tillandsia* genus are common in this ecosystem, since they possess morphological and anatomical traits such as narrow leaves, abundant trichomes, and crassulacean acid metabolism, avoiding water loss by transpiration [27,28]. Species such as *T. recurvata* L., *T. schiedeana* Steud., and *T. pringlei* S. Watson have also been reported in submontane scrub, although they are the least abundant species [29].

Vascular epiphytes show vertical distribution patterns on trees, indicating their ranges of tolerance to light, moisture, and other microclimatic factors [30]. Usually, epiphytes are more abundant in the intermediate zone on host trees and least abundant in the upper crown [10,31,32], where the prevalence of high radiation levels and less water availability are restrictive for some species [33,34]. The bark type of host trees plays an important role in successful dispersal, since rough bark is better at catching seeds than smooth bark, from which seeds can slip to the ground [35–37]. Additionally, host size is a determinant factor for the establishment of epiphytes, since taller individuals are older and have a greater probability of increasing epiphytic colonization [38].

The vertical distribution of vascular epiphytes on host trees can vary according to plant size. It has been suggested that adult (or larger) plants occupy the thickest branches because they can support more weight, while seedlings (or smaller plants) are distributed on the periphery or in the upper crown [39–41]. However, the vertical distribution of plant size in modified or disturbed ecosystems is poorly studied. Therefore, the success of epiphytes is related to the size and position they occupy in the host tree, among other factors. We propose the following questions: (a) How do the abundance, species richness, and vertical distribution of epiphyte bromeliads vary in forests with different degrees of conservation? (b) How does the vertical distribution of plant size of *Tillandsia baileyi* Rose ex Small differ in gallery forest? We hypothesize that (i) abundance and species richness are higher in gallery forest due to the frequency of high humidity, despite being a modified ecosystem, and (ii) the distribution of adult plants of *T. baileyi* is higher in the middle zone of host trees, while seedlings and juvenile plants occupy the upper and lower strata.

2. Results

2.1. Abundance, Species Richness, and Vertical Distribution of Bromeliads at Two Study Sites

Species richness of bromeliads was two-fold in the gallery forest (eight species) in comparison to submontane scrub (four species). Additionally, the abundance of epiphytic bromeliads was higher in the gallery forest (5316 individuals) than in submontane scrub (1360 individuals) ($p < 0.05$; Figure 1). *Tillandsia baileyi* was the most abundant in the gallery forest (3467 individuals), followed by *T. ionantha* Planch. (811 individuals) and *T. usneoides* L. (688 individuals). In contrast, *T. usneoides* was the most abundant (1170 individuals) in submontane scrub, and *T. baileyi* was the least abundant (19 individuals) (Figure 2). In the gallery forest, *Taxodium mucronatum* had the highest relative abundance, coinciding with its high relative dominance and IVI (Table 1), while in submontane scrub, two host species (*Ehretia anacua* (Terán & Berland.) I. M. Johnst and *Ocotea tampicensis* (Meisn.) Hemsl.) had the highest relative abundance, relative dominance, and IVI. Despite *Quercus* sp. being fourth in IVI, it hosts half of the epiphyte bromeliads in submontane scrub (Table 2). We found one host species of each type of bark in the gallery forest, but in submontane scrub, 61.5% of the trees had smooth bark, followed by 30.8% with rugose bark and 7.6% with semi-rugose bark.

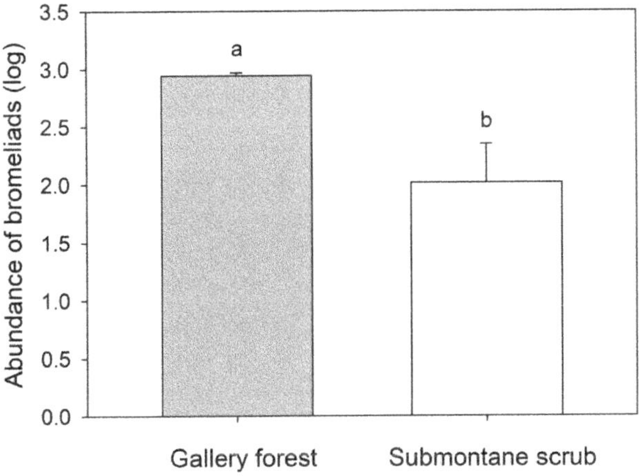

Figure 1. Abundance of epiphytic bromeliads in gallery forest and submontane scrub in Tamaulipas, Mexico. Each bar represents the mean ± SE. Different letters indicate significant differences with a p value of <0.05.

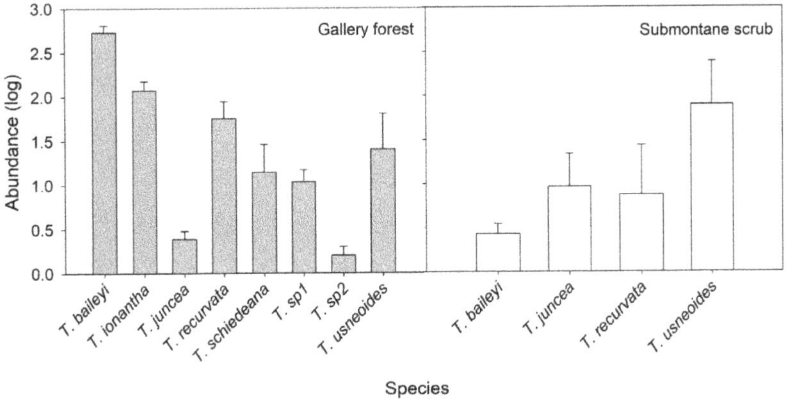

Figure 2. Abundance of Bromeliaceae species in gallery forest and submontane scrub in Tamaulipas, Mexico. Each bar represents the mean ± SE.

Table 1. Host-tree traits from the gallery forest (DBH ≥ 10 cm), including bark type, relative abundance (%), relative frequency, relative dominance (%), importance value index (IVI), and epiphyte abundance (%).

Tree Species	Family	Bark Type	Relative Abundance (%)	Relative Frequency (%)	Relative Dominance (%)	IVI 100%	Epiphyte Abundance (%)
Taxodium mucronatum	Cupressaceae	Rugose	88.88	60	80.08	76.34	97.31
Ficus sp.	Moraceae	Smooth	8.35	30	19.89	19.40	2.60
Tabernaemontana alba Mill.	Apocynaceae	Semi-rugose	2.77	10	0.03	4.26	0.09
	Total		100	100	100	100	100

Study sites and Johansson zones influenced the abundance of epiphytic bromeliads ($F_{3,43} = 15.1$, $p < 0.001$). JZ2, 3, and 4 contributed positively to the model, while submontane scrub had a negative effect on the abundance of epiphytic bromeliads (Table 3). Differences were found only between JZ1 and other zones (JZ2, JZ3, and JZ4; $p < 0.001$; Table S1; Figure S2). Moreover, the abundance of epiphytic bromeliads differed between sites ($p < 0.001$).

2.2. Distribution of T. baileyi in Two Forests and on Host Tree Species

The abundance of *T. baileyi* significatively differed among size categories ($F_{2,32} = 3.37$, $p < 0.05$) and between study sites ($F_{1,32} = 270.15$, $p < 0.001$) (Figure 3). Submontane scrub was the only variable that significantly contributed to the model, although it was negative ($p < 0.001$) (Table S2). The only size-pair category that was significant was juvenile vs. seedling *T. baileyi* ($p < 0.05$) (Table S3). The abundance of each size category of *T. baileyi* was significantly differed among JZs in the gallery forest ($X^2 \leq 494.99$, gl = 3, $p < 0.001$; Figure 4). However, in submontane scrub, only the juvenile category showed differences among the JZs ($X^2 \leq 19.143$, gl = 3, $p < 0.005$; Figure 4).

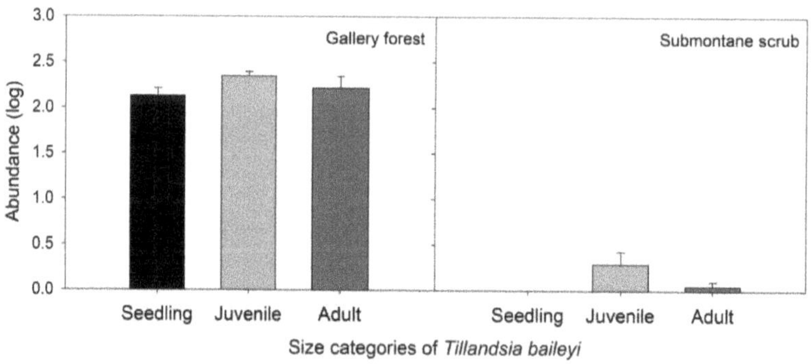

Figure 3. Abundance of *Tillandsia baileyi* based on size categories in gallery forest and submontane scrub in Tamaulipas, Mexico. Each bar represents the mean ± SE.

Contingency table analysis showed that the distribution of *T. baileyi* by category size is not homogeneous in the Johansson zones of host trees in the gallery forest ($X^2 = 87.5$, df = 6, $p < 0.001$, Table S4). Standard residual analysis indicated that seedlings of *T. baileyi* were significantly more abundant than expected by chance in Johansson zones 1 and 4, where seedlings have a 34% chance of occurrence. Moreover, juvenile plants of this species were more abundant in Johansson zone 3, with 43% of occurrence on trees in the same forest (Figure 5).

Table 2. Host-tree traits in submontane scrub (DBH ≥ 10 cm), including bark type, relative abundance (%), relative frequency, relative dominance (%), importance value index (IVI), and epiphyte abundance (%).

Tree Species	Family	Bark Type	Relative Abundance (%)	Relative Frequency (%)	Relative Dominance (%)	IVI 100%	Epiphyte Abundance (%)
Ehretia anacua	Boraginaceae	Rugose	20.00	12.90	30.61	21.17	25.12
Ocotea tampicensis	Lauraceae	Smooth	24.71	12.90	15.63	17.75	2.29
Zanthophyllum fagara (L.) Sarg.	Rutaceae	Rugose	11.76	9.68	8.97	10.14	7.74
Quercus sp.	Fagaceae	Rugose	5.88	9.68	14.78	10.11	49.85
Havardia pallens (Benth.) Britton & Rose.	Fabaceae	Smooth	8.24	6.45	4.99	6.56	2.07
Sebastiana pavonia Muell.	Euohorbiaceae	Semi-rugose	7.06	9.68	2.39	6.37	1.69
Pithecellobium flexicaule (Benth.) Coult.	Fabaceae	Rugose	5.88	6.45	5.94	6.09	0.61
Casimiroa greggi (S.Watson) F.Chiang	Rutaceae	Smooth	3.53	9.68	4.68	5.96	0
Ugnadia speciosa Endl.	Sapindaceae	Smooth	3.53	6.45	3.09	4.36	7.73
Randia obcordata S. Watson	Rubiaceae	Smooth	3.53	3.23	5.30	4.02	0.15
Acacia coulteri Benth.	Fabaceae	Smooth	3.53	6.45	1.46	3.81	0
Persea liebmannii Mez	Lauraceae	Smooth	1.18	3.23	1.83	2.08	2.75
Robinsonella discolor Rose & Baker f. ex Rose	Malvaceae	Smooth	1.18	3.23	0.34	1.58	0
Total			100	100	100	100	100

Table 3. Generalized linear model with quasi-Poisson distribution where the variables of Johansson zone (JZ2, JZ3, and JZ4) and the submontane scrub were evaluated. Variable JZ1 and the gallery forest are not presented as coefficients because they were taken as reference points. p value < 0.001.

	Estimate	Standard Error	t Value	Pr(>|t|)
Intercept	3.730	0.394	9.45	0.001
JZ2	1.759	0.416	4.22	0.001
JZ3	2.116	0.411	5.14	0.001
JZ4	1.673	0.422	3.96	0.001
Submontane scrub	−1.324	0.197	−6.70	0.001

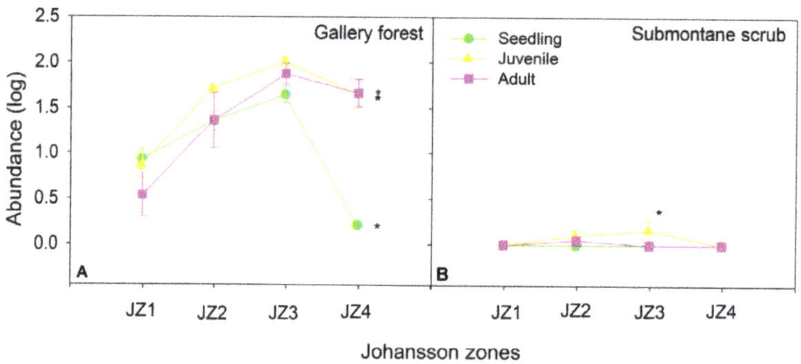

Figure 4. Size categories (seedling, juvenile, and adult) of *Tillandsia baileyi* based on Johansson zone in the gallery forest and submontane scrub in Tamaulipas, Mexico. Each point represents the mean ± SE; asterisk indicate significant differences among Johansson zones for each size category with $p < 0.05$.

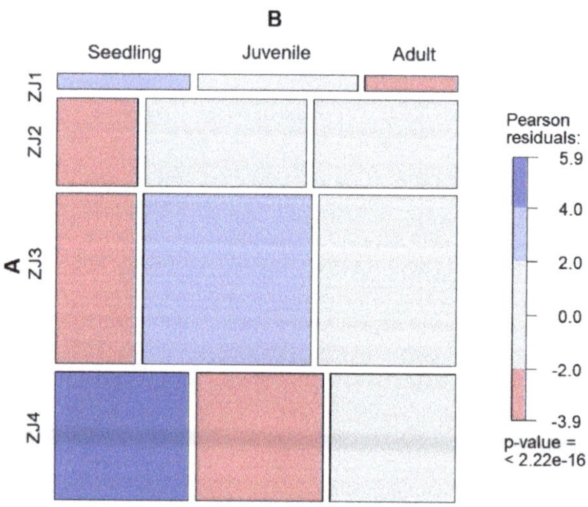

Figure 5. Mosaic display of the standardized residuals of (**A**) Johansson zones (JZ1, JZ2, JZ3, and JZ4) and (**B**) size categories of *Tillandsia baileyi* in the gallery forest. Blue boxes indicate a significant association between epiphyte species and hosts ($p < 0.05$). Association was considered positive when standardized residual values were >2 and negative when they were <−2 [42].

3. Discussion

Epiphytes are generally more abundant and species-rich in primary forests than in disturbed or modified landscapes, a pattern observed in diverse ecosystems [1,7,43–46]. For instance, 178 species of vascular epiphytes were recorded in primary forest, compared with only 81 species found in secondary forest in the Venezuelan Andes [43]. However, our results did not fulfill this expectation, as both the abundance and species richness of bromeliads were higher in the gallery forest than the disturbed forest. This unexpected result could be explained by the ability of certain epiphyte species to thrive in disturbed environments or on isolated trees due to their drought-adaptive traits [45,47–49]. For example, some ecosystems, such as wetlands, are characterized by high humidity and support greater epiphyte biodiversity, despite human disturbance being less pronounced in such environments [46,50,51].

Although the gallery forest represents a disturbed site, the species richness of bromeliads was comparable to that found in drier, more conserved habitats, such as the tropical dry forest and the submontane scrub in the Biosphere Reserve of "El Cielo" [29], where species richness is relatively lower. Despite its disturbance, the gallery forest supports a relatively high abundance of specific bromeliad species, while generally known for hosting only 8% of Mexico's total epiphyte species (144 species) [21]. In our study, atmospheric bromeliads like *Tillandsia baileyi*, *T. ionantha*, and *T. usneoides* were most abundant in the gallery forest. Similarly, *T. usneoides* and *T. recurvata* were dominant in the submontane scrub. These species cataloged as atmospheric epiphytes have morphological adaptations such as narrow leaves and abundant trichomes, which enhance their ability to capture atmospheric water and nutrients, facilitating their survival in disturbed or drier environments [52]. In contrast, less abundant tank bromeliads such as *T.* sp1 and *T.* sp2 exhibit adaptations like broad, flat leaves that form water-holding chambers, which allow them to capture and store water from fog or dew. This water retention capability likely increases resilience in the species in drier and sun-exposed environments [53]. These ecological adaptations offer a potential explanation for the success of bromeliads in both disturbed and conserved sites, highlighting the complex relationship between disturbance and epiphyte diversity.

On the other hand, low abundance and species richness in the submontane scrub coincides with previous findings with respect to the same vegetation type within the "El Cielo" Biosphere Reserve, where epiphytes were also found to be less abundant than semideciduous and tropical montane cloud forest [29]. This may be attributed to the environmental conditions in submontane scrub characterized by high irradiance and drought [26], where trees and shrubs have broad crowns that expose epiphytes to increased light levels. Only species adapted to these stressful conditions, such as drought-tolerant bromeliads, can survive under the intense radiation in this habitat. For instance, *T. usneoides*, one of the few species present, has small leaves (43 mm leaf length; [52]) covered by abundant trichomes that aid in water absorption and uses crassulacean acid metabolism (CAM) to minimize water loss through transpiration [52]. Another plausible factor contributing to low epiphyte abundance is the prevalence of trees with smooth bark; in submontane scrub, approximately 70% of trees have smooth bark, which lacks the surface irregularities necessary to anchor fragile coma hairs.

In contrast *T. baileyi*, an atmospheric epiphyte, was highly abundant in the gallery forest, constituting 97% of all bromeliads and colonizing 89% of *T. mucronatum* trees, which were the primary host in this ecosystem (Table 1). While the abundance of *T. baileyi* was notable in our study, its overall distribution in Mexico has been declining. Previously recorded across southern Mexico [54], recent reports suggest its current distribution is limited to the northeast [21], likely due to the extensive transformation of tropical dry forest caused by agricultural expansion and human disturbance [55]. The tropical dry forest, once a suitable habitat for *T. baileyi*, [54], may no longer support its populations effectively.

Our study indicates that *T. mucronatum* trees, which grow in disturbed gallery forest areas surrounded by agricultural land, play a critical role in supporting epiphytic bromeliads. These trees are large and old and have extensive, well-branched canopies (the widest tree has a DBH of 228 cm)—traits that enhance their suitability as phorophytes. Additionally, their crowns provide a heterogeneous microenvironment for epiphytes, remaining leafy for most of the year, except from December to January, when sunlight is lowest [56]. This structural complexity is crucial for bromeliad colonization, as tree height and branch circumference have been positively correlated with the abundance of epiphytic bromeliads in different forest types, including the Atlantic Rainforest [57] and tropical forests [58]. Larger trees offer multiple shaded habitats, which are favorable for the growth of epiphytes [40,59,60].

Moreover, *T. mucronatum* may function as remanent tree species in gallery forest, providing essential support structures for epiphytic bromeliads. Studies have shown that isolated trees can harbor a significant number of epiphyte species; for example, one tree was found to support 34 epiphytic species, compared to a maximum of 66 species in primary forest [43]. This underscores the importance of remanent hosts for the conservation of epiphytes. It is also unlikely that seed dispersal limitations explain the abundance patterns observed in the gallery forest, as epiphyte seeds have been shown to disperse over distances of up to 2 km from their source [61]. Despite being surrounding by agricultural land, isolated patches of tropical dry deciduous forest may serve as a viable seed source for epiphytic colonization in the gallery forest.

Additionally, cypress (*Taxodium* sp.) host trees exhibit rough bark that facilitates the anchoring of bromeliad seeds, as similarly observed with *Taxodium ascendens* Brongn in Florida, USA; this rough bark hosts abundant *Tillandsia circinatta* Schlecht [62]. The rough bark of these trees presents fissures and cavities that can accumulate dust, moisture, and nutrients, providing an advantageous microenvironment for epiphytes [35,63]. In this study, cypresses growing along creek edges were found to benefit from the permanent availability of water, which contributes to a humid microclimate that mitigates the harsh conditions typically found in epiphytic habitats. This is analogous to findings in a tropical dry forest, where bromeliad abundance was found to decrease with distance from cenotes (water holes), as sites closer to cenotes experience higher nighttime humidity, which contributes to the maintenance of the bromeliad community [64].

Our results indicate that epiphytic bromeliads were more abundant in Johansson zones 2, 3, and 4 at both study sites, where host trees exhibit well-branched canopies. This pattern aligns with other studies suggesting that intermediate zones (JZ2 and 3) on the host trees provide more colonization area, favorable microclimates, and increased survival chances for epiphytes [10,31,34,65].

The abundance of *T. baileyi* sizes varied significantly between seedlings and juveniles, with fewer seedlings indicating low recruitment rates. While other clonal species also showed low seedling recruitment despite high-speed production [66–68], it is generally observed that smaller plants are more susceptible to desiccation during the dry season, which is a critical phase for epiphyte establishment. Seedlings are particularly vulnerable in the epiphytic habitat, which poses a significant challenge for their survival [68]. Understanding population structure is crucial when implementing conservation strategies, especially when detailed demographic data are lacking [69].

Tillandsia baileyi displayed differential colonization patterns on *T. mucronatum* host trees, preferring juvenile and adult individuals in the middle canopy (JZ3), while seedlings were more frequently found in the upper canopy or twigs (JZ 4) (Figure 5, Table S4). The higher-than-expected number of seedlings in the upper canopy of the gallery forest suggests that the dense foliage of *T. mucronatum* may mitigate harsh microclimatic conditions, such as intense solar radiation and low humidity [70], thereby facilitating colonization in these zones. This is similar to *Tillandsia schiedeana* in tropical dry forests, which colonizes the outer crown of *Bursera simaruba* (L.) Sarg., showing drought adaptations throughout its morphophysiology [37]. The preference of juvenile *T. baileyi* for the middle canopy (JZ3), which offers more stable microhabitats, further supports the notion that well-structured host trees are essential for epiphytic colonization [10].

4. Materials and Methods

4.1. Study Site

This study was carried out at two sites in Tamaulipas. (1) The first site was ejido La Cabecera in the municipality of Aldama, which is a gallery forest with anthropic disturbance, which was considered a disturbed forest. Gallery forest comprises groups of trees that develop along more or less permanent water courses. The vegetation in gallery forests is heterogeneous, with tree species that can reach 4 to 40 m in height, including numerous climbers and epiphytes. Some of the representative genera are *Salix, Taxodium, Acer, Inga, Carya, Fraxinus,* and *Alnus* [17]. (2) The second site was ejido Carricitos in the municipality of San Nicolas, which is characterized by generally unarmed, 3 to 5 m high and dense submontane scrub; this site was called primary forest. While dominant tree species vary from one region to another, the most frequent are *Helietta parvifolia* (A. Gray ex Hemsl.) Benth, *Neopringlea integrifolia* (Hemsl.) S. Watson, *Gochnatia hypoleuca* (DC.) A. Gray, *Pithecellobium brevifolium* Benth, *Quercus fursiformis* Small, and *Cordia boissieri* A.DC [17]. At both sites, a semi-warm, sub-humid climate predominates, with summer rains, presenting an average annual temperature higher than 18 °C and annual rainfall ranging from 900 to 1000 mm (UNAFOR 2803 and 2801, 2010).

4.2. Data Collection

Six 50 × 5 m linear transects were established at each study site (1.5 ha per site). In each transect, the tree species were identified. Subsequently, each tree was measured for diameter at breast height (DBH) and height using a tape measure and a distance meter (D210; Leica, Wetzlar, Germany), respectively. Only trees with a DBH greater than 10 cm were considered, since they offer a great diversity of ecological niches due to their variety of sizes and positions of their branches, facilitating colonization by epiphytes [71]. Identification of the trees at the species level was carried out using taxonomic keys and with the help of a specialist from the Institute of Applied Ecology of the Autonomous University of Tamaulipas. Tree bark was classified into the following three categories [72]: rugose (with deep grooves and ridges), semi-rugose (with small grooves and ridges), and smooth (lacking peeling or cracks).

Individuals of all epiphytic bromeliads at both study sites were recorded by direct observation with the aid of binoculars (EO-D102; Eagle optics, Middleton, WI,, USA). When the epiphytic individuals were at a distance greater than 5 m from the ground, a simple rope technique [73] was used to ascend to the canopy. The height of the bromeliads on the host was measured using a distance meter (D210, Leica, Wetzlar, Germany). Clones from rhizomatous plants or clumps of plants were considered one individual.

Since *T. baileyi* was the most abundant species in the gallery forest, the size categories of this species were analyzed to determine its structure and vertical distribution. *T. baileyi* individuals were classified into the following three categories according to size: seedling, juvenile, and adult. The seedling individuals measured between 1 and 10 cm in height without inflorescence, juveniles were between 11 and 25 cm tall and could present inflorescence or not, and adult plants measured more than 26 cm in height and presented inflorescence (Figure S1). Plant height was considered from the base to the longest leaf of the plant (Figure S1).

Vertical distribution of epiphytic bromeliads in the host was established through the following categories proposed in [63] and modified in this work: Zone 1 (JZ 1) corresponds to the total portion of the trunk, Zone 2 (JZ 2) is the lower canopy or first branches, Zone 3 (JZ 3) concerns the middle canopy of secondary branches, and Zone 4 (JZ 4) represents the upper canopy or twigs.

4.3. Data Analysis

Data were analyzed with R software version 4.0.4 [74]. The relative abundance of species was calculated as the number of individuals of each species/total abundance × 100. We assessed the importance value index (IVI) to ascertain the dominant tree species in each forest [75]. For details of IVI calculation, see [37]. If the IVI value for different species is similar, they contribute similarly to the composition, structure, site quality, and dynamics of the forest [76]. A one-way ANOVA was used to analyze the differences in abundance between study sites.

The effects of the Johansson zones (JZ1, JZ2, JZ3, and JZ4) and sites on the abundance of epiphytic bromeliads and the effects of Johansson zones and the size categories (seedling, juvenile, and adult) of *T. baileyi* on the abundance of *T. baileyi* were evaluated using generalized linear models. To correct overdispersion, a quasi-Poisson distribution and a logit link function were used [77]. An analysis of variance (ANOVA) was applied to determine the significance of the following factors: Johansson zone, size category of *T. baileyi*, and site [78]. Normality was checked using the Shapiro–Wilk test. Subsequently, multiple means comparisons (Holm, $p < 0.05$) were performed to evaluate possible differences between sites, Johansson zones, and size categories for *T. baileyi*. Holm's method is a *p*-value adjustment used in multiple-hypothesis testing to avoid false statistical inferences [79]. An χ^2 analysis was performed to evaluate the possible significant differences among the Johansson zones for each studied species. A contingency table with three size categories (rows) and four Johansson zones (columns) was structured. The association of size categories with Johansson zones of the gallery forest was evaluated with an χ^2 test [80]. The frequency of the expected abundance of bromeliads was obtained with the product of the row and column totals divided by the total number of observations. When the $\chi 2$ test was significant ($p \leq 0.05$), a standardized residual analysis was performed [42]. The association between the size categories and Johansson zones, on the one hand, was considered positive when the observed values were higher than the expected values and the values of the standardized residuals were >2. A negative association was considered when the observed values were lower than the expected values and standardized residual values were <-2 [42].

5. Conclusions

The gallery forest, despite being surrounded by agricultural land, is an important reservoir for epiphytic bromeliads. This means that large and old trees of *T. mucronatum* offer a great microsite to epiphytic bromeliads are crucial for their conservation and the maintenance of the species richness of the site. *T. baileyi* was the most abundant species in the gallery forest, and seedling was the least abundant category size, abundantly colonizing the upper crown, maybe in response to the dry and bright conditions prevalent in the outer crown. Our results represent a tool to propose conservation initiatives and management programs in the gallery forest, especially for *T. baileyi*, a species that has reduced its distribution in recent years [21]. According to NatureServe, *T. baileyi* is considered an endangered species in the USA [81], but in Mexico, its ecological and physiological requirements are unknown. Therefore, this study establishes a baseline for the conservation of this species. Research in gallery forest is essential and should be extended to wetland sites to better understand the role it plays as a habitat for epiphytes and the dynamics and ecological processes that occur in such habitats.

Supplementary Materials: The following supporting information can be downloaded at: https://www.mdpi.com/article/10.3390/plants13192754/s1, Figure S1. Size categories of bromeliad *T. baileyi*: (A) seedling; (B) juvenile; (C) adult. The diameter of the coin is 2.5 cm. Figure S2. Abundance of epiphytic bromeliads based on Johansson zone in gallery forest and submontane scrub in Tamaulipas, Mexico. Each bar represents the mean ± SE. Table S1. Paired comparisons between the Johansson zones of the gallery forest and submontane scrub in Tamaulipas, Mexico. JZ = Johansson zone. Table S2. Generalized linear model with quasi-Poisson distribution where the size categories of *Tillandsia baileyi* were evaluated. The adult category and the gallery forest were taken as reference points ($p < 0.001$). Table S3. Paired comparisons for size categories of *Tillandsia baileyi* in the gallery forest and the submontane scrub in Tamaulipas, Mexico. Table S4. Observed values and standardized residuals showing the association between size categories of *Tillandsia baileyi* and the Johansson zones of trees from the gallery forest. Association was considered positive when standardized residual values were >2, indicating that individuals of *T. baileyi* were more abundant than expected by chance and negative when residuals were <−2, suggesting that individuals of this bromeliad were less abundant than expected by chance [42].

Author Contributions: Conceptualization, E.d.l.R.-M. and S.S.S.T.; Methodology, E.d.l.R.-M., L.U.A.-M. and S.S.S.T.; Formal Analysis, E.d.l.R.-M., L.U.A.-M. and S.S.S.T.; Writing—original draft preparation, E.d.l.R.-M., L.U.A.-M. and S.S.S.T.; Writing—review and editing, E.d.l.R.-M., S.S.S.T., L.U.A.-M., K.M.A.-D., J.G.M.Á. and M.C.J.A. All authors have read and agreed to the published version of the manuscript.

Funding: This research was funded by CONAHCYT (CB 2015–259248). The authors extend their gratitude to the COTACYT, for funding the publishing of this research work.

Data Availability Statement: In accordance with Open Science communication practices, the authors declare that all data are available within the manuscript.

Conflicts of Interest: The authors declare no conflicts of interest.

References

1. Turner, I.M.; Tan, H.T.W.; Wee, Y.C.; Ibrahim, A.B.; Chew, P.T.; Corlett, R.T. A study of plant species extinction in Singapore: Lessons for the conservation of tropical biodiversity. *Conserv. Biol.* **1994**, *8*, 705–712. [CrossRef]
2. Wolf, J.H.D. The response of epiphytes to anthropogenic disturbance of pine-oak forest in the highlands of Chiapas, México. *Forest Ecol. Manag.* **2005**, *212*, 376–393. [CrossRef]
3. Pérez-Peña, A.; Krömer, T. ¿Qué pueden aportar los acahuales y las plantaciones de cítricos a la conservación de las epífitas vasculares en Los Tuxtlas, Veracruz? In *Avances y Perspectivas en la Investigación de los Bosques Tropicales y sus Alrededores: La Región de Los Tuxtlas*; Reynoso, V.H., Coates, R.I., Vázquez-Cruz, M.L., Eds.; Instituto de Biología, Universidad Nacional Autonoma de México: Ciudad de México, Mexico, 2017; pp. 569–580.
4. Köster, N.; Friedrich, K.; Nieder, J.; Barthlott, W. Conservation of epiphyte diversity in an Andean landscape transformed by human land use. *Conserv. Biol.* **2009**, *23*, 911–919. [CrossRef] [PubMed]
5. Hietz-Seifert, U.; Hietz, P.; Guevara, S. Epiphyte vegetation and diversity on remnant trees after forest clearance in southern Veracruz. *Biol. Conserv.* **1996**, *75*, 103–111. [CrossRef]
6. Larrea, M.L.; Werner, F.A. Response of vascular epiphyte diversity to different land-use intensities in a neotropical montane wet forest. *Forest Ecol. Manag.* **2010**, *260*, 1950–1955. [CrossRef]
7. Krömer, T.; García-Franco, J.G.; Toledo-Aceves, T. Epífitas vasculares como bioindicadoras de la calidad forestal: Impacto antrópico sobre su diversidad y composición. In *Bioindicadores: Guardianes de Nuestro Futuro Ambiental*; González, C.A., Vallarino, A., Pérez, J.C., Low, A.M., Eds.; Instituto Nacional de Ecología y Cambio Climático (INECC), El Colegio de la Frontera Sur (ECOSUR): Mexico City, Mexico, 2014; pp. 605–623.
8. Zotz, G.; Weigelt, P.; Kessler, M.; Kreft, H.; Taylor, A. EpiList 1.0: A Global Checklist of Vascular Epiphytes. *Ecology* **2021**, *102*, e0332. [CrossRef]
9. Nadkarni, N.M. Epiphyte biomass and nutrient capital of a neotropical elfin forest. *Biotropica* **1984**, *16*, 249–256. [CrossRef]
10. Benzing, D.H. *Vascular Epiphytes. General Biology and Related Biota*; Cambridge University Press: Cambridge, UK, 1990; p. 346.
11. Hietz, P. Fern adaptations to xeric environments. In *Fern Ecology*; Mehltreter, K., Walker, L., Sharpe, J., Eds.; Cambridge University Press: Cambridge, UK, 2010; pp. 140–176.
12. Sutton, S.L.; Whitmore, T.C.; Chadwich, A.C. *Tropical Rain Forest: Ecology and Management*; Blackwell Scientific Publication: Oxford, UK, 1983; p. 498.
13. Zotz, G. The systematic distribution of vascular epiphytes-a critical update. *Bot. J. Linn. Soc.* **2013**, *171*, 453–481. [CrossRef]
14. Benzing, D.H. *Bromeliaceae. Profile of an Adaptive Radiation*; Cambridge University Press: Cambridge, UK, 2000; p. 708.
15. Rosas-Mejía, M.; Siaz-Torres, S.S.; Arellano-Méndez, L.U.; Torres-Castillo, J.A.; de la Rosa-Manzano, E. *Azteca instabilis*: First Report of the Ant in the Tropical Dry Forest at Tamaulipas, Mexico. *Southwest. Entomol.* **2022**, *47*, 515–518. [CrossRef]
16. 16 Rusch, A.; Chaplin-Kramer, R.; Gardiner, M.M.; Hawro, V.; Holland, J.; Landis, D.; Thies, C.; Tscharntke, T.; Weisser, W.W.; Winqvist, C.; et al. Agricultural landscape simplification reduces natural pest control: A quantitative synthesis. *Agric. Ecosyst. Environ.* **2016**, *221*, 198–204. [CrossRef]
17. Rzedowski, J. *Vegetación de México*, 1st ed.; Comisión Nacional para el Conocimiento y uso de la Biodiversidad: Tlalpan, Mexico, 2006; p. 504.
18. González-Elizondo, J. Vegetación de la reserva de la biósfera "La Michilia", Durango, México. *Acta Bot. Mex.* **1993**, *22*, 1–104. [CrossRef]
19. Carter, V.; Gammon, P.T.; Garret, M. Ecotone dynamics and boundary determination in the Great Dismal Swamp. *Ecol. Appl.* **1994**, *4*, 189–203. [CrossRef]
20. Banner, A.; MacKenzie, M. *Riparian Areas: Providing Landscape Habitat Diversity, Part 5 of 7*; British Columbia Ministry of Forests: Victoria, BC, Canada, 1998; p. 8.
21. Espejo-Serna, A.; López-Ferrari, A.R.; Mendoza-Ruiz, A.; García-Cruz, J.; Ceja-Romero, J.; Pérez-García, B. Mexican vascular epiphytes: Richness and distribution. *Phytotaxa* **2021**, *503*, 1–124. [CrossRef]
22. Enríquez-Peña, E.G.; Suzán-Azpiri, H. Estructura poblacional de *Taxodium mucronatum* en condiciones contrastantes de perturbación en el estado de Querétaro, México. *Rev. Mex. Biodivers.* **2011**, *82*, 153–167. [CrossRef]

23. Canizales-Velázquez, P.A.; Alanís-Rodríguez, E.; García-García, S.A.; Holguín-Estrada, V.A.; Collantes-Chávez-Costa, A. Estructura y diversidad arbórea de un bosque de galería urbano en el río Camachito, Noreste de México. *Polibotánica* **2021**, *51*, 91–105. [CrossRef]
24. Vergara-Rodríguez, D.; Mathieu, G.; Samain, M.-S.; Armenta-Montero, S.; Krömer, T. Diversity, distribution and conservation status of Peperomia (Piperaceae) in the state of Veracruz, Mexico. *Trop. Conserv. Sci.* **2017**, *10*, 1–28. [CrossRef]
25. Grella, C.; Renshaw, A.; Wright, I. Invasive weeds in urban riparian zones: The influence of catchment imperviousness and soil chemistry across an urbanization gradient. *Urban Ecosyt.* **2018**, *21*, 505–517. [CrossRef]
26. Estrada-Castillón, E.; Villarreal-Quintanilla, J.A.; Jurado-Ybarra, E.; Cantú-Ayala, C.; García-Aranda, M.A.; Sánchez-Salas, J.; Jiménez-Pérez, J.; Pando-Moreno, M. Classification, structure and diversity of the piedmont scrub adjacent to the Northern Gulf coastal plain in Northeastern Mexico. *Bot. Sci.* **2012**, *90*, 37–52. [CrossRef]
27. Martin, C.E.; Adams, W.W. Crassulacean acid metabolism, CO_2-recycling, and tissue desiccation in the Mexican epiphyte *Tillandsia schiedeana* Steud (Bromeliaceae). *Photosynth. Res.* **1987**, *11*, 237–244. [CrossRef]
28. Loeschen, V.; Martin, C.; Smith, M.; Eder, S. Leaf anatomy and CO_2 recycling during Crassulacean acid metabolism in twelve epiphytic species of *Tillandsia* (Bromeliaceae). *Int. J. Plant Sci.* **1993**, *154*, 100–106. [CrossRef]
29. de la Rosa-Manzano, E.; Mendieta-Leiva, G.; Guerra-Pérez, A.; Aguilar-Dorantes, K.; Arellano-Méndez, L.U.; Torres-Castillo, J.A. Vascular epiphytic diversity in a Neotropical transition zone is driven by environmental and structural heterogeneity. *Trop. Conserv. Sci.* **2019**, *12*, 1–16. [CrossRef]
30. Krömer, T.; Kessler, M.; Gradstein, S.R. Vertical stratification of vascular epiphytes in submontane and montane forest of the Bolivian Andes: The importance of the understory. *Plant Ecol.* **2007**, *189*, 261–278. [CrossRef]
31. Nieder, J.; Engwald, S.; Klawun, S.; Barthlott, W. Spatial distribution of vascular epiphytes (including hemiepiphytes) in a lowland Amazonian rain forest (Surumoni Crane Plot) of southern Venezuela. *Biotropica* **2000**, *32*, 385–396. [CrossRef]
32. Cervantes, S.E.; Graham, E.A.; Andrade, J.L. Light microhabitats, growth and photosynthesis of an epiphytic bromeliad in a tropical dry forest. *Plant Ecol.* **2005**, *179*, 107–118. [CrossRef]
33. Zotz, G. Substrate use of three epiphytic bromeliads. *Ecography* **1997**, *20*, 264–270. [CrossRef]
34. de la Rosa-Manzano, E.; Andrade, J.L.; Zotz, G.; Reyes-García, C. Epiphytic orchids in tropical dry forests of Yucatan, Mexico—Species occurrence, abundance and correlations with host tree characteristics and environmental conditions. *Flora* **2014**, *209*, 100–109. [CrossRef]
35. Callaway, R.M.; Reinhart, K.; Moore, G.W.; Pennings, S. Epiphyte host preferences and host traits: Mechanisms for species-specific interactions. *Oecologia* **2002**, *132*, 221–230. [CrossRef]
36. Hietz, P.; Winkler, M.; Scheffknecht, S.; Hülber, K. Germination of epiphytic bromeliads in forests and coffee plantations: Microclimate and substrate effects. *Biotropica* **2012**, *44*, 197–204. [CrossRef]
37. Siaz-Torres, S.S.; Mora-Olivo, A.; Arellano-Méndez, L.U.; Vanoye-Eligio, V.; Flores-Rivas, J.; de la Rosa-Manzano, E. Contribution of peeling host for epiphyte abundance in two tropical dry forests in the "El Cielo Biosphere Reserve", Mexico. *Plant Species Biol.* **2021**, *36*, 269–283. [CrossRef]
38. Wagner, K.; Zotz, G. Including dynamics in the equation: Tree growth rates and host specificity of vascular epiphytes. *J. Ecol.* **2020**, *108*, 761–773. [CrossRef]
39. ter Steege, H.; Cornelissen, J.H.C. Distribution and ecology of vascular epiphytes in lowland rain forest of Guyana. *Biotropica* **1989**, *21*, 331–339. [CrossRef]
40. Hietz, P.; Hietz-Seifert, U. Composition and ecology of vascular epiphyte communities along an altitudinal gradient in central Veracruz, Mexico. *J. Veg. Sci.* **1995**, *6*, 487–498. [CrossRef]
41. Zotz, G.; Vollrath, B. The epiphyte vegetation of the palm *Socratea exorrhiza*-correlations with tree size, tree age and bryophyte cover. *J. Trop. Ecol.* **2003**, *19*, 81–90. [CrossRef]
42. Agresti, A. *An Introduction to Categorical Data Analysis*, 2nd ed.; Wiley-Interscience: New York, NY, USA, 1996; p. 357.
43. Barthlott, W.; Schmit-Neuerburg, V.; Nieder, J.; Engwald, S. Diversity and abundance of vascular epiphytes: A comparison of secondary vegetation and primary montane rain forest in the Venezuelan Andes. *Plant Ecol.* **2001**, *152*, 145–156. [CrossRef]
44. Kubota, Y.; Katsuda, K.; Kikuzawa, K. Secondary succession and effects of clear-logging on diversity in the subtropical forests on Okinawa Island, southern Japan. *Biodivers. Conserv.* **2005**, *14*, 879–901. [CrossRef]
45. García-González, A.; Pérez, R. La comunidad orquideológica en la Reserva de la Biosfera Sierra del Rosario, Cuba. *Rev. Biol. Trop.* **2011**, *59*, 1805–1812. [CrossRef]
46. Pincheira-Ulbrich, J.; Hernández, C.E.; Saldaña, A. Consequences of swamp forest fragmentation on assemblages of vascular epiphytes and climbing plants: Evaluation of the metacommunity structure. *Ecol. Evol.* **2018**, *8*, 11785–11798. [CrossRef] [PubMed]
47. Solís-Montero, L.; Flores-Palacios, A.; Cruz-Angon, A. Shade-coffee plantations as refuges for tropical wild orchids in Central Veracruz, Mexico. *Conserv. Biol.* **2005**, *19*, 908–916. [CrossRef]
48. Hietz, P.; Buchberger, G.; Winkler, M. Effect of forest disturbance on abundance and distribution of epiphytic bromeliads and orchids. *Ecotropica* **2006**, *12*, 103–112.
49. Flores-Palacios, A.; García-Franco, J.G. Habitat isolation changes the beta diversity of the vascular epiphyte community in lower montane forest, Veracruz, Mexico. *Biodivers. Conserv.* **2008**, *17*, 191–207. [CrossRef]
50. Zimmerman, J.K.; Olmsted, I.C. Host tree utilization by vascular epiphytes in a seasonally inundated forest (Tintal) in Mexico. *Biotropica* **1992**, *24*, 402–407. [CrossRef]

51. García-González, A.; Plasencia-Vázquez, A.H.; Serrano-Rodríguez, A.; Riverón-Giró, F.B.; Ferrer-Sánchez, Y.; Zamora-Crescencio, P. Assemblage of vascular epiphytes associated to seasonally inundated forest in the Southeastern Mexico: Challenges for its conservation. *Glob. Ecol. Conserv.* **2021**, *25*, e01404. [CrossRef]
52. Martin, C.E.; McLeod, K.W.; Eades, C.A.; Pitzer, A.F. Morphological and physiological responses to irradiance in the CAM epiphyte *Tillandsia usneoides* L. (Bromeliaceae). *Bot. Gaz.* **1985**, *146*, 489–494. [CrossRef]
53. Adams, W.W., III; Martin, C.E. Morphological changes accompanying the transition from juvenile (atmospheric) to adult (tank) form in the Mexican epiphyte *Tillandsia deppeana* (Bromeliaceae). *Am. J. Bot.* **1986**, *73*, 1207–1214.
54. Diego-Escobar, M.V.; Flores-Cruz, M.; Koch, S.D. *Tillandsia* L. (Bromeliaceae). In *Flora de Guerrero*; Universidad Nacional Autónoma de Mexico, Facultad de Ciencias: Ciudad de Mexico, Mexico, 2013; p. 122.
55. Trejo, I.; Dirzo, R. Deforestation of seasonally dry tropical forest: A national and local analysis in Mexico. *Biol. Conserv.* **2000**, *94*, 133–142. [CrossRef]
56. Comision Nacional Forestal. Mexico. 2023. Available online: https://www.gob.mx/conafor (accessed on 12 September 2023).
57. Bonnet, A.; Queiroz, M.H.; Lavoranti, O.J. Relações de bromélias epifíticas com características dos forófitos em diferentes estádios sucessionais da floresta ombrófila densa, Santa Catarina, Brasil. *Floresta* **2007**, *37*, 83–94. [CrossRef]
58. Ingram, S.; Nadkarni, N. Composition and distribution of epiphytic organic matter in a Neotropical cloud forest, Costa Rica. *Biotropica* **1993**, *25*, 370–383. [CrossRef]
59. Burns, K.C. Meta-community structure of vascular epiphytes in a temperate rainforest. *Botany* **2008**, *86*, 1252–1259. [CrossRef]
60. Woods, C.L.; Cardelús, C.L.; Dewalt, S.J. Microhabitat associations of vascular epiphytes in a wet tropical forest canopy. *J. Ecol.* **2015**, *103*, 421–430. [CrossRef]
61. Werner, F.A.; Gradstein, S.R. Diversity of dry forest epiphytes along a gradient of human disturbance in the tropical. *Andes. J. Veg. Sci.* **2009**, *20*, 59–68. [CrossRef]
62. Benzing, D.H. The life history profile of *Tillandsia circinnata* (Bromeliaceae) and the rarity of extreme epiphytism among the angiosperm. *Selbyana* **1978**, *2*, 325–337.
63. Johansson, D. Ecology of vascular epiphytes in West African rain forest. *Acta Phytogeogr. Suec.* **1974**, *59*, 1–136.
64. Chilpa-Galván, N.; Tamayo-Chim, M.; Andrade, J.L.; Reyes-García, C. Water table depth may influence the asymmetric arrangement of epiphytic bromeliads in a tropical dry forest. *Plant Ecol.* **2013**, *214*, 1037–1048. [CrossRef]
65. Zotz, G. Johansson revisited: The spatial structure of epiphyte assemblages. *J. Veg. Sci.* **2007**, *18*, 123–130. [CrossRef]
66. Hernández-Apolinar, M. Dinámica Poblacional de *Laelia speciosa* (H.B.K.) Schltr. (Orchidaceae). Bachelor's Thesis, Facultad de Ciencias, UNAM, Ciudad de México, Mexico, 1992; p. 86.
67. Larson, R.J. Population dynamics of *Encyclia tampensis* in Florida. *Selbyana* **1992**, *13*, 50–56.
68. Zotz, G. Demography of the epiphytic orchid, *Dimerandra emarginata*. *J. Trop. Ecol.* **1998**, *14*, 725–741. [CrossRef]
69. Wang, B.; Ma, Y.; Chen, G.; Li, C.; Dao, Z.; Sun, W. Rescuing *Magnolia sinica* (Magnoliaceae), a Critically Endangered species endemic to Yunnan, China. *Oryx* **2016**, *50*, 446–449. [CrossRef]
70. Cardelús, C.L.; Chazdon, R.L. Inner-crown microenvironments of two emergent tree species in a lowland wet forest. *Biotropica* **2005**, *37*, 238–244. [CrossRef]
71. Arévalo, R.; Betancur, J. Vertical distribution of vascular epiphytes in four forest types of the Serranía de Chiribiquete, Colombian Guayana. *Selbyana* **2006**, *27*, 175–185.
72. Pennington, T.D.; Sarukhán, J. *Árboles Tropicales de México: Manual para la Identificación de las Principales Especies*, 1st ed.; Fondo de Cultura Económica, Universidad Nacional Autónoma de México: Ciudad de México, Mexico, 2005; p. 523.
73. Lowman, M.D.; Schowalter, T.D.; Franklin, J.F. *Methods in Forest Canopy Research*; University of California Press: Berkeley, CA, USA, 2012; p. 221.
74. R Core Team. *R: A Language and Environment for Statistical Computing*; R Foundation for Statistical Computing: Vienna, Austria, 2021. Available online: https://www.r-project.org. (accessed on 10 September 2023).
75. Martínez-Bernal, A.; Vasquez-Velasco, B.; Ramírez-Arriaga, E.; Zárate-Hernández, M.; Martínez-Hernández, E.; Téllez-Valdés, O. Composition, structure and diversity of tree and shrub strata in a tropical deciduous forest at Tehuacán Valley, Mexico. *Rev. Mex. Biodivers.* **2021**, *92*, 1–16. [CrossRef]
76. Mishra, A.K.; Behera, S.K.; Singh, K.; Sahu, N.; Bajpai, O. Relation of forest structure and soil properties in natural, rehabilitated and degraded forest. *J. Biodivers. Manag. For.* **2013**, *8*, 27–29. [CrossRef]
77. Crawley, M.J. The R Book. *Stat. Pap.* **2009**, *50*, 445–446. [CrossRef]
78. Dunn, P.K.; Smyth, G.K. Generalized Linear Models: Inference. In *Generalized Linear Models with Examples in R*; Springer: New York, NY, USA, 2018; pp. 265–296.
79. Holm, S. A simple sequentially rejective multiple test procedure. *Scand. J. Stat.* **1979**, *6*, 65–70.
80. Vergara-Torres, C.A.; Pacheco-Álvarez, M.C.; Flores-Palacios, A. Host preference and host limitation of vascular epiphytes in a tropical dry forest of central Mexico. *J. Trop. Ecol.* **2010**, *26*, 563–570. [CrossRef]
81. NatureServe. US. 2023. Available online: www.natureserve.org (accessed on 21 October 2023).

plants

MDPI

Article

The Impact of Changing Climate on an Endangered Epiphytic Orchid (*Pleione formosana*) in a Montane Cloud Forest and the Conservation Challenge Ahead

Rebecca C.-C. Hsu [1,*], Yi-Chiann Chen [2] and Chienyu Lin [1]

[1] Forest Ecology Division, Taiwan Forestry Research Institute, Taipei 100051, Taiwan; cylin@tfri.gov.tw
[2] Silviculture Division, Taiwan Forestry Research Institute, Taipei 100051, Taiwan; chiann@tfri.gov.tw
* Correspondence: ecogarden@tfri.gov.tw

Abstract: *Pleione formosana* Hayata is an endemic orchid that was once widely distributed across the mid-elevations of Taiwan. However, populations of this orchid have steadily shrunk due to orchid poaching in most of its habitats. By correlating data from micrometeorological stations that we installed in the cloud forest canopy at the study site, Yuanyang Lake (YYL) from 2017, we discovered the critical role of spring rainfall in triggering flowering and summer rainfall in promoting the growth of new bulbs. We found that rising temperatures and frequent drought events threaten orchid growth, potentially leading to pathogen infections. We climbed old-growth yellow cypresses to collect seed capsules of *P. formosana* for in vitro germination at YYL beginning in the autumn of 2018. Orchid plantlets were reintroduced to the study site in mid-August of 2022. However, the seedlings failed to survive the summer of 2023. This study is the first persistent monitoring of this rare orchid in the forest canopy of this old-growth cloud forest. Based on the result, we propose conservation strategies and directions for protecting this orchid on a regional scale. Our study highlights the mounting challenge to conservation efforts posed by global climate change.

Keywords: ex situ conservation; in vitro germination; montane cloud forest; reintroduction biology; vascular epiphyte

1. Introduction

Montane cloud forests (MCFs) are recognized as one of the most valuable ecosystems threatened by climate change from many perspectives [1–4]. First, the seasonal and diurnal changes in the microclimates of MCFs are generally small due to their constant envelopment in daily clouds, which moderate temperature fluctuations and sustain high humidity [5,6]. Therefore, many species inhabiting MCFs are adapted to relatively stable microclimates and sensitive to climate changes [7]. Second, as the climate warms, the altitude of cloud formation rises [2], and species, especially plants, may have difficulty migrating to keep pace with these changes. The cloud belt occurring at a specific elevation also isolates the species that grow within it, hindering interpopulation connection and promoting endemic specification [8]. Therefore, many MCF species are relatively rare, with fragmented distributions, which increase their vulnerability in the face of climate change [9].

Astride the Tropic of Cancer in the southwest Pacific, more than half of Taiwan's land is over 1000 m above sea level [10], where moist ocean winds induce cloud formation and nurture a vast expanse of montane cloud forests [7]. Our previous study revealed that the montane cloud forest ecosystem of NE Taiwan may be the most severely impacted under various climate change scenarios compared to other forest types, threatening the suitable habitats of many epiphytic plants of this ecosystem [11]. Our species distribution models (SDMs) for the year 2100 found that cypress forests and many associated epiphytic plants will lose their suitable habitats. The forests of Chilan were one of the areas found to be in need of monitoring in our previous study [11]. The area is located in the Central Mountain

Range in northern Taiwan, ranging in altitude from 1650 to 2444 m. It is a typical montane cloud forest ecosystem comprising pristine old-growth forests of mainly yellow cypress (*Chamaecyparis obtusa* var. *formosana* (Hayata)) (Figure 1). The area, including Yuanyang Lake (YYL), has been declared a nature reserve and is a long-term ecosystem research (LTER) site. The epiphyte inventory shows that YYL has the most abundant and diverse epiphytic community in Chilan [12].

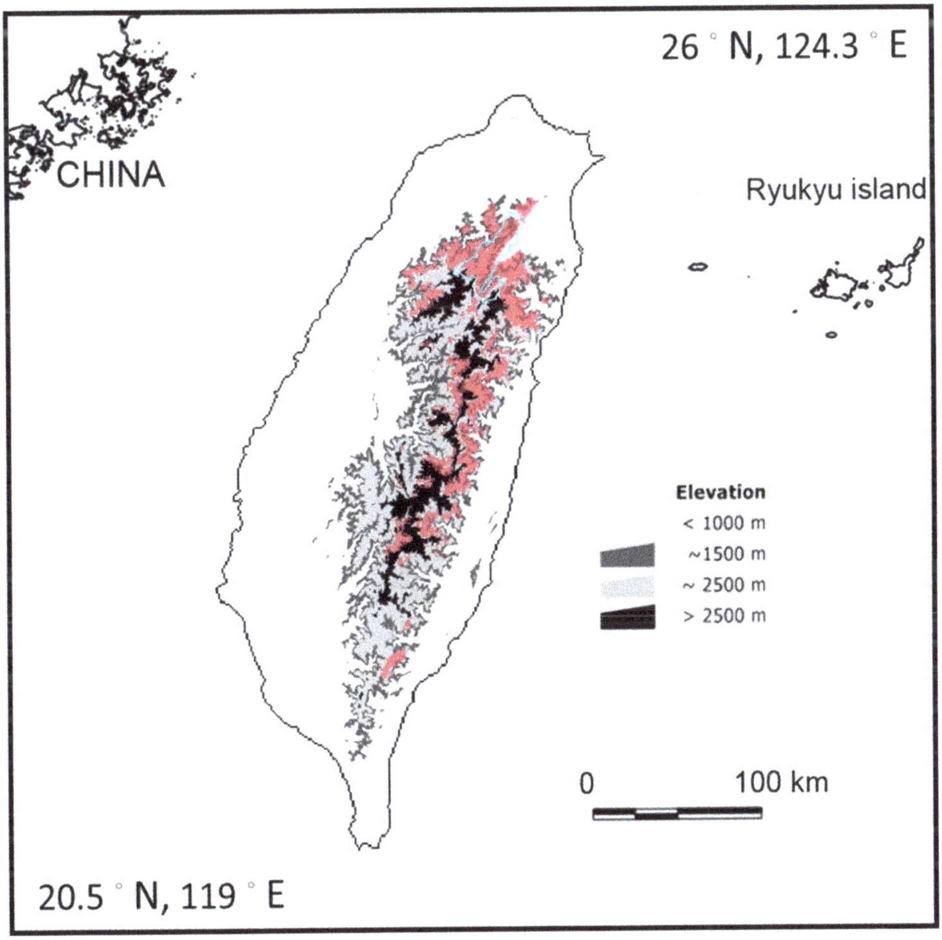

Figure 1. Distribution of cypress forests (red areas, MCFs) in Taiwan. Chilan, located in NE Taiwan (cyan boundary), comprises a vast area of cypress forests.

YYL is one of the most intact habitats of *Pleione formosana* Hayata, hosting abundant populations in northern Taiwan (Figure 2). The epiphytic orchid *P. formosana* is endemic to Taiwan, growing on peat moss-covered branches or rocks, distributed between 1200 and 2500 m elevation in montane cloud forests throughout the island. The species was discovered and named in 1911 [13]. The white flower variety was initially described as a new species in 1933, but this was later corrected [14]. Although commercial nurseries cultivate the orchid for ornamental use, the wild population continues to shrink due to orchid poaching [15]. Moreover, populations of *P. formosana* are predicted to diminish dramatically under the future effects of climate change [16,17]. In 2017, we established a study site in YYL, recording the microclimate and monitoring populations of *P. formosana*.

2.3. In Vitro Germination and Reintroduction

We have climbed yellow cypress trees to collect seed capsules of *P. formosana* every autumn since 2018 and brought intact seed capsules back to the lab for asymbiotic seed germination (Table 1). In the first three seasons, we germinated seeds at an orchid nursery in Yilan near the study site. Beginning in 2021, the capsules collected were brought back to the lab of the Taiwan Forestry Research Institute (TFRI) for asymbiotic seed germination (Figure 4). Capsules were surface-sterilized using 70% (*v*/*v*) ethanol for 1 min and 1% (*v*/*v*) sodium hypochlorite solution (NaOCl) with Tween 20 for 5 min under ultrasonication. Finally, the capsules were rinsed with autoclaved distilled water three times in a laminar flow cabinet, and moisture on the capsule surface was absorbed using sterilized filter paper. The capsules were cut longitudinally, and the seeds were equally distributed in bottles containing sowing medium (1/4 strength Murashige and Skoog (MS) medium [24], with 20 g L^{-1} sucrose, 2 g L^{-1} active charcoal, and 8 g L^{-1} agar, adjusted to pH 5.6 prior to autoclaving for 15 min at 121 °C). Seed-derived protocorms were transferred to seedling multiplication medium (1/2 MS, supplemented with 20 g L^{-1} sucrose, 8 g L^{-1} agar, 2 g L^{-1} activated charcoal, and 3 g L^{-1} Hyponex No. 1, 1 g L^{-1} tryptone, adjusted to pH 5.6 prior to autoclaving) to promote seedling growth and development [25]. Bottle seedlings were cultured in a growth chamber with the temperature controlled at 21 (±2) °C under a 12 h dark and 12 h light cycle at TFRI. In the summer of 2022, aggregate seedlings with 3 cm-long leaves and mini bulbs were reintroduced to natural peat moss at YYL.

Figure 4. The adult *P. formosana* with a seed capsule and a new bulb formation (**a**), fully matured seed capsules were collected in late September before dehiscing (**b**), capsules were cut longitudinally after surface sterilized (**c**), and a healthy capsule contains many dusty seeds (**d**).

Table 1. Phenology of *P. formosana* observed from 2018 to 2023 at YYL. Bold text indicates extreme values during the study period. Canopy temperature records in the summer of 2020 were lost due to mechanical problems.

Year	2018	2019	2020	2021	2022	2023
Flowering Date [1]	+1.1	−0.5	0	0	0	+0.7
Flowers	Average	Average	Few	Few	Average	Few
Fruiting Date [1]	+3.3	−2	0	0	0	+1.1
Fruits	Average	Average	Few	Few	Abv. average	Few
Spring rainfall [2] temperature [3]	210 −1.7/21.2	**397** 1.6/**25.5**	236 −0.5/21.3	**197** 2.2/21.2	355 −0.8/21.4	154 −**1.8**/24.2
Summer rainfall [2] temperature [3]	761 11.1/24.6	715 13.1/25.4	711 -	**1178** 12.3/25.6	**216** 12/26.5	755 12.9/**26.8**
Extreme Weather Event	Snow event				Drought	Foehn wind
Seed Collection Date	4 7 November	1 10 September	2 * 9 October	6 ** 23 September	6 *** 21 June and 22 September	0

* One capsule dehisced before germination, ** two capsules dehisced before germination, *** capsules collected in June did not germinate. [1] Days/year compared with 2009, [2] mm, [3] min/max °C.

2.4. Data Analysis

We calculated vapor pressure deficit (VPD) using a simple, accurate formula proposed by Huang [25], described in our previous study [12]. The formula takes into account both temperature and relative humidity (RH). Statistical analyses were conducted using R version 4.1.0 [26]. The mean climatic factors among different years were compared using an ANOVA test, and the correlation between visibility and temperature was assessed using Pearson correlation preceded by a Shapiro–Wilk normality test.

3. Results

3.1. Climate Recorded on Site

3.1.1. Temperature and Rainfall from 2018

Summaries of rainfall and mean temperature during the spring (March and April) and summer (July and August) at YYL recorded by the CWA local meteorological station from 2018 are shown in Figures 5 and 6. Spring rainfall ranged from 155 to 355 mm, with an average of ca. 250 mm. Rainfall for 2023 was only half of that of 2022 due to abnormal drought. Since 2018, there have been fewer typhoons across Taiwan, especially across NE Taiwan, which historically is most frequently affected by typhoons, so summer rainfall was significantly reduced at YYL over the last six years. In October of 2022, torrential rain brought by Tropical Storm Nesat, along with the NE monsoon, caused serious damage to the roads around YYL, so we stopped visiting study sites until March of 2023. In August of 2023, foehn winds brought by Typhoon Khanun caused dry leaves for most *P. formosana* individuals in YYL. Although the overall summer precipitation remained relatively unchanged (Table 1), the three-day mean temperature (from 5 to 7 August 2023) rose to 20.8 °C, with a maximum of 26.8 °C recorded, and the mean relative humidity dropped to a very low 76%. These conditions are stressful for the survival of *P. formosana*. Additionally, we noted a trend of reduced low-cloud events at YYL consistent with observations at other MCFs [3]. Our visibility records also showed a significant positive correlation (r = 0.59, $p < 0.001$) between daily mean temperature and visibility (i.e., fewer cloudy days, Figure 7).

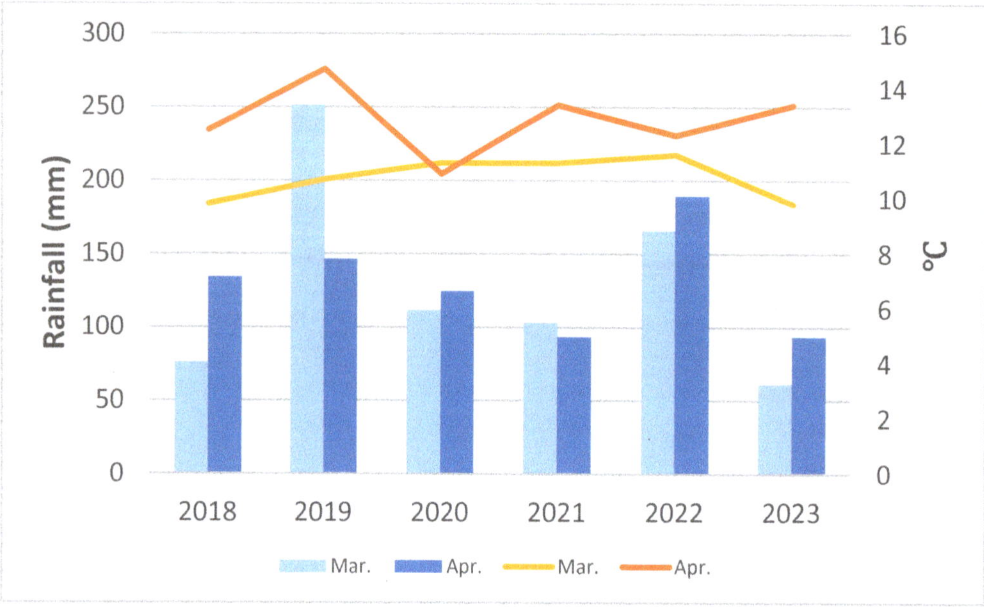

Figure 5. Spring rainfall (bars) and average temperature (lines) in YYL from 2018 to 2023 (CWA records).

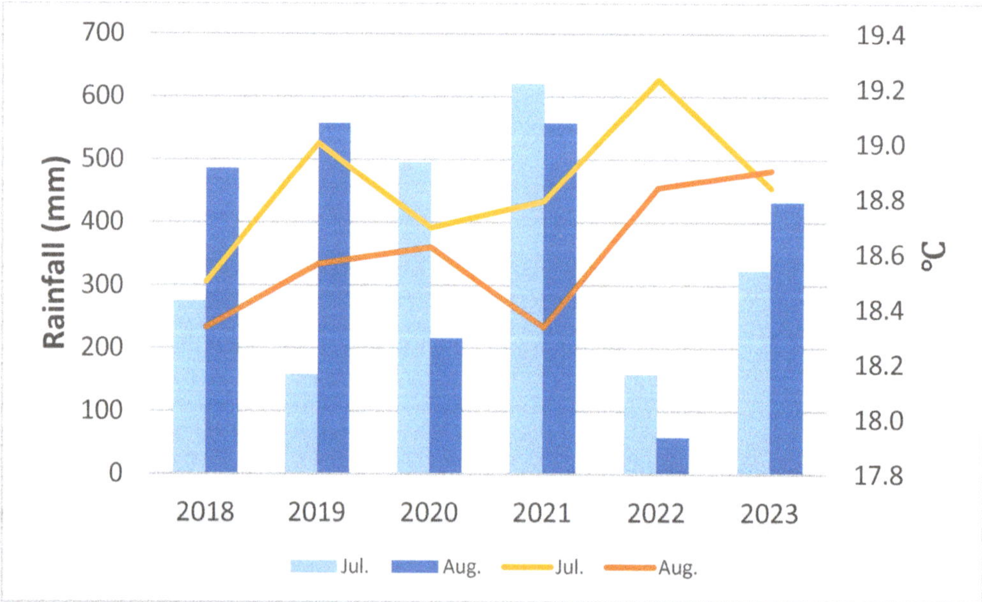

Figure 6. Summer rainfall (bars) and average temperature (lines) in YYL from 2018 to 2023 (CWA records).

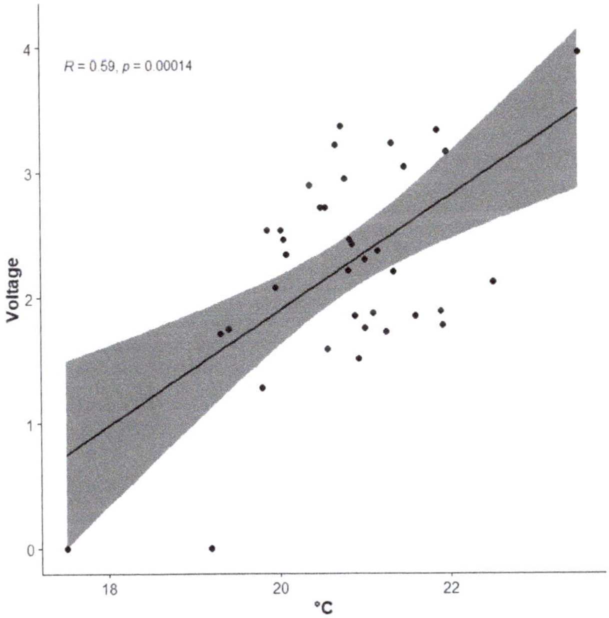

Figure 7. Pearson correlation between daily mean temperature (°C) and visibility (voltage) in 2020 summer. Data excluded rainy days with relative humidity (RH) over 95 percent.

3.1.2. Canopy Sensors

We calculated the 24 h VPD and temperature differences for 2018 and the summer (July and August) of 2022 (Figure 8). Temperature changes are presented in line charts, and compared with 2018, the average summer temperature rose about 1.5 °C by noon and dropped 0.35 °C before dawn in 2022. The VPD (bar chart) increased more through the day and night in 2022 than in 2018, with the greatest increase (0.23) at 10 AM.

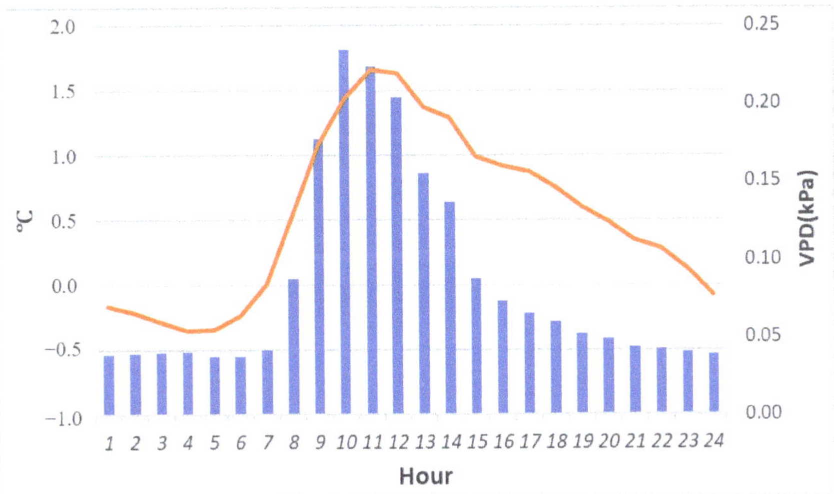

Figure 8. VPD (bar chart, kPa) and 24 h temperature (line chart, °C) difference between 2018 and 2022 summers (July and August).

3.2. Observations on P. formosana Phenology and Field Collection

Beginning in 2018, we observed six flowering seasons and collected five seasons of seed capsules (Table 1, Figure 9). Seed capsule maturation dates varied from early September to early November, and the flowering dates determined the dates of mature fruits. We found fluctuated flowering and fruit mature dates rather than delayed or advanced trends. The flowering date was decided by the spring mean temperature and rainfall. In 2018 and 2023, a cold spring caused postponed flowering, while *P. formosana* flowered about five days earlier in 2019 than in 2009. Extreme weather and a heavy snow event in February 2018 delayed the flowering of *P. formosana* until May (usually in late March), and abnormally high spring temperatures brought by El Niño in 2019 accelerated fruiting by one month to August (normally in late September), so we only collected one capsule in 2019. Insufficient rainfall in March 2020, 2021, and 2023 resulted in poor flowering conditions in those years. In general, over the last 6 years, only 2022 was a rewarding season for fruit collection. Flowering was delayed in the spring of 2023, probably due to a cold and dry spring (Figure 5), compounded by the previous dry summer. In 2023, no mature fruits were observed in the field, making it the only year that we did not collect any capsules.

Figure 9. Delayed flowering was observed on 22 March 2018 (**b**) and 23 March 2023 (**c**). Compared with 31 March 2009 (**a**), flowering individuals have declined sharply over the last five years, as observed on 31 March 2021 (**d**) and 26 March 2022 (**e**), especially for the white variety, as observed on 26 March 2022 (**f**). Photo credit: Rebecca Hsu.

3.3. Growth of Seedlings and Reintroduction

We collected six capsules on 23 September 2021, while two capsules dehisced before germination. The average length (±SE.) of capsules was 2.82 (0.53) cm, and the average width was 0.85 (0.12) cm. In comparison with terrestrial orchids, we found that epiphytic *P. formosana* was relatively easy to germinate asymbiotically. All capsules collected in June 2022 failed to germinate (Table 1), suggesting that mature seeds were necessary for the successful germination of *P. formosana*. There was an absence of seed dormancy of *P. formosana*, and within 50 to 80 days of inoculation, mature seeds began to germinate and started to show healthy growth (Figure 10). We speculated that the bottle seedlings of the first 3 years (2018 to 2020) died gradually in summer due to lowland heat. Therefore, we cultured embryos in a growth chamber with the temperature controlled at 21 °C at the TFRI in 2021 and 2022, and the young grew quickly. On 17 August 2022, we introduced the plantlets germinated in 2021 to YYL (Figure 11). The young plants survived until 3 May 2023 but waned after the summer of 2023. The 2023 summer was particularly dry and hot due to extreme foehn winds brought by Typhoon Khanun.

Figure 10. In vitro germination of *P. formosana* after 75 days (**a**,**b**), seedlings grew densely after nine months (**c**), and seedlings were ready to transplant after 14 months (**d**).

Figure 11. Plantlets were reintroduced to YYL in August of 2022 (**a**) but disappeared after the summer of 2023. September 2022 (**b**), March 2023 (**c**), May 2023 (**d**).

4. Discussion

4.1. Key Climate Factors for P. formosana Survival

There is no constant change in direction from our six-season phenology observation, especially the flowering date of *P. formosana* (Table 1). In comparison with 2009, peak flowering dates shifted back and forward during the study period. Climatic records suggest that spring (March and April) rainfall is crucial for the flowering condition of the orchid. Summer rainfall helps the new bulbs grow, which is important for the next year's flowering. Lower summer rainfall may have reduced the mature size of *P. formosana*'s bulbs, which we observed in the autumn of 2022. The following spring (2023) was the worst flowering season during the study period due to very low spring rainfall and small bulbs.

Table 1 shows the trend of rising maximum summer temperatures during the study period. Compared to 2018, the summer mean temperature increased about 1.5 °C at noon in 2022. There were 10 days in 2022 with daily maximum temperatures exceeding 25 °C, while there were no records above 25 °C in 2018. The extreme heat event was critically unfavorable to *P. formosana* survival; the ideal cultivation temperature at the orchid nursery is under 25 °C [27]. From our germination experiment, we also found that temperature is a decisive factor in the survival of *P. formosana*. Rising temperatures are especially harmful to seedling growth and will reduce the population of *P. formosana* in the future. We also noted an increased diurnal range of temperatures in 2022 (Figure 8). This is probably due to radiation cooling before dawn in the forest. Nights with clear skies result in steep temperature declines [7]. Although the R squared value of 0.35 is not particularly high, Figure 7 highlights the significant relationship between reduced clouds and rising air temperature. This suggests a more variable temperature regime of this cloud forest ecosystem in the future.

Our previous study found that the VPD is a decisive factor for epiphyte distribution in the studied forest [12]. The VPD takes into account both relative humidity (RH) and air temperature; therefore, it is an informative indicator of the key microclimate conditions for epiphytic plants [28]. Our results suggest stable VPD favors epiphyte diversity and abundance; however, there were also significantly rising mean and extreme values of VPD over the five years (Figure 12, ANOVA, $p < 0.001$). The mean VPD and standard deviations dramatically increased both in the forest canopy and on the forest floor, suggesting a trend of drought stress and unstable microclimates for *P. formosana* in the future.

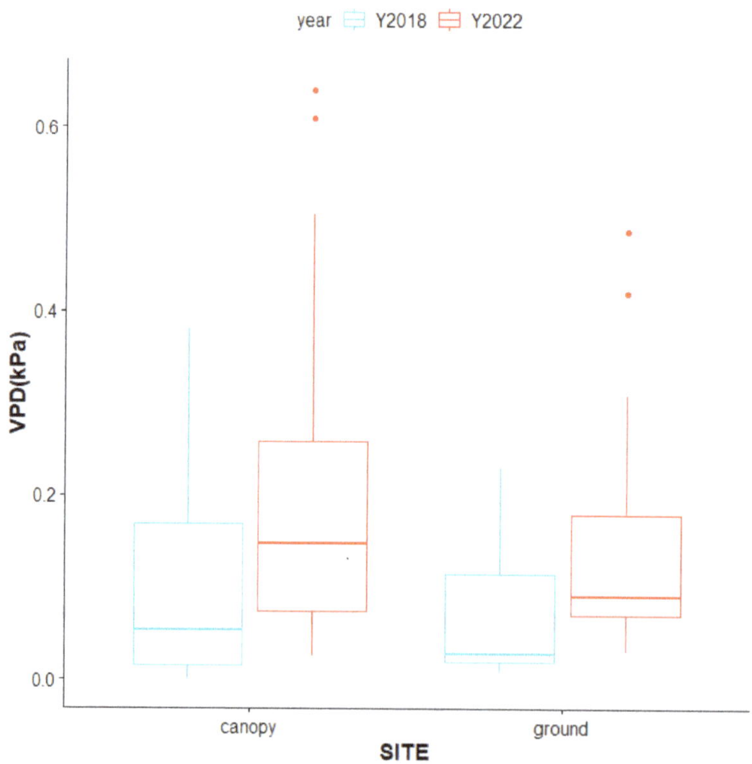

Figure 12. Boxplots (range with median line in the box) of mean VPD in 2018 (blue) and 2022 (red) in the forest canopy and on the forest floor (differences: sites, $p < 0.05$; year, $p < 0.001$).

4.2. Potential Threats from Changing Climate to *P. formosana* Survival

During the study period, we observed a decline in the population size of *P. formosana* in situ. From photos taken during the study period, we found many populations had disappeared from branches or trunks where they once grew. Since YYL is a conservation site, illegal poaching is unlikely to be the main reason for population decline. We also observed that populations of *P. formosana* near the forest floor sharply decreased. Two years ago, there were many individuals growing near the forest floor, but these have nearly all disappeared in recent years (field observation). Our data reveal that the VPD and mean temperature at the forest floor were lower than in the canopy, although both vertical forest strata exhibited an increasing trend (Figure 12, Appendix A Figure A1). If rising temperatures and the VPD were not the main reason for the disappearance of *P. formosana* near the forest floor, we can consider other influential microclimatic factors. Aside from light intensity, wind flow is one of the major differences between the forest canopy and floor. Raised garden beds are usually recommended for *P. formosana* nurseries [27] because elevated cultivation promotes air circulation and prevents pathogen infection in the orchid. Recent studies suggest that rising temperatures and high humidity may reduce disease resistance for native orchids [29]. A possible pathogen (Trichosphaeriaceae, unpublished data) that causes leaf spot in *P. formosana* was detected in the canopy soil at YYL [30]. However, strong airflow increases evapotranspiration in the forest canopy (Video, https://photos.app.goo.gl/dRFNvFVzsfjkHZQSA, accessed on 2 April 2024) and is probably unfavorable to the pathogen growth that infects *P. formosana*. We also found that individuals of the white flower variety have gradually disappeared at YYL. Further research is needed to determine whether the habitat niche of the white variety is more limited than that of pink flower individuals.

4.3. Implications for Re-Establishment of *P. formosana*

The *P. formosana* plantlets were reintroduced to YYL in mid-August of 2022. The onset of the NE monsoon (the first drop of air temperature below 10 °C) in 2022 was on September 8. The NE monsoon usually brings abundant rainfall to YYL, which enhances the survival of *P. formosana* seedlings. However, the reintroduced seedlings did not survive the following summer of 2023 and eventually shriveled. Snow events occur regularly at YYL, but the preliminary outcome of the reintroduction experiment suggests that summer heat rather than winter cold is probably the limiting factor for the survival of the species. The most suitable reintroduction season for YYL may be autumn rather than spring because the constant light rain brought by the NE monsoon provides the necessary moisture for young plants [30]. Moreover, to improve the survival rate, it is best to acclimatize the plantlets for a few months in the nursery on sphagnum moss after they are taken out from aseptic culture bottles [31].

5. Conclusions

This conservation project of *P. formosana* at YYL highlights the importance of the long-term monitoring of sensitive ecosystems in the face of anthropogenic climate change. Our six-year climate record reveals that climate change is not constant in direction, and it fluctuates greatly; therefore, it is difficult for forest organisms to adapt. Extreme weather events, such as drought or torrential rain, are becoming more likely with climate change and can have a greater impact on forest ecosystems than gradual global warming. The interaction between organisms of the forest ecosystem (e.g., microbial connections) may change with rising temperatures [32]. Preserving rare species in the face of global climate change is a mounting challenge to modern conservationists. An intact old-growth forest, such as the one at YYL, provides a micro-site with relatively stable habitats for sensitive species in a changing climate. Our report presents practical conservation measures to preserve the population of *P. formosana*, and our goal going forward is to focus on the management of local healthy populations of *P. formosana*.

Author Contributions: Conceptualization, R.C.-C.H.; methodology, R.C.-C.H. and C.L.; investigation, R.C.-C.H. and C.L.; In vitro germination, Y.-C.C.; formal analysis, R.C.-C.H.; writing—review and editing, R.C.-C.H. All authors have read and agreed to the published version of the manuscript.

Funding: This research was sponsored by the Ministry of Agriculture of Taiwan (112AS-7.1.2-FI-G3).

Data Availability Statement: The data presented in this study are available on request from the corresponding author. The data are not publicly available due to regulatory restrictions of the funding organization.

Acknowledgments: We thank Mei-Chun Lu of the Miaoli District Agricultural Research and Extension Station for advice on *P. formosana* cultivation. We also express our appreciation to the Forestry and Nature Conservation Agency for their kind assistance in Chilan.

Conflicts of Interest: The authors declare no conflicts of interest.

Appendix A

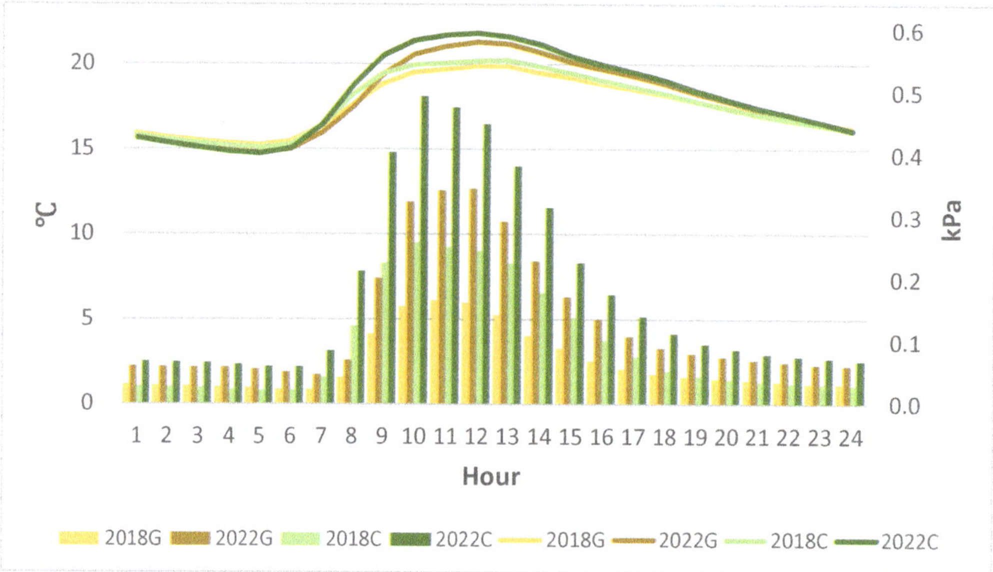

Figure A1. Comparison of 24 h temperature (lines chart, °C) and VPD (bar chart, kPa) in summers of 2018 and 2022 in the forest canopy (C) and on the forest floor (G).

References

1. Foster, P. The Potential Negative Impacts of Global Climate Change on Tropical Montane Cloud Forests. *Earth Sci. Rev.* **2001**, *55*, 73–106. [CrossRef]
2. Still, C.J.; Foster, P.N.; Schneider, S.H. Simulating the Effects of Climate Change on Tropical Montane Cloud Forests. *Nature* **1999**, *398*, 608–610. [CrossRef]
3. Guzmán, Q.J.A.; Hamann, H.F.; Sánchez-Azofeifa, G.A. Multi-Decadal Trends of Low-Clouds at the Tropical Montane Cloud Forests. *Ecol. Indic.* **2024**, *158*, 111599. [CrossRef]
4. Hu, T.; Liu, W.Y.; Wen, H.D.; Song, L.; Zhang, T.T.; Chen, Q.; Liu, S. Vascular Epiphyte Populations with Higher Leaf Nutrient Concentrations Showed Weaker Resilience to an Extreme Drought in a Montane Cloud Forest. *Plant Biol.* **2023**, *25*, 215–225. [CrossRef] [PubMed]
5. Hsu, R.C.; Oostermeijer, J.G.; Wolf, J.D. Adaptation of a Widespread Epiphytic Fern to Simulated Climate Change Conditions. *Plant Ecol.* **2014**, *215*, 889–897. [CrossRef]
6. El-Madany, T.S.; Walk, J.B.; Deventer, M.J.; Degefie, D.T.; Chang, S.; Juang, J.; Griessbaum, F.; Klemm, O. Canopy-atmosphere Interactions under Foggy Condition—Size-resolved Fog Droplet Fluxes and Their Implications. *J. Geophys. Res. Biogeosci.* **2016**, *121*, 796–808. [CrossRef]

7. Jang, Y.; Shen, S.; Juang, J.; Huang, C.; Lo, M. Discontinuity of Diurnal Temperature Range along Elevated Regions. *Geophys. Res. Lett.* **2022**, *49*, e2021GL097551. [CrossRef]
8. Hamilton, L.S.; Juvik, J.O.; Scatena, F.N. *Tropical Montane Cloud Forests*; Springer: New York, NY, USA, 1995; ISBN 0387943234.
9. Benzing, D.H. Vulnerabilities of Tropical Forests to Climate Change: The Significance of Resident Epiphytes. *Clim. Change* **1998**, *39*, 519–540. [CrossRef]
10. Hsu, R.C.C.; Wolf, J.H.D.; Tamis, W.L.M. Regional and Elevational Patterns in Vascular Epiphyte Richness on an East Asian Island. *Biotropica* **2014**, *46*, 549–555. [CrossRef]
11. Hsu, R.C.C.; Tamis, W.L.M.; Raes, N.; de Snoo, G.R.; Wolf, J.H.D.; Oostermeijer, G.; Lin, S.-H. Simulating Climate Change Impacts on Forests and Associated Vascular Epiphytes in a Subtropical Island of East Asia. *Divers. Distrib.* **2012**, *18*, 334–347. [CrossRef]
12. Hsu, R.C.-C.; Lin, C.; Chen, C. Topography-Induced Local Climatic Variations as the Decisive Factor in the Shaping of Epiphyte Distributions in Chilan, Northeastern Taiwan. *Forests* **2023**, *14*, 358. [CrossRef]
13. Hayata, B. Materials for a Flora of Formosa: Supplementary Notes to the Enumeratio Plantarum Formosanarum and Flora Montana Formosae. *J. Coll. Sci. Imp. Univ. Tokyo Jpn.* **1911**, *30*, 1–471.
14. Fukuyama, N. Neue Orchideen von Formosa. *Trans. Nat. Hist. Soc. Formosa* **1933**, *22*, 413–416.
15. Chao, W.-C.; Liu, Y.-C.; Jiang, M.-T.; Wu, S.-S.; Fang, C.-L.; Ho, J.-F.; Huang, C.-L. Genetic Diversity, Population Genetic Structure and Conservation Strategies for Pleione Formosana (Orchideace). *Taiwania* **2021**, *66*, 20–30.
16. Hsu, R.C.-C. Modelling Spatial Patterns of Rare Orchids for Conservation Priority in Taiwan. In Proceedings of the 20th World Orchid Conference, Singapore, 13–20 November 2011.
17. Hsu, R.C.-C. Using Spatially Autocorrelated Environmental Conditions in Habitats to Project Potential Distributions of Rare Orchids. *Taiwan J. For. Sci.* **2015**, *30*, 97–107.
18. Lai, I.-L.; Chang, S.-C.; Lin, P.-H.; Chou, C.-H.; Wu, J.-T. Climatic Characteristics of the Subtropical Mountainous Cloud Forest at the Yuanyang Lake Long-Term Ecological Research Site, Taiwan. *Taiwania* **2006**, *51*, 317–329.
19. Chang, S.-C.; Lai, I.-L.; Wu, J.-T. Estimation of Fog Deposition on Epiphytic Bryophytes in a Subtropical Montane Forest Ecosystem in Northeastern Taiwan. *Atmos. Res.* **2002**, *64*, 159–167. [CrossRef]
20. Jiang, M.-T.; Chao, W.-C.; Huang, C.-L.; Lan, S.-R.; Liu, Z.-J.; Wu, S.-S. The Complete Chloroplast Genome of *Pleione Formosana* (Orchidaceae). *Mitochondrial DNA Part B* **2019**, *4*, 1044–1046. [CrossRef]
21. Chiang, Y.L.; Chen, Y.-R. Observations on *Pleione formosana* Hayata. *Taiwania* **1968**, *14*, 271–301.
22. Lu, M.-C. High Frequency Plant Regeneration from Callus Culture of *Pleione Formosana* Hayata. *Plant Cell Tissue Organ. Cult.* **2004**, *78*, 93–96. [CrossRef]
23. Lu, M.-C. Tissue Culture of *Pleione Formosana* Hayata. *Seed Nurs.* **2004**, *6*, 60–71. (In Chinese)
24. Murashige, T.; Skoog, F. A Revised Medium for Rapid Growth and Bio Assays with Tobacco Tissue Cultures. *Physiol. Plant* **1962**, *15*, 473. [CrossRef]
25. Huang, J. A Simple Accurate Formula for Calculating Saturation Vapor Pressure of Water and Ice. *J. Appl. Meteorol. Clim.* **2018**, *57*, 1265–1272. [CrossRef]
26. R Core Team. *R: A Language and Environment for Statistical Computing*; R Foundation for Statistical Computing: Vienna, Austria, 2021.
27. Chang, S.-T. *The cultivation of Pleione formosana*; Bulletin of Taoyuan District Agricultural Research and Extension Station: Taoyuan City, Taiwan, 2001; pp. 1–8. (In Chinese)
28. Gotsch, S.G.; Davidson, K.; Murray, J.G.; Duarte, V.J.; Draguljić, D. Vapor Pressure Deficit Predicts Epiphyte Abundance across an Elevational Gradient in a Tropical Montane Region. *Am. J. Bot.* **2017**, *104*, 1790–1801. [CrossRef]
29. Mangelsdorff, R.; Piepenbring, M.; Perdomo-Sánchez, O. Correlation of Diversity of Rust Fungi and Their Host Plants with Disturbance and Conservation of Vegetation in Western Panama: A Case Study in Western Panama Focused on Orchidaceae and Pteridophytes as Host Plants. *Biodivers. Conserv.* **2012**, *21*, 2323–2339. [CrossRef]
30. Jain, A.; Sarsaiya, S.; Chen, J.; Wu, Q.; Lu, Y.; Shi, J. Changes in Global Orchidaceae Disease Geographical Research Trends: Recent Incidences, Distributions, Treatment, and Challenges. *Bioengineered* **2021**, *12*, 13–29. [CrossRef]
31. Kang, H.; Kang, K.W.; Kim, D.H.; Sivanesan, I. In Vitro Propagation of *Gastrochilus matsuran* (Makino) Schltr., an Endangered Epiphytic Orchid. *Plants* **2020**, *9*, 524. [CrossRef]
32. Gautam, H.R.; Bhardwaj, M.L.; Kumar, R. Climate Change and Its Impact on Plant Diseases. *Curr. Sci.* **2013**, *105*, 1685–1691.

plants

MDPI

Article

Variation in the Floral Morphology of *Prosthechea karwinskii* (Orchidaceae), a Mexican Endemic Orchid at Risk

María Hipólita Santos-Escamilla [1], Gabriela Cruz-Lustre [2,*], Manuel Cuéllar-Martínez [3], Luicita Lagunez-Rivera [4] and Rodolfo Solano [3,*]

1 Instituto Tecnológico del Valle de Oaxaca, Ex Hacienda Nazareno, Santa Cruz Xoxocotlán 71233, Oaxaca, Mexico; mhsantosescamilla55@gmail.com
2 Departamento de Botânica, Instituto de Ciências Biológicas, Universidade Federal de Minas Gerais, Avenida Antônio Carlos 6627, Pampulha, Caixa Postal 486, Belo Horizonte 31270-910, Minas Gerais, Brazil
3 Centro Interdisciplinario de Investigación para el Desarrollo Integral Regional Unidad Oaxaca, Instituto Politécnico Nacional, Santa Cruz Xoxocotlán 71230, Oaxaca, Mexico; manuel.cuellarm@gmail.com
4 Laboratorio de Extracción y Análisis de Productos Naturales Vegetales, Centro Interdisciplinario de Investigación para el Desarrollo Integral Regional Unidad Oaxaca, Instituto Politécnico Nacional, Santa Cruz Xoxocotlán 71230, Oaxaca, Mexico; llagunez@ipn.mx
* Correspondence: gcruzlustre@gmail.com (G.C.-L.); asolanog@ipn.mx (R.S.)

Abstract: *Prosthechea karwinskii* is an orchid endemic to Mexico, threatened by the destruction of its habitat and the extraction of specimens to meet its demand for ornamental and religious use. Most of its populations, including the most locally abundant ones, are found in Oaxaca state. Variations in some floral traits have been observed in these populations. We implemented a morphometric analysis to assess their floral variation and identify the most significant characters in the morphological patterns of this orchid. Floral samples were collected from 17 populations of *P. karwinskii* in Oaxaca, as well as from specimens used as ornaments during Easter in an Oaxacan community (Zaachila), whose origin is unknown. Sampling of natural populations covered the environmental, geographic, and morphological variation of the species. We performed an analysis of variance (ANOVA), principal component analysis (PCA), canonical variate analysis (CVA), and cluster analysis, including 185 individuals and 45 variables (12 of them were discarded in the multivariate analyses due to high correlation). Characters of the column, lateral sepal, and labellum were most informative for the observed morphological patterns. Albarradas showed the greatest morphological differentiation, mainly due to the column. In general, individuals from the same locality tended to overlap more, especially the populations of Jaltianguis and Yahuiche, which were different from the geographically close population of Etla. Teposcolula presented the highest values in perianth characters, unlike Sola_Rancho Viejo. The specimens recovered from religious ornaments were morphologically more similar to those from Yanhuitlan and Etla. This morphometric analysis identified characters as potential taxonomic markers for *P. karwinskii* and related species, showing its potential to associate specimens of unknown origin with their probable geographical region. Our work encourages working on collaborative conservation strategies to ensure the long-term permanence of both the species and its traditional uses.

Keywords: endangered species; epiphyte; floral variation; intraspecific variation; morphometrics; orchids; ornamental/useful species

1. Introduction

Prosthechea karwinskii (Mart.) J.M.H. Shaw is an orchid endemic to western and southern Mexico, inhabiting mountainous regions where it grows as a hanging epiphyte in oak or oak-pine forests subject to a well-marked seasonal drought (Figure 1A). The inflorescence is a two-flowered (exceptionally three flowers) raceme arising from the last developed pseudobulb (Figure 1B). The taxon was described in 1830, but its taxonomic history has been linked to that of *Prosthechea citrina* (Lex.) W.E. Higgins, or any of its nomenclatural synonyms, a very similar species, with which it becomes sympatric in some locations in Guerrero and Michoacán. In the past, the information available for *P. karwinskii* was invariably attributed to *P. citrina*; only in recent years have both species come to be considered different [1,2]. This species is one of the most distinctive orchids in the Mexican flora, valued for the ornamental beauty and pleasant aroma of its flowers, as well as having cultural significance since pre-Hispanic times. It was one of the orchids from which mucilage was obtained and used as an adhesive in feather art in pre-Hispanic times and during the colonial period [3]. In traditional medicine, it has been used to soothe coughs, heal wounds and burns, treat diabetes, prevent miscarriage, and aid in childbirth [4–7]. The flowers are used as decorations in homes, commercial stands, and temples during Easter commemorations [8,9]. Additionally, due to the beauty, color, and aroma of its flowers, this plant is cultivated in a rustic manner in orchards in many communities in Oaxaca.

Figure 1. (**A**) Habitat of *Prosthechea karwinskii*, an oak forest in San Pedro y San Pablo Teposcolula, Oaxaca. (**B**) *Prosthechea karwinskii* growing in situ as a hanging epiphyte on *Quercus* sp. in Santo Domingo Yanhuitlan, Oaxaca. Photographs by R. Solano.

In most of the localities where *P. karwinskii* grows, it is locally scarce and survives in forest fragments surrounded by a matrix of environments modified by anthropogenic causes (conversion into crop fields, human settlements, opening of new roads, goat farming). In Oaxaca, the most abundant populations of the species occur, but they face the risk of extraction for temporary adornment in local trade [8–10], for religious purposes [8], and to a lesser extent, for medicinal use [4]. This practice occurs annually, mainly during the species' flowering season, leading to its inclusion in the list of species of Mexican wild flora at risk [11]. For other epiphytic orchids growing in the mountain forests of Mexico, it has been demonstrated that extraction for local trade has effects on species subjected to this practice, such as reductions in population size and rates of fertility and recruitment, as well as loss of genetic diversity [12–15].

In the localities of *P. karwinskii* in Oaxaca, variation has been observed in some floral traits, such as flower size and coloration, the shape of the labellum, and the shape of the apical teeth of the column. This leads to the assumption of the existence of intraspecific variation, at least among populations in the state, which has not been analyzed either through the use of morphological or molecular markers. The analysis of this variation will be important for identifying morphotypes with ornamental potential and desirable in a management program, recognizing phytogenetic diversity present in the species, identifying forms or subspecies within the orchid, as well as determining a possible geographical pattern associated with morphological variation.

Morphometric studies have been employed in some species of Orchidaceae, primarily analyzing variation associated with floral morphology [16–22], although vegetative morphology has also been considered, including attributes of leaf anatomy [17,22]. Studies in this regard seek to find intraspecific differences [18,23] and interspecific differentiation to delimit similar taxa considered as cryptic species [16,22], recognize taxa of hybrid origin [24], or identify morphotypes with phytogenetic value [19,20]. The quantitative analysis of characters identified in the labellum has been valuable for recognizing and characterizing intraspecific variation in orchids [19,20], although the use of traits present in other floral structures has also been useful [18]. Morphometry has also been used to trace the geographical origin of samples of unknown origin, particularly for species or products of economic importance [25–30]. Although various sources of information (e.g., genetic and chemical) and analytical tools can address this issue, the use of multivariate methods with morphological characters offers the advantage of low cost [27–30] and relatively easy data collection for a large number of individuals [26,27]. However, this topic is analytically complex and requires caution in its implementation and interpretation due to the need for a robust reference and the requirements of the analyses [26,28].

The objective of this study was to analyze the variation among populations from different localities in Oaxaca, Mexico, and to identify variables with taxonomic potential, through a morphometric analysis. Additionally, an attempt was made to determine if this set of characters allows relating specimens extracted from their habitats and whose origin is unknown, which were recovered after having had a religious use. Predictions for this study are as follows: (1) the floral morphology of *P. karwinskii*, analyzed using morphometric methods, will allow us to recognize the interpopulation variation of the species; and (2) the floral traits of *P. karwinskii* could serve as a morphological marker to associate the geographical origin of specimens from unknown localities.

2. Results

2.1. ANOVA and Kruskal–Wallis Tests

Most of the 40 floral variables analyzed with ANOVA showed significant differences among the localities of *P. karwinski* (Table 1). However, the characters lengths between maximum width and apex (SlLa) and angle at the apex (SlAa) of the lateral sepal, width at 1/3 (SdA1), length between maximum width and apex (SdLa), and angle at the apex (SdAa) of the dorsal sepal, and width at the base of the middle lobe of the labellum (LaAbm) did not show significant differences among localities. The specimens from Teposcolula exhibited the highest values for the length and width of floral segments, which determine flower size, such as total length (SlLt) and maximum width (SlAm) of the lateral sepal, total length (SdLt) and maximum width of the dorsal sepal (SdAm), total length (PeLt) and maximum width (PeAm) of the petal, total length (LaLt) and maximum width (LaAm) of the labellum, and total length of the column (CoLt) (see Table S1). In contrast, the specimens from Sola_Rancho Viejo appeared to have the smallest flowers, as characters like total length (SdLt) and maximum width (SdAm) of the dorsal sepal, total length (PeLt) and maximum width (PeAm) of the petal, total length (LaLt), and maximum width (LaAm) of the labellum showed the lowest values (see Table S1). Additionally, the locality of Albarradas showed the highest values in two characters of the labellum (LaAml, LaA1l), in the separation between the teeth of the column (AnDlDm) and in the width of the cuniculus near the base (CuA1). However, it presented the lowest values for one character of the petal (PeAa), two in the labellum (LaAul, LaAlm) and three in the teeth of the column (DlAl, DlAn, DmAl).

The Kruskal–Wallis test revealed significant differences among localities (df = 11, p-value < 0.05) for the thickness at the middle part (CoGm, $X^2 = 39.759$) and the anther level of the column (CoGa, $X^2 = 40.521$), and width of its middle tooth (DmAn, $X^2 = 27.5236$). However, there were no differences among localities (df = 11, p-value > 0.05) for the apex angle of the middle lobe of the labellum (LaAam, $X^2 = 18.374$) and width of the column at the stigma level (CoAe, $X^2 = 17.823$). The specimens from Teposcolula and Sola_El Lazo recorded the highest values for the thickness of the column at the anther level (CoGa), while the individuals from Zaachila and Albarradas had the lowest values for this character

(Table S1). For thickness at the middle part of the column (CoGm), the individuals from Teposcolula had the highest value. For the width of the middle tooth of the column (DmAn), the individuals from Sola_El Lazo had the highest value, while those from Albarradas had the lowest value.

Table 1. Results of the analysis of variance (ANOVA) for 40 of the 45 floral characters that met the assumption of normality recorded in 185 individuals from 12 expanded populations of *Prosthechea karwinskii*, including Zaachila. df = degrees of freedom, ss = sum of squares, ms = mean square. Differences are significant at $p \leq 0.05$, number of asterisks indicates increasing levels of significance, a dot indicates that the test value was not significant.

Variable	df	ss	ms	F-Value	*p*-Value	Significance
SlLt	11	1546	140.51	3.214	<0.001	***
SlAm	11	120.4	10.944	2.213	0.015	*
SlA1	11	127.1	11.552	2.096	0.022	*
SlA2	11	135.1	12.278	2.209	0.022	*
SlLa	11	463	42.13	1.35	0.201	.
SlAa	11	1846	167.86	1.702	0.076	.
SdLt	11	2005	182.27	4.338	<0.001	***
SdAm	11	142	12.913	2.262	0.013	*
SdA1	11	121.9	11.084	1.72	0.072	.
SdA2	11	137.1	12.462	2.177	0.017	*
SdLa	11	467	42.47	1.718	0.072	.
SdAa	11	2178	198	1.513	0.130	.
PeLt	11	1777	161.52	4.154	<0.001	***
PeAm	11	516.8	46.98	3.065	<0.001	***
PeA1	11	473.5	43.05	3.222	<0.001	***
PeA2	11	351.4	31.95	2.592	0.004	**
PeLa	11	515.5	46.86	2.643	0.003	**
PeAa	11	8608	862.5	4.436	<0.001	***
LaLt	11	2123	193	4.351	<0.001	***
LaAm	11	677	61.55	2.497	0.006	**
LaLl	11	678	61.63	2.99	0.001	**
LaAml	11	3.817	0.347	4.113	<0.001	***
LaA1l	11	172.9	15.714	3.881	<0.001	***
LaAul	11	14463	1314.8	9.014	<0.001	***
LaAlm	11	3.817	0.347	4.113	<0.001	***
LaLm	11	5.668	0.5152	4.874	<0.001	***
LaAbm	11	172.6	15.695	1.839	0.050	.
LaAmm	11	169.7	15.423	1.816	0.054	.
CoLt	11	262.2	23.836	6.174	<0.001	***
CoA1	11	46.92	4.266	2.917	0.001	**
RoAl	11	48.02	4.366	3.471	<0.001	***
RoAn	11	42.36	3.85	2.119	0.016	*
DlAl	11	95.07	8.642	14.04	<0.001	***
DlAn	11	27.24	2.4768	3.08	0.008	**
DmAl	11	109.69	9.972	18.58	<0.001	***
AnDlDm	11	12.45	1.1315	6.602	<0.001	***
CuLt	11	112.9	10.264	2.263	0.013	*
CuAm	11	12.1	1.1	3.215	<0.001	***
CuA1	11	1.27	0.11542	3.472	<0.001	***
CuA2	11	7.75	0.7046	2.245	0.014	*

2.2. Multivariate Analyses (PCA, CVA, and Cluster Analysis)

Both the PCA and CVA produced similar results whether the samples obtained in Zaachila were excluded or included. Therefore, for both methods, only the results of the analyses including the floral samples obtained from this community in Oaxaca are presented. For the final analyses (PCA and CVA), 185 individuals and 33 morphological variables of *P. karwinskii* were included.

The PCA showed that eight principal components had eigenvalues > 1.0, which together accounted for 71.91% of the total variance (Table 2). Among the eight components that retained the highest percentage of variance, the first one explained a third of it (33.51%) and was more correlated with variables related to floral size. Specifically, the maximum width of the labellum (LaAm), the petal (PeAm), and the lateral sepal (SlAm) were the most important variables in the morphological patterns observed in this axis. This analysis showed a high overlap among individuals from different populations; however, some individuals from Teposcolula and the populations of Yanhuitlan, Etla, and Sola_Rancho Viejo appeared at the extremes (Figure 2).

Table 2. Contribution between variables and axes with eigenvalues > 1.00 of PCA (18 populations, including Zaachila) and CVA (12 expanded populations) of *Prosthechea karwinskii* from Oaxaca, Mexico. Values in bold correspond to the three highest for each axis (A).

Variable	PCA (n =185)								CVA (n =185)		
	A1	A2	A3	A4	A5	A6	A7	A8	A1	A2	A3
SlLt	−0.835	0.367	0.015	0.063	0.114	0.132	0.030	0.031	−0.128	**−0.941**	−0.066
SlAm	**−0.852**	−0.089	−0.055	0.070	−0.247	0.085	0.067	−0.019	**−0.551**	0.230	−0.146
SlLa	−0.671	0.369	−0.209	0.087	−0.050	0.187	−0.059	−0.117	0.435	−0.159	−0.004
SlAa	−0.281	**−0.701**	0.027	0.372	−0.134	0.054	−0.001	−0.050	0.411	0.241	0.222
SdLa	−0.657	0.403	−0.166	0.028	−0.041	0.161	−0.051	−0.242	−0.230	0.006	0.112
SdAa	−0.174	**−0.696**	−0.001	0.407	−0.121	−0.033	0.056	0.161	−0.242	−0.121	−0.024
PeAm	**−0.853**	0.104	−0.045	0.114	−0.221	0.025	0.059	0.092	0.062	0.352	−0.007
PeLa	−0.706	0.407	−0.128	0.124	0.002	0.043	0.032	−0.164	0.082	−0.188	0.205
PeAa	−0.176	**−0.632**	0.114	**0.457**	0.059	0.189	0.016	0.129	0.341	−0.104	0.191
LaLt	−0.827	0.323	0.106	0.103	0.121	0.063	−0.067	0.153	0.158	**1.015**	0.088
LaAm	**−0.861**	0.069	0.069	0.236	−0.244	−0.081	−0.007	0.081	0.291	−0.420	**0.480**
LaLl	−0.804	0.201	0.147	0.205	−0.006	−0.021	−0.040	0.150	0.143	0.241	−0.436
LaAml	−0.630	0.199	0.145	0.204	−0.376	−0.276	−0.052	0.184	**−0.533**	0.040	−0.050
LaAul	0.251	−0.043	**−0.518**	−0.069	0.206	0.137	0.018	−0.242	0.082	−0.230	**0.628**
LaAlm	−0.074	0.020	−0.022	−0.010	**0.474**	**0.441**	**0.375**	**0.404**	0.034	−0.040	0.157
LaLm	−0.604	0.329	0.096	0.049	0.147	0.192	−0.094	−0.031	0.227	−0.273	−0.064
LaAmm	−0.661	−0.047	−0.018	0.131	−0.121	0.270	0.028	−0.166	−0.432	−0.050	−0.139
LaAam	0.016	−0.220	0.070	0.026	−0.102	**0.458**	**−0.717**	−0.011	0.049	−0.104	−0.053
CoLt	−0.622	0.168	−0.107	0.016	0.249	−0.313	−0.037	0.161	−0.152	0.557	−0.050
CoAe	−0.729	−0.390	−0.297	−0.277	−0.039	−0.125	0.024	0.078	−0.497	−0.334	0.034
CoA1	−0.711	−0.311	−0.327	−0.208	−0.065	0.094	0.204	−0.065	0.031	−0.074	−0.076
CoGa	−0.750	−0.274	−0.345	−0.195	−0.002	−0.090	0.233	−0.026	0.081	0.142	**0.484**
RoAl	−0.394	−0.147	−0.248	**−0.405**	0.094	−0.156	**−0.488**	0.207	−0.327	−0.198	0.354
RoAn	−0.588	−0.372	−0.350	−0.404	0.009	−0.062	−0.121	0.145	0.285	0.219	−0.269
DlAl	−0.339	0.008	0.067	0.189	**0.580**	−0.345	−0.163	0.260	0.162	**0.594**	0.339
DlAn	−0.543	−0.199	−0.204	0.021	0.137	−0.316	0.028	**−0.297**	−0.240	−0.393	0.309
DmAl	−0.252	−0.253	0.140	0.261	**0.505**	0.237	−0.057	−0.193	**0.950**	−0.243	−0.160
DmAn	−0.494	−0.329	−0.096	−0.040	0.262	0.044	−0.071	−0.283	0.108	−0.213	−0.342
AnDlDm	−0.089	−0.043	0.019	**−0.489**	−0.278	**0.463**	0.081	**0.314**	−0.311	−0.402	−0.360
CuLt	−0.466	−0.051	0.303	−0.273	0.306	0.073	−0.035	−0.019	−0.098	0.211	−0.088
CuAm	−0.424	−0.186	**0.651**	−0.201	−0.088	−0.131	−0.102	−0.133	0.144	0.206	−0.107
CuA1	−0.486	−0.126	**0.647**	−0.315	0.108	0.013	0.117	−0.088	0.217	0.221	−0.202
CuA2	−0.484	−0.218	0.622	−0.356	−0.026	−0.042	0.146	−0.169	−0.096	−0.088	0.016
Eigenvalue	11.06	3.16	2.35	1.93	1.65	1.42	1.13	1.03	3.29	1.96	1.27
Accumulated variance (%)	33.51	43.10	50.23	56.09	61.08	65.37	68.80	71.91	33.56	53.52	66.46

In the CVA, three canonical axes had eigenvalues > 1.0 and together explained 66.46% of the total variance (Table 2). The first canonical axis explained 33.56% of the variance and reflected a greater contribution of the height of the middle tooth of the column (DmAl), the maximum width of both the lateral sepal (SlAm) and the lateral lobe of the labellum (LaAml). On this axis, there was a clear separation of individuals from the Albarradas population from the other populations (Figure 3A). The samples from Jaltianguis and Yahuiche showed less dispersion along the first three canonical axes and overlapped with each other. Along axis 1, individuals from these two populations were completely separated from Etla and Zaachila (except for one individual from Jaltianguis). Axis 1 also showed a complete separation of the samples from Zaachila from those from Juquila and the populations of Sola (El Lazo and Lachixío), with only a marginal overlap with the remaining Sola population (Rancho Viejo) and Tlaxiaco. Except for one individual, Teposcolula was also almost completely separated from Zaachila and Etla.

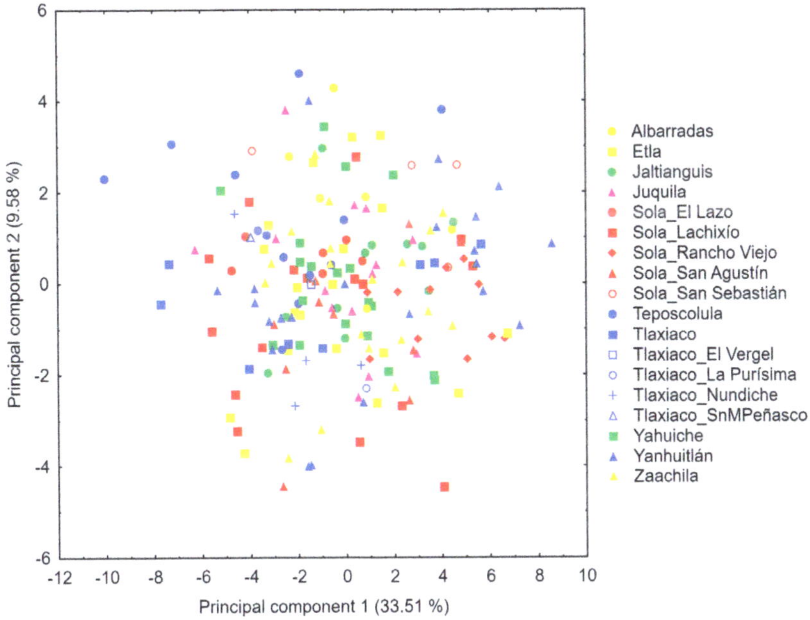

Figure 2. Representation of axes 1 and 2 of the PCA resulting from the variation of 33 floral variables in 185 individuals of *Prosthechea karwinskii* from 18 localities (including Zaachila) in Oaxaca, Mexico.

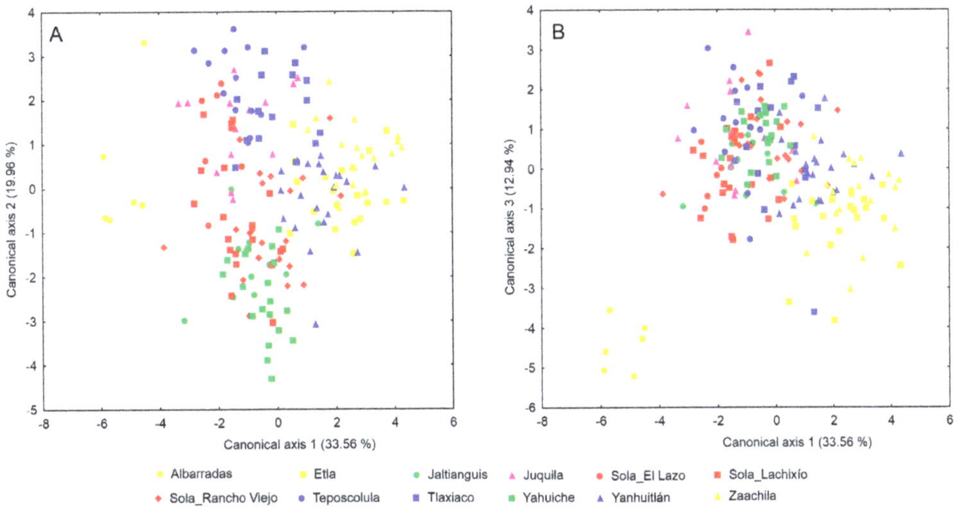

Figure 3. Representation of the axes 1–2 (**A**) and 1–3 (**B**) of the CVA resulting from the variation of 33 floral variables in 185 individuals of *Prosthechea karwinskii* from 12 expanded populations in Oaxaca, Mexico. See Section 4 to see how the expanded populations were integrated.

The second axis explained 19.96% of the variance and showed a greater contribution from the total length of both the labellum (LaLt) and the lateral sepal (SlLt), as well as the height of the lateral tooth of the column (DlAl). On axis 2, individuals from the Yahuiche and Jaltianguis populations were completely separated from Tlaxiaco and Teposcolula (Figure 3A) and present a marginal overlap with Juquila and Sola_El Lazo. On this axis, Yanhuitlan is completely separate from Teposcolula and partially overlaps with Sola_Rancho

Viejo. The third axis accounted for 12.94% of the variance and had as the most important variables the angle between the lateral lobe and the claw of the labellum (LaAul), the thickness of the column at the level of the anther (CoGa) and the maximum width of the labellum (LaAm). On this axis, the separation of individuals from Albarradas from all other populations was again highlighted (Figure 3B), as seen on axis 1. Along the first three axes, the samples from Zaachila showed greater overlap with those from the populations of Yanhuitlan and Etla (Figure 3).

The cluster analysis also revealed the high morphological divergence of Albarradas population (Figure 4). In the UPGMA dendrogram, this population was externally linked to the group containing the remaining populations. This latter group is divided into three subgroups: the first formed by intermixed populations from the southern regions of Oaxaca (Sola_El Lazo and Juquila) and two populations from the Mixteca (Teposcolula and Tlaxiaco); the second includes the remaining populations from Sola de Vega (Sola_Rancho Viejo and Sola_Lachixío) nested between those of the Sierra Norte (Jaltianguis and Yahuiche); and the third subgroup comprises Zaachila linked to Yanhuitlan, with Etla also joined to them.

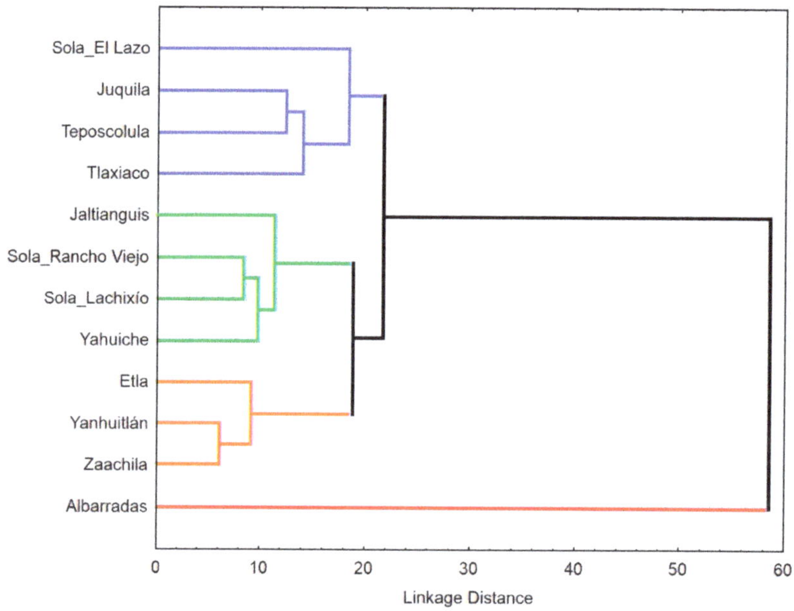

Figure 4. Phenogram showing the relationships between 12 expanded populations of *Prosthechea karwinskii* from Oaxaca, Mexico. The dendrogram was constructed using the UPGMA clustering algorithm and the squared Mahalanobis distances between the population centroids, calculated from 33 floral variables.

3. Discussion

3.1. Morphological Variation in Natural Populations of *Prosthechea karwinskii*

Plants have the ability to modify their phenotype in response to environmental conditions. However, the variation within a species due to the environment is expected to be smaller and more limited in floral characters compared to vegetative ones, since the former are related to reproductive success and must maintain their function [31,32]. The variation in floral morphology can be interpreted as an adaptation to selection by different pollinators [33,34]. Interactions between plant-pollinators and climatic influence can explain the variation in floral traits, suggesting that the variation expressed on them is a product of an adaptive response [35]. However, other processes can produce divergence among geographically separated populations of flowering plants, such as random genetic drift, isolation, indirect selection, and genetic factors [36,37]. Because floral traits are considered phenotypically more stable than vegetative ones, their variation tends to be less within populations, making them valuable for recognizing infraspecific variation [19,23]. Studies evaluating infraspecific morphological variation in orchids have so far been conducted using floral traits and have generally been useful in identifying such variation [16–25].

Of the floral characters evaluated in the ANOVA, most (38 out of 40) showed significant variation among the populations of *P. karwinskii* analyzed here, as demonstrated by the ANOVA and Kruskal–Wallis tests. Of the other five floral characters that did not show normality and were evaluated with the Kruskal–Wallis test, three of them (corresponding to the column) showed significant differences among populations (thickness in the middle part, thickness at the level of the anther, and width of the middle tooth). These tests identified Teposcolula (Mixteca region) as the population in Oaxaca that includes individuals with the largest flowers (showing the highest values for sepal, petal, and labellum length and width), while those from Sola_Rancho Viejo (Sierra Sur region) include individuals with the smallest flowers. Among orchids of ornamental value, specimens with larger flowers are preferred for cultivation and are selected as mother plants or as parents for artificial hybrids. Since *P. karwinskii* is an orchid appreciated for its ornamental value, the Teposcolula population holds greater importance in horticulture compared to other localities in Oaxaca.

The PCA conducted with the set of floral characters for *P. karwinskii* showed a high overlap between individuals from the 17 sampled locations and those obtained from Zaachila. Thus, it seems that the traits of floral morphology do not allow for detecting differences between populations, or their number is not sufficient to discriminate the infraspecific variation of *P. karwinskii*. Nevertheless, along the axis that accumulated the highest percentage of variance, several individuals from Teposcolula and Yanhuitlan appear at the extremes. The characters associated with this pattern were related to the width of the perianth segments (labellum, petal and lateral sepal). Among all the populations analyzed, Teposcolula has the widest segments, while Yanhuitlan, Etla, and Sola_Rancho Viejo have the narrowest segments, as also revealed by the univariate analyses. Ibáñez [27] suggested that complex patterns revealed through multivariate analyses of morphological data may be associated with the life history of the organisms. The localities of Teposcolula and Yanhuitlan host two of the largest populations of the species in Oaxaca, where the highest percentage of individuals are reproductive. Possibly, these two populations present high levels of genetic variation, which could be related to how their individuals are dispersed in the PCA graph. Other studies that have evaluated interspecific [17,22] or intraspecific variation [23] in orchids have also not found differentiation between populations that are geographically separated when analyzing floral characters with PCA. Such results are common given the method's assumptions, particularly the lack of a priori categorization into groups that would typically minimize intragroup variance and maximize intergroup variance [38]. However, this multivariate method has been useful for recognizing morphotypes in orchid species represented by wild specimens [20] or cultivated ones [19,25] in Mexico.

The analysis of *P. karwinskii*'s floral morphology with the CVA and cluster analysis were informative for discriminating infraspecific variation. These analyses showed that individuals from Alabarradas are well separated from the rest of the localities. Differences in the teeth of the column discriminate these individuals from those from other localities. Additionally, individuals from Jaltianguis and Yahuiche showed low dispersion and tended to overlap with each other. This morphological pattern is congruent with geography since these two locations are very close to each other in the region known as Sierra Norte of Oaxaca. Interestingly, individuals from Etla, the locality closest to the previous two, tend to show low dispersion among themselves and separate from those of Jaltianguis-Yahuiche. The morphological differentiation that is present with respect to Teposcolula is more expected. The CVA showed that the variables with the greatest contribution are traits of the column (height of the middle and the lateral teeth, and thickness at the level of the anther), lateral sepal (maximum width, total length), labellum (maximum width of the lateral lobe, total length, angle between the lateral lobe and claw, and maximum width). Due to the nature of the characters revealed by the analyses as the most important for the morphological patterns of the species, these traits could be related to the attraction of pollinators, as has been hypothesized in other studies analyzing morphological variation in other plants [22,39]. We recommend paying attention to the most important variables presented in this work, as they can be useful as taxonomic markers at the intraspecific level and possibly at the level of the *Prosthechea citrina* complex, the species group to which *P. karwinskii* belongs.

The population of Albarradas turned out to be the most differentiated among those of *P. karwinskii* from Oaxaca. The individuals from this locality exhibit the lowest height of the middle and lateral teeth of the column, the smaller angle between the lateral lobe and claw of the labellum and the thinnest thickness of the column at the level of the anther. Additionally, they have the greatest maximum width of the lateral lobe of the labellum. This population could be recognized as a variety or geographic form of the species, geographically isolated in the central part of the state of Oaxaca. Moreover, the locality of Albarradas is a priority for the conservation of the species, as it represents a unique morphological variant. Unfortunately, the forest where this form grows in Albarradas hosts one of the least numerous populations of this orchid in Oaxaca. It will be interesting to verify if the morphological differentiation of the Albarradas population is related to genetic differentiation, as has been corroborated for other orchid species using vegetative characters [17].

3.2. *Morphological Patterns of Zaachila Flowers of Unknown Origin*

Among the natural populations of *P. karwinskii* used as reference in this study to compare flowers obtained from Easter celebrations in Zaachila (2017–2019), morphometric analyses revealed greater overlap and morphological similarity between the Zaachila material and populations from Yanhuitlan and Etla. Flowers from Zaachila are distinct from those in Albarradas, Jaltianguis and Yahuiche (Sierra Norte), Teposcolula (Mixteca), Juquila and Sola_El Lazo (Southern mountains of Oaxaca). Conversely, Zaachila specimens exhibit varying degrees of overlap with individuals from the Sierra Sur (Sola_Lachixío and Sola_Rancho Viejo) and Mixteca (Tlaxiaco) regions. While the species is relatively common across various forests in Oaxaca, our inference suggests that individuals rescued from Zaachila could come from the Mixteca or vicinity of Etla. However, the available material spans more than one year of festivities, indicating participants likely gathered flowers from multiple locations over three years. The findings presented are specific to the study period, and we caution that they may vary annually depending on where extraction occurs.

Morphometric methods have been successfully employed to determine the unknown origins of organisms in various animal groups, utilizing live specimens, preserved in museums or commercialized [26–30]. However, they are relatively unexplored in plant species for this purpose. This study represents an initial attempt to infer the origins of Zaachila individuals based on morphological data. We point out the challenges (even with other types of data, e.g., genetics), including potential variance due to collection from multiple localities during the sampling period, gaps in samples from other distribution sites [26,28], and varying population sizes [40], which are intrinsic to orchids and other epiphytes [41].

Several populations studied here are situated in areas where specimens are harvested for local markets and religious use [3,8,9], impacting genetic variability and effective population sizes, as well increasing the chances of experiencing inbreeding depression and bottleneck events [13–15]. Genetic analysis could help elucidate the causes of floral morphological variation in *P. karwinskii* and its biogeographical patterns, though correlations between genetic and morphological variability are not always straightforward.

Mexico's use of plants to satisfy aspects of the cultural and social life of local communities is vast, given the high cultural and biological diversity [42–44] Among them, ceremonial uses constitute a cultural element of the people, making their transmission important. However, the extraction of wild plants used in such ceremonies often has a negative impact on their populations [45,46]. The conservation and use of wildlife are controversial [47], but these topics must be addressed through collective strategies [45,48]. There are some preliminary initiatives for *P. karwinskii* [10,49]. These efforts motivate us to develop collaborative conservation strategies that ensure the maintenance of the morphological and genetic variability of the species in regions most susceptible to flower extraction for traditional uses, as well as in the communities where these extracted specimens are destined. This will help guarantee the long-term preservation of both the species and its cultural significance.

4. Materials and Methods

4.1. Biological Material

During the flowering season of *P. karwinskii* (March–April), localities (populations) representing its distribution in Oaxaca were visited between 2015 and 2021. The geographical information of the sampled localities is presented in Table 3, while Figure 5 shows their geographical distribution on a map of Oaxaca. In each visited locality, one flower per individual was collected, ensuring they were on different host trees to avoid collecting ramets from the same individual; the proximal flower on the inflorescence was chosen when it had more than one. The sample size of each locality depended on the population abundance and is indicated in Table 3. A voucher specimen (herbarium or spirit) was prepared from each locality, which was deposited in the Herbarium OAX (acronym according to [50]. Additionally, flowers rescued from specimens used as decorations in Catholic temples during the Easter celebration in the Villa de Zaachila, Oaxaca (2017–2019) were obtained. These samples, whose original locality is unknown, were obtained with permission from the organizing committee of this commemoration in the community in 2019, once they were removed from the temples. Both the flowers collected in the field and the rescued ones were preserved in a fixing solution of water (78%), ethanol 96% (21%), lactic acid 85% (6%), benzoic acid (0.5% *w*/*v*), and glycerin (5%), and then deposited in the Laboratory of Extraction and Analysis of Natural Products (CIIDIR Oaxaca, Instituto Politecnico Nacional).

Table 3. Geographic information for the localities of *Prosthechea karwinskii* in Oaxaca, Mexico. See Section 4.2 to know how the expanded populations were integrated. QF = *Quercus* forest, QPF = *Quercus-Pinus* forest. *n* = number of individuals represented in the sample size per population. The locality/population numbers correspond to those shown in the map of Figure 5. The species is an orchid protected by Mexican environmental legislation; therefore, the coordinates of the localities were omitted. NA = the data does not apply.

Number	Locality/Population	*n*	Expanded Population (*n*)	Elevation	Vegetation
1	Amialtepec, Santa Catarina Juquila	14	Juquila (14)	2105 m	QF
2	El Lazo, Sola de Vega	6	Sola_El Lazo (6)	1840 m	QPF
3	El Vergel, Tlaxiaco	1	Tlaxiaco (15)	1900 m	QPF
4	La Purisima, Tlaxiaco	1		2160 m	QPF
5	San Mateo Peñasco, Tlaxiaco	1		2150 m	QF
6	Tlaxiaco	8		2160 m	QPF
7	Santiago Nundiche, Tlaxiaco	4		2200 m	QPF
8	Rancho Viejo, Sola de Vega	11	Sola_Rancho Viejo (21)	1838 m	QF
9	San Agustin, Sola de Vega	10		2091 m	QPF

Table 3. *Cont.*

Number	Locality/Population	*n*	Expanded Population (*n*)	Elevation	Vegetation
10	San Agustin, Etla	19	Etla (19)	1950 m	QPF
11	San Pedro y San Pablo Teposcolula	13	Teposcolula (13)	2410 m	QPF
12	El Vado-San Sebastian de las Grutas, Sola de Vega	4	Sola_Lachixio (20)	2100 m	QF
13	San Vicente Lachixio, Sola de Vega	16		2240 m	QPF
14	Santa Maria Jaltianguis	11	Jaltianguis (11)	2164 m	QPF
15	San Lorenzo Albarradas	6	Albarradas (6)	2230 m	QPF
16	Santo Domingo Yanhuitlan	22	Yanhuitlan (22)	2400 m	QPF
17	Yahuiche, Ixtlan de Juarez	19	Yahuiche (19)	2017 m	QF
18	Zaachila	19	Zaachila (19)	NA	NA

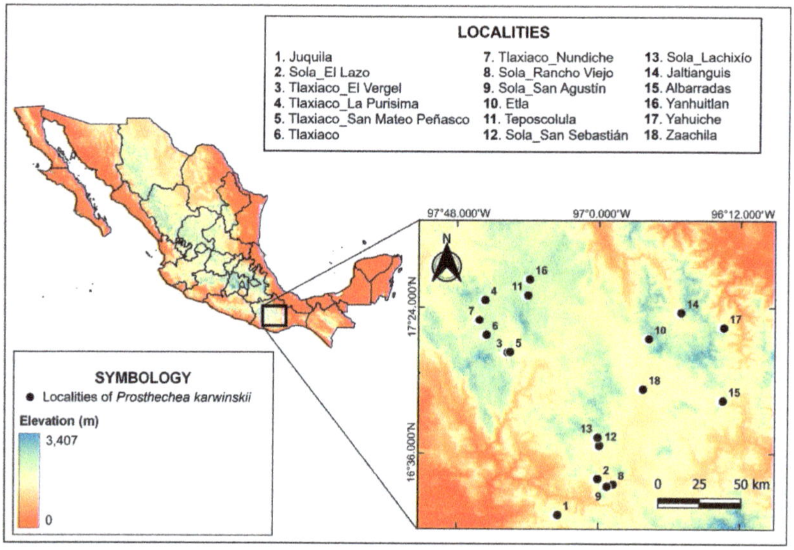

Figure 5. Localities where individuals of *Prosthechea karwinskii* were sampled in Oaxaca, Mexico. See Table 3 for additional information of the localities.

4.2. Selection of Floral Characters

Each flower of *P. karwinskii* was dissected into sepals, petals, labellum, and column to select the characters used as variables in the morphometric analyses, according to Borba et al. [18] and da Cruz et al. [23] with some modifications. Photographs (Canon Rebel camera) were taken of the sepals, petals, and labellum, which were spread out as shown in Figure 6A. The column was separated from the rest of the perianth for recording the characters on this structure, as shown in Figure 6B. The ovary and column were longitudinally sectioned to show the nectary or cuniculus and to record their characters, as shown in Figure 6C. Since the flower of *P. karwinskii* has bilateral symmetry, only the right-side sepal, petal, lateral lobe of the labellum, and lateral tooth of the column were considered. A total of 45 characters were selected from these structures, of which 39 were linear measurements and 6 were angles (Table 4, Figure 6). Each linear measurement was taken with a digital caliper, and the angles were measured with a protractor. Missing data in some individuals for certain characters, due to herbivory, were replaced with the population average for that variable. The set of these continuous characters was recorded in a total of 185 individuals of *P. karwinskii*. Specimens rescued from Zaachila were considered as a separate population. Specimens morphologically similar from populations very close to each other (<5 km apart) were integrated as a single population (expanded population); when a locality was represented by fewer than five specimens, they were also integrated with those from the nearest locality, thus avoiding the effect of a non-representative population. Due to this, the CVA and cluster analysis included 12 expanded populations, as shown in Table 3.

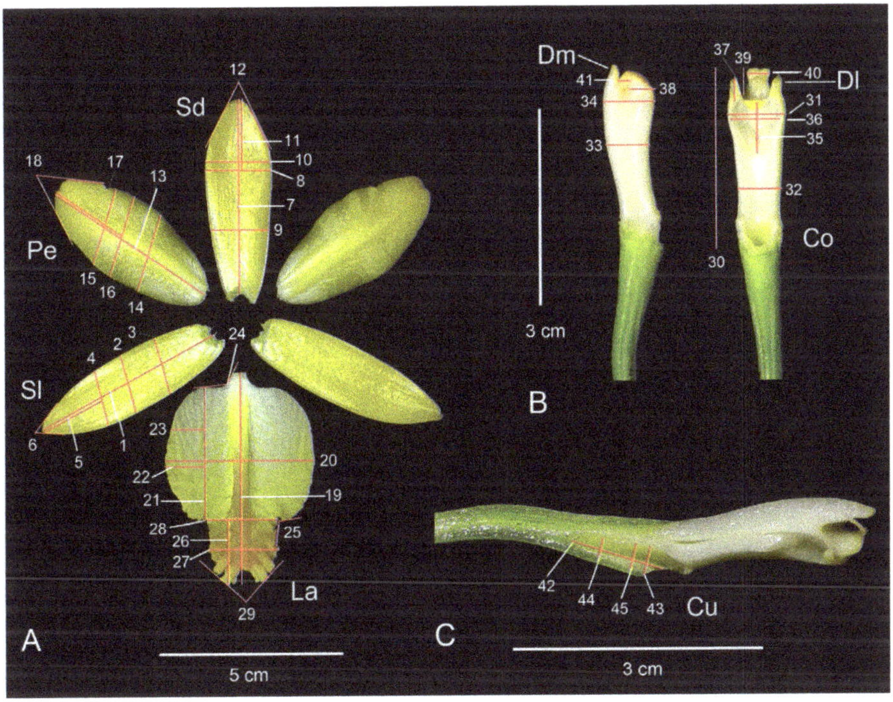

Figure 6. Floral structures showing the variables evaluated in the morphometric analyses of the 18 populations (including Zaachila) of *Prosthechea karwinskii* from Oaxaca, Mexico. (**A**) Flower dissection. (**B**) Lateral and ventral views of the column (**C**) Longitudinal section of the column and ovary, showing the cuniculus. Sl = lateral sepal, Sd = dorsal sepal, Pe = petal, La = labellum, Co = column, Dl = lateral tooth of the column, Dm = median tooth of the column, Cu = cuniculus. See Table 4 for the names of the variables. Photographs by R. Solano.

Table 4. Floral morphologic variables of *Prosthechea karwinskii* and their coding used in the present study. The number and code of these variables correspond to those shown in Tables 1 and 2, as well as in Figure 6.

Floral Structure	Variable	Number	Code
Lateral sepal	Total length	1	SlLt
	Maximum width	2	SlAm
	Width at 1/3	3	SlA1
	Width at 2/3	4	SlA2
	Length between maximum width and apex	5	SlLa
	Angle at the apex	6	SlAa

Table 4. *Cont.*

Floral Structure	Variable	Number	Code
Dorsal sepal	Total length	7	SdLt
	Maximum width	8	SdAm
	Width at 1/3	9	SdA1
	Width at 2/3	10	SdA2
	Length between maximum width and apex	11	SdLa
	Angle at the apex	12	SdAa
Petal	Total length	13	PeLt
	Maximum width	14	PeAm
	Width at 1/3	11	PeA1
	Width at 2/3	16	PeA2
	Length between maximum width and apex	17	PeLa
	Angle at the apex	18	PeAa
Labellum	Total length	19	LaLt
	Maximum width	20	LaAm
	Length of the lateral lobe	21	LaLl
	maximum width of the lateral lobe	22	LaAml
	Width at 1/3 of lateral lobe	23	LaA1l
	Angle between lateral lobe and claw	24	LaAul
	Angle between lateral and middle lobes	25	LaAlm
	Length of middle lobe	26	LaLm
	Maximum width of middle lobe	27	LaAmm
	Width at the base of the middle lobe	28	LaAbm
	Angle at the apex of the middle lobe	29	LaAam
Column	Total length	30	CoLt
	Width at the stigma level	31	CoAe
	Width at 1/3	32	CoA1
	Thickness in the middle part	33	CoGm
	Thickness at the level of the anther	34	CoGa
Rostellum	Height of the stigmatic cavity	35	RoAl
	Width of the stigmatic cavity	36	RoAn
Teeth of the column	Height of the lateral tooth	37	DlAl
	Width of the lateral tooth	38	DlAn
	Height of the middle tooth	39	DmAl
	Width of the middle tooth	40	DmAn
	Width of the gap between the middle and the lateral teeth	41	AnDlDm
Cuniculus	Total length	42	CuLt
	Maximum width	43	CuAm
	Width at 1/3	44	CuA1
	Width at 2/3	45	CuA2

4.3. Statistical Analyses

An analysis of variance (ANOVA) and a Tukey test as a post-hoc analysis were conducted to find significant differences among each of the 45 floral characters and the origin locality of the individuals. Out of the 45 floral traits, 5 did not meet all the assumptions for applying an ANOVA, including normality. To assess if there are differences between populations for these five characters (angle at the apex of the middle lobe, thickness at the level of the anther, thickness in the middle part, width at the stigma level, and width of the middle tooth), a Kruskal–Wallis non-parametric test was applied, followed by Dunn's test with Bonferroni correction as a post-hoc analysis. These analyses were performed using R 2023.12.0 [51] in the stats package.

Multivariate analyses were implemented in Statistica 10 [52]. First, a correlation test was conducted between all possible pairs of variables. Out of the 45 selected variables, 12 showed a correlation greater than 0.9 with each other and were therefore discarded for further analyses. By floral structure, the discarded variables were from the lateral sepal (width at 1/3 of its length, width at 2/3 of its length), dorsal sepal (total length, maximum width, width at 1/3 of its length, width at 2/3 of its length), petal (total length, width at 1/3 of its length, width at 2/3 of its length), labellum (width at 1/3 of its length, width of the base of the middle lobe), and column (thickness in the middle part). The remaining 33 characters were subjected to ordination analysis to evaluate morphological variation among populations, which allowed identifying the most significant variables for morphological patterns. To explore the data structure and detect possible outliers, a principal component analysis (PCA) based on the correlation matrix (correlated variables excluded) was performed. Since this exploratory analysis did not show the presence of outliers, no individuals were removed from the analyses. Subsequently, a canonical variate analysis (CVA) was performed using the expanded populations as categorical variables. We used the standardized coefficients for canonical variables to identify the most important in the observed patterns [18]. Both the correlation test, PCA, and CVA were conducted including and excluding samples from Zaachila (of unknown origin). We calculated a matrix of squared Mahalanobis distances between the expanded population centroids (including Zaachila), using the expanded populations as the group variable and the 33 floral variables. This matrix was used to perform a cluster analysis using UPGMA (unweighted pair-group method) with arithmetic averages as the clustering algorithm.

5. Conclusions

Through morphometric analysis, we examined the intraspecific variation in the floral morphology of *P. karwinskii*, identifying characters as potential taxonomic markers for the species related to the *P. citrina* complex. Traits from lateral sepal, labellum, and column, are useful for this purpose and they are related to the morphological patterns of *P. karwinskii*. The majority of floral characters analyzed here varied significantly among populations of *P. karwinskii*. The CVA was informative for discriminating the infraspecific variation in this orchid, but the PCA was not. Albarradas harbors the most differentiated population of *P. karwinskii* in Oaxaca, and it might be recognized as a variety or geographical form of the species and thus considered a priority for conservation. The Teposcolula population has value for horticultural management due to its individuals having the largest flowers in this species.

The results show that the floral characters have the potential to associate specimens of unknown origin with their probable geographical region. The individuals rescued in the Zaachila group with those from Etla and Yanhuitlan suggest that the origin of the former might be assigned to these two localities or others very close to them. Additional studies are recommended to compare morphological variation with genetic and geographical variation of the populations studied here. Finally, this study constitutes an initial attempt to determine the unknown geographical origin of an orchid extracted for religious use by Mexican communities. The method employed here is low-cost, allows the analysis of a large number of individuals, and could be applied to other orchid species. However, the results must be interpreted with caution, as the accuracy of traceability depends on a good reference and a careful analysis.

Supplementary Materials: The following supporting information can be downloaded at: https://www.mdpi.com/article/10.3390/plants13141984/s1, Table S1: Results of the ANOVA and Tukey tests for the floral characters recorded in 12 expanded populations of *Prosthechea karwinskii* from Oaxaca, including individuals rescued from Zaachila. The values in parentheses show the mean ± standard error (in mm) for each floral variable recorded in each population. The populations did not show significant differences among themselves when the value of the variable shared the same script letter; when the script letter was different, there were significant differences between populations (with $p < 0.05$). See Table 4 for variable names.

Author Contributions: Conceptualization, R.S. and G.C.-L.; methodology, M.H.S.-E., G.C.-L. and R.S.; software, M.H.S.-E., G.C.-L. and M.C.-M.; validation, M.H.S.-E., G.C.-L., M.C.-M., L.L.-R. and R.S.; formal analysis, M.H.S.-E., G.C.-L. and M.C.-M.; investigation, M.H.S.-E., G.C.-L. and R.S.; resources, R.S. and L.L.-R.; data curation, M.H.S.-E., G.C.-L. and R.S.; writing—original draft preparation, M.H.S.-E., G.C.-L., M.C.-M., L.L.-R. and R.S.; writing—review and editing, M.H.S.-E., G.C.-L., M.C.-M., L.L.-R. and R.S.; visualization, M.H.S.-E., G.C.-L. and R.S.; supervision, G.C.-L. and R.S.; project administration, L.L.-R. and R.S.; funding acquisition, L.L.-R. and R.S. All authors have read and agreed to the published version of the manuscript.

Funding: This research was partially funded by Instituto Politécnico Nacional, through grant number SIP-20220893.

Data Availability Statement: The authors assure that, by mutual agreement, we have shared data and information included in this study for its potential publication. We note that access to the Supplementary Material accompanying the manuscript should be restricted until publication is finalized.

Acknowledgments: We appreciate the permission granted by the municipal and community authorities of the visited localities where the floral samples were collected. Special thanks to Clarita Ibarra-Contreras, Gabriela Cruz-García, and Beatriz Pérez-López who provided floral material or assisted in their collection. We are grateful to the reviewers whose constructive comments improved the manuscript. We thank Brenda Solis-García for her assistance in maintaining the collection of flowers used in the study.

Conflicts of Interest: The authors declare no conflicts of interest.

References

1. Solano, R.; Salazar-Chávez, G.A.; Jiménez-Machorro, R. New combinations in Orchidaceae of Mexico. *Act. Bot. Mex.* **2011**, *97*, 49–56. [CrossRef]
2. Solano, R.; Salazar-Chávez, G.A.; Jiménez-Machorro, R.; Hágsater, E.; Cruz-García, G. Actualización del catálogo de autoridades taxonómicas de Orchidaceae de México. Instituto Politécnico Nacional. Centro Interdisciplinario de Investigación para el Desarrollo Integral Regional Unidad Oaxaca. Informe Final, SNIB-CONABIO, Proyecto No. KT005. Mexico City. 2020. Available online: http://www.conabio.gob.mx/institucion/proyectos/resultados/KT005_Anexo_listado_taxonomico.pdf (accessed on 10 June 2024).
3. Hágsater, E.; Soto-Arenas, M.A.; Salazar-Chávez, G.A.; Jiménez-Machorro, R.; López-Rosas, M.A.; Dressler, R.L. *Las Orquídeas de México*; Instituto Chinoin: Mexico City, Mexico, 2007; 302p.
4. Cruz-García, G.; Solano, R.; Lagunez-Rivera, L. Documentation of the medicinal knowledge of *Prosthechea karwinskii* (Orchidaceae) in a Mixtec community in Mexico. *Rev. Bras. Farmacogn.* **2014**, *24*, 153–158. [CrossRef]
5. Barragán-Zarate, G.S.; Lagunez-Rivera, L.; Solano, R.; Pineda-Peña, E.A.; Landa-Juárez, A.Y.; Chávez-Piña, A.E.; Carranza-Álvarez, C.; Hernández-Benavides, D.M. *Prosthechea karwinskii*, an orchid used as traditional medicine, exerts anti-inflammatory activity and inhibits ROS. *J. Ethnopharmacol.* **2020**, *253*, 112632. [CrossRef] [PubMed]
6. Barragán-Zarate, G.S.; Lagunez-Rivera, L.; Solano, R.; Carranza-Álvarez, C.; Hernández-Benavides, D.M.; Belmonte-Jiménez, S.I.; Vilarem, G. UPLC-ESI-qTOF-MS/MS characterization of bioactive constituents and ROS inhibition in *Prosthechea karwinskii* leaves, pseudobulbs, and flowers. *Heliyon* **2022**, *8*, e09867. [CrossRef]
7. Lagunez-Rivera, L.; Barragan-Zarate, G.S.; Solano, R.; Alexander-Aguilera, A.; Chávez-Piña, A.E. Mexican orchid (*Prosthechea karwinskii*) and use in cardiovascular protection cellular and physiological aspects. In *Ancient and Traditional Foods, Plants, Herbs and Spices*; Rajendram, R., Patel, V.R., Patel, V.B., Eds.; Taylor & Francis: Boca Raton, FL, USA, 2023; pp. 259–279. [CrossRef]
8. Solano, R.; Cruz-Lustre, G.; Martínez-Feria, A.; Lagunez-Rivera, A. Plantas utilizadas en la celebración de la Semana Santa en Zaachila, Oaxaca. *Polibotánica* **2010**, *29*, 263–279.

9. Cruz-García, G.; Lagunez-Rivera, L.; Chávez-Angeles, M.G.; Solano, R. The wild orchid trade in a Mexican local market: Diversity and economics. *Econ. Bot.* **2015**, *69*, 291–305. [CrossRef]
10. Dutra, D. Demography, Wild Harvest Patterns and Trade of Culturally Important Species: Priorities for Management and Conservation. Ph.D. Dissertation, The University of Hawaii, Honolulu, HI, USA, 2014; 76p.
11. SEMARNAT. Norma Oficial Mexicana NOM-059-SEMARNAT-2010, Protección Ambiental—Especies Nativas de México de Flora y Fauna Silvestres—Categorías de Riesgo y Especificaciones Para su Inclusión, Exclusión o Cambio—Lista de Especies en Riesgo. Modificación del Anexo Normativo III. Diario Oficial de la Federación, 14 November 2019. Available online: https://www.dof.gob.mx/nota_detalle.php?codigo=5578808&fecha=14/11/2019 (accessed on 16 March 2024).
12. Coutiño-Cortés, A.G.; Bertolini, V.; Archila-Morales, F.; Valle-Mora, J.; Iracheta-Donjuan, J.; García-Bautista, M.; Ruiz-Montoya, L. El uso ornamental de *Guarianthe skinneri* (Orchidaceae), en Chiapas y Guatemala, determina parcialmente su diversidad y estructura genética. *Act. Bot. Mex.* **2018**, *124*, 35–48. [CrossRef]
13. Emeterio-Lara, A.; García-Franco, J.; Hernández-Apolinar, M.; Mora-Herrera, M.; Toledo-Hernández, V.; Valencia-Díaz, S.; Flores-Palacios, A. Endogamy cost and reproductive biology of *Laelia autumnalis*, an endemic orchid of Mexico. *Plant Ecol.* **2018**, *219*, 1423–1434. [CrossRef]
14. Emeterio-Lara, A.; García-Franco, J.; Hernández-Apolinar, M.; Mora-Herrera, M.; Toledo-Hernández, V.; Valencia-Díaz, S.; Flores-Palacios, A. Is pseudobulb harvest a sustainable management strategy in wild orchid populations? An experiment with *Laelia autumnalis*. *Forest Ecol. Manag.* **2021**, *491*, 119205. [CrossRef]
15. Rojas-Méndez, K.J.; Peñaloza-Ramírez, J.M.; Rocha-Ramírez, V.; Cortés-Palomec, A.; McCauley, R.A.; Oyama, K. Massive extraction of the orchid *Laelia speciosa* (HBK) Schltr. for trading in local markets affects its population genetic structure in a fragmented landscape in central Mexico. *Trop. Conserv. Sci.* **2020**, *10*, 1940082917693235. [CrossRef]
16. Catling, P.M. *Malaxis salazarii*, a new species from Mexico and Northern Mesoamerica. *Orquidea* **1990**, *12*, 93–104.
17. Borba, E.L.; Shepher, G.J.; van den Berg, C.; Semir, J. Floral and vegetative morphometrics of five *Pleurothallis* (Orchidaceae) species: Correlation with taxonomy, phylogeny, genetic variability and pollination systems. *Ann. Bot.* **2002**, *90*, 219–230. [CrossRef] [PubMed]
18. Borba, E.L.; Funch, R.R.; Ribeiro, P.L.; Smidt, E.C.; Silva, P.V. Demography, and genetic and morphological variability of the endangered *Sophronitis sincorana* (Orchidaceae) in the Chapada Diamantina, Brazil. *Plant Syst. Evol.* **2007**, *267*, 129–146. [CrossRef]
19. Salazar-Rojas, V.M.; Herrera-Cabrera, B.E.; Soto-Arenas, M.A.; Castillo-González, F. Morphological variation in *Laelia anceps* subsp. dawsonii f. chilapensis Soto-Arenas Orchidaceae in traditional home gardens of Chilapa, Guerrero, México. *Genet. Resour. Crop Evol.* **2010**, *57*, 543–552. [CrossRef]
20. Hernández-Ruíz, J.; Herrera-Cabrera, B.E.; Delgado-Alvarado, A. Variación morfológica del labelo de *Vanilla pompona* (Orchidaceae) en Oaxaca, México. *Rev. Mex. Biodivers.* **2019**, *90*, 2–90. [CrossRef]
21. Melo, C.M.; Borba, E.L. Morphological variability in rupicolous species of the *Acianthera prolifera* complex (Orchidaceae) occurring in southeastern Brazil. *Plant Syst. Evol.* **2011**, *293*, 135–145. [CrossRef]
22. Cruz-Lustre, G.; Batista, J.A.N.; Radins, A.J.; González, A.; Borba, E.L. Morphometric analysis of the *Habenaria parviflora* complex (Orchidaceae). *Plant Syst. Evol.* **2020**, *306*, 37. [CrossRef]
23. da Cruz, D.T.; Selbach-Schnadelbach, A.; Mota-Lambert, S.; Ribeiro, P.L.; Borba, E.L. Genetic and morphological variability in *Cattleya elongata* Barb. Rodr. (Orchidaceae), endemic to the campo rupestre vegetation in northeastern Brazil. *Plant Syst. Evol.* **2011**, *294*, 87–98. [CrossRef]
24. Bateman, R.M.; Smith, R.J.; Fay, M.F. Morphometric and population genetic analyses elucidate the origin, evolutionary significance and conservation implications *of Orchis angusticruris (O. purpurea x O. simia)*, a hybrid orchid new to Britain. *Bot. J. Linn. Soc.* **2008**, *157*, 687–711. [CrossRef]
25. Lima-Morales, M.; Herrera-Cabrera, B.E.; Delgado-Alvarado, A. Intraspecific variation of *Vanilla planifolia* (Orchidaceae) in the Huasteca region, San Luis Potosi, Mexico: Morphometry of floral labellum. *Plant Syst. Evol.* **2021**, *307*, 40. [CrossRef]
26. Chiari, Y.; Claude, J. Morphometric identification of individuals when there are more shape variables than reference specimens: A case study in Galápagos tortoises. *Comptes Rendus Biol.* **2008**, *335*, 62–68. [CrossRef] [PubMed]
27. Ibánez, A.L. Fish traceability: Guessing the origin of fish from a seafood market using fish scale shape. *Fish. Res.* **2015**, *170*, 82–88. [CrossRef]
28. Trivellini, M.M.; Van der Molen, S.; Filun, L.; Márquez, F. Can shell shape be used to find the origin of South American mussels? *Mar. Biol. Res.* **2021**, *17*, 215–222. [CrossRef]
29. Zheng, C.; Jiang, T.; Luo, R.; Chen, X.; Liu, H.; Yang, J. Geometric morphometric analysis of the Chinese mitten crab *Eriocheir sinensis*: A potential approach for geographical origin authentication. *N. Am. J. Fish. Manag.* **2021**, *41*, 891–903. [CrossRef]
30. Oleksa, A.; Căuia, E.; Siceanu, A.; Puškadija, Z.; Kovačić, M.; Pinto, M.A.; Tofilski, A. Honeybee (*Apis mellifera*) wing images: A tool for identification and conservation. *GigaScience* **2023**, *12*, giad019. [CrossRef]
31. Pélabon, C.; Armbruster, W.S.; Hansen, T.F. Experimental evidence for the Berg hypothesis: Vegetative traits are more sensitive than pollination traits to environmental variation. *Funct. Ecol.* **2011**, *25*, 247–257. [CrossRef]
32. Romero-Bravo, A.; Castellanos, M.C. Nectar and floral morphology differ in evolutionary potential in novel pollination environments. *New Phytol.* **2024**, *243*, 753–764. [CrossRef]
33. Clegg, M.T.; Durbin, M.L. Flower color variation: A model for the experimental study of evolution. *Proc. Natl. Acad. Sci. USA* **2000**, *97*, 7016–7023. [CrossRef]

34. Mascó, M.; Noy-Meir, I.; Sersic, A.N. Geographic variation in flower color patterns within *Calceolaria uniflora* Lam. in Southern Patagonia. *Plant Syst. Evol.* **2004**, *244*, 77–91. [CrossRef]
35. Weber, U.K.; Nuismer, S.L.; Espíndola, A. Patterns of floral morphology in relation to climate and floral visitors. *Ann. Bot.* **2019**, *125*, 433–445. [CrossRef]
36. Podolsky, R.H.; Holtsford, T.P. Population structure of morphological traits in *Clarkia, Dudleyana* I, comparison of Fst between allozymes and morphological traits. *Genetic* **1995**, *140*, 733–744. [CrossRef] [PubMed]
37. Schemske, D.W.; Bierzychudek, P. Perspective: Evolution of flower color in the desert annual *Linanthus parryae*: Wright revisited. *Evolution* **2001**, *55*, 1269–1282. [PubMed]
38. Palacio, F.X.; Apodaca, M.J.; Crisci, J.V. *Análisis Multivariado Para Datos Biológicos: Teoría y su Aplicación Utilizando el Lenguaje R*; Fundación de Hstoria Natural Félix de Azara: Buenos Aires, Argentina, 2020; 265p.
39. Benítez-Vieyra, S.; Pérez-Alquicira, J.; Sazatornil, F.D.; Domínguez, C.A.; Boege, K.; Pérez-Ishiwara, R.; Fornoni, J. Evolutionary transition between bee pollination and hummingbird pollination in *Salvia*: Comparing means, variances and covariances of corolla traits. *J. Evol. Biol.* **2019**, *32*, 783–793. [CrossRef] [PubMed]
40. Congiu, L.; Striebel-Greiter, B.; Gessner, J.; Boscari, E.; Boner, M.; Jahrl, J.; Della-Porte, S.; Ludwig, A. Identification and tracking of sturgeons and paddlefish products in trade: Implications for trade control and biodiversity management. *Aquaculture* **2023**, *574*, 739708. [CrossRef]
41. Mondragón, D.M.; Valverde-Valdés, M.T.; Hernández-Apolinar, M. Population ecology of epiphytic angiosperms: A review. *Trop. Ecol.* **2015**, *56*, 01–39. [CrossRef]
42. Beltrán-Rodríguez, L.A.; Martínez-Rivera, B.; Maya, A.P. Etnoecología de la flor de catarina-*Laelia autumnalis* (La Llave & Lex.) Lindl.)-(Orchidaceae) en una comunidad campesina al sur del estado de Morelos, México: Conservando un recurso y preservando saberes populares. *Etnobiología* **2012**, *10*, 1–17.
43. Casas, A.; Blancas, J.; Lira, R. *Mexican Ethnobotany: Interactions of People and Plants in Mesoamerica*; Ethnobotany of Mexico: Interactions of People and Plants in Mesoamerica; Springer: New York, NY, USA, 2016.
44. Briseño-Tellez, J.M.; Pulido-Silva, M.T.; Bautista, K.; García-Mera, A.; Larios-Lozano, O.; López-Gutiérrez, B.N.; Zepeda-Hernández, Z.K. Palm Sunday in central Mexico: Among sellers, palms and syncretism. *J. Ethnobiol. Ethnomed.* **2023**, *19*, 22. [CrossRef]
45. Ticktin, T. The ecological implications of harvesting non-timber forest products. *J. Appl. Ecol.* **2024**, *41*, 11–21. [CrossRef]
46. Chapagain, D.J.; Meilby, H.; Baniya, C.B.; Budha-Magar, S.; Ghimire, S.K. Illegal harvesting and livestock grazing threaten the endangered orchid *Dactylorhiza hatagirea* (D. Don) Soó in Nepalese Himalaya. *Ecol. Evol.* **2021**, *11*, 6672–6687. [CrossRef]
47. Chan, K.M.; Pringle, R.M.; Ranganathan, J.A.I.; Boggs, C.L.; Chan, Y.L.; Ehrlich, P.R.; Macmynowski, D.P. When agendas collide: Human welfare and biological conservation. *Conserv. Biol.* **2007**, *21*, 59–68. [CrossRef]
48. Robinson, J.G. Ethical pluralism, pragmatism, and sustainability in conservation practice. *Biol. Conserv.* **2011**, *144*, 958–965. [CrossRef]
49. Douglas, J. (University of Hawaii, Mānoa, HI, USA); Hernández-Apolinar, M. (Universidad Nacional Autónoma de Mexico, Mexico City, Mexico); Ticktin, T. (University of Hawaii, Mānoa, HI, USA). Personal Communication, 2024.
50. Thiers, B. Index Herbariorum: A Global Directory of Public Herbaria and Associated Staff. New York Botanical Garden's Virtual Herbarium. Available online: http://sweetgum.nybg.org/science/ih/ (accessed on 19 April 2024).
51. R Core Team. *R: A Language and Environment for Statistical Computing*; R Foundation for Statistical Computing: Vienna, Austria, 2023. Available online: https://www.R-project.org (accessed on 15 June 2024).
52. StatSoft. *Statistica 10.0*; TIBCO Software Inc.: Palo Alto, CA, USA, 2010. Available online: https://statistica.software.informer.com/10.0/ (accessed on 25 April 2024).

plants

MDPI

Article

From Phenotypes to Genotypes: Enhancing the Identification of *Cymbidium* Species with DNA Barcoding

Yaonan Peng [1,2], Yao Chen [1,2], Hongfan Ding [1,2], Xiangdong Liu [1,2], Fuxiang Cao [1,2,3,*] and Lu Xu [1,2,3,*]

[1] College of Horticulture, Hunan Agricultural University, Changsha 410128, China; pengyaonan2022@163.com (Y.P.); m13201018828@163.com (Y.C.); 19896267791@163.com (H.D.); m15116158531_1@163.com (X.L.)
[2] Hunan Mid-Subtropical Quality Plant Breeding and Utilization Engineering Technology Research Center, Changsha 410128, China
[3] Yuelushan Laboratory, Changsha 410128, China
* Correspondence: csfucao@163.com (F.C.); xulu@hunau.edu.cn (L.X.)

Abstract: The genus *Cymbidium*, with its intricate floral elements, pronounced endemicity, and patchy distribution, evolves a rich diversity of morphological forms and a wide variety of species while causing an indistinctness in the classification of its species. To elucidate the phylogenetic relationships among *Cymbidium* species and enhance their taxonomic classification by DNA barcoding, this study conducted amplification and sequence results of nuclear (ITS) and chloroplast genes (*matK*, *rbcL*, *trnL-F*, *psbA-trnH*) with phenotypic genetic diversity analysis, genetic distance analysis, and phylogenetic analysis from 48 samples of *Cymbidium* species. The comparison of genetic distance variations showed that *psbA-trnH*, ITS + *psbA-trnH*, and ITS + *matK* + *psbA-trnH* exhibit minimal overlap and significant genetic variation within *Cymbidium* species. The phylogenetic analysis indicated that the combination, ITS + *matK* + *psbA-trnH*, has the highest identification rate. Notably, both the phylogenetic analysis and the genetic diversity analysis of phenotypic traits consistently indicated a clear divergence between epiphytic and terrestrial orchids, with epiphytic orchids forming a distinct clade. This provides reference evidence for studying the ecological adaptations and evolutionary differences between epiphytic and terrestrial orchids, as well as a scientific basis for the classification and identification, germplasm conservation, resource utilization, and phylogenetic evolution of orchids.

Keywords: *Cymbidium*; phenotypic diversity; DNA barcoding; classification identification; affiliation

1. Introduction

Cymbidium Sw., a member of the Orchidaceae family within the Epidendroideae subfamily, includes perennial herbs predominantly found in the tropical and subtropical regions of Asia, extending southward to Australia [1]. It stands as one of the most prolific flowering families globally. The genus *Cymbidium* occupies a leading position in orchid plants, with its scientific, economic, cultural, and social values being of paramount importance. Notably, the terrestrial species (such as *C. goeringii*, *C. faberi*, *C. ensifolium*, *C. kanran*, and *C. sinense*), colloquially known as 'Guolan' in China, have been cultivated in the country for over two millennia [2]. Since 1981, scholars have initiated the establishment of orchid germplasm conservation nurseries in Zhejiang Province, with subsequent centers and nature reserves being set up in various provinces such as Guangxi, Jiangsu, and Guangdong, emphasizing the importance of enhancing and safeguarding orchid germplasm resources [3]. This indicates that the preservation and utilization of orchid germplasm resources in China have officially commenced. However, the widespread natural hybridization phenomenon of *Cymbidium* species enhances the diversity of morphological features but makes its classification ambiguous. Moreover, the nomenclature system for *Cymbidium* resources lacks standardization, leading to frequent issues with synonyms and homonyms [4]. The confusion between Latin and Chinese names, coupled with the outdated methods of information resource management, impedes the conservation, exchange, utilization, and innovation of *Cymbidium* germplasm resources. These factors exacerbate the difficulties in research classification and the study of genetic diversity within *Cymbidium* species [5].

The infrageneric classification of *Cymbidium* was initially based on the classification system proposed by Schlechter, which divided the genus into eight sections [6]. Scholars initially categorized the genus *Cymbidium* into three subgenera, such as subgenus *Cymbidium*, subgenus *Cyperorchis*, and subgenus *Jensoa* [7]. Subsequently, Puy and Cribb [8] and Liu et al. [9] made further discoveries regarding the classification of the genus and conducted supplementary research and refinements. However, morphological characteristics and their statistical descriptions are easily influenced by environmental variability, making it challenging to distinguish *Cymbidium* species solely based on phenotypic traits. Therefore, utilizing more stable and concise methods for the identification and classification of *Cymbidium* species is extremely necessary.

DNA barcoding is a method of molecular identification that employs short, standardized DNA sequences to swiftly ascertain the species of biological specimens [10–13]. In 2003, Hebert first formally proposed the concept of DNA barcoding at the first International Conference on the Barcoding of Life (CBOL), introducing molecular biology techniques into the broader realm of biological classification [14] and announcing it as an essential tool for the identification of species across the globe. The Sloan Gene Society (2004) established a consortium for the CBOL to develop a standard barcode process and a comprehensive DNA barcode database and gradually extend DNA barcoding research to unknown species and to develop a global standard for species identification [15]. At the Third International Conference on DNA Barcoding in 2009, *rbcL* and *matK* were proposed as core barcodes for terrestrial plants, with the rapidly evolving ITS and *psbA-trnH* sequences suggested as supplementary barcodes for plant identification [16]. The Chinese Plant Barcoding Consortium assessed the identification ability of *psbA-trnH*, ITS/ITS2 and *rbcL* + *matK* sequences or combinations in 2011. The results showed that the ITS/ITS2 sequence could serve as the core barcode for seed plants, with *psbA-trnH* emerging as a candidate barcode sequence [11]. In addition, previous studies have shown that a complete chloroplast genome can serve as an effective tool for identifying *Cymbidium* species and resolving their phylogenetic relationships [17–19]. Despite some progress in these studies, the phylogenetic relationships among many *Cymbidium* species remain controversial [20–22]. Therefore, more effective molecular techniques are needed for further research on orchid species. As a form of digital information, DNA barcoding technology holds significant value in taxonomy with its accuracy, richness, and unique repeatability [12,23,24]. With the rapid advancement of biotechnology, sequencing reactions are becoming more accessible and cost-effective, facilitating the comprehensive construction of public sequence databases. Consequently, DNA barcoding, with its speed and efficiency, is becoming increasingly practical [25]. Its applications extend beyond taxonomy to include evolutionary ecology, food quality and safety, forensic evidence, pharmacology, and other fields of identification.

The application of DNA barcoding technology has greatly enriched the genetic information of *Cymbidium* species, aiding in the exploration of plant morphological evolution caused by genetic mutations and variations. Therefore, in order to resolve the taxonomic ambiguities caused by the rich and diverse phenotypic variations, genetic variations, and taxonomic confusion in *Cymbidium*, this study employed five single-DNA-fragment barcodes and four combined barcodes to conduct phylogenetic analysis on 48 plant materials of *Cymbidium*. Concurrently, combined with morphological classification, we aim to identify superior and more precise DNA sequence markers for the classification of *Cymbidium* species and comprehensively explore the affiliation and potential divergence of closely related *Cymbidium* species across different life forms. This study aspires to tackle theoretical challenges in the classification of *Cymbidium* species, lay a foundation for clarifying the genetic variation of *Cymbidium*, and provide basic technical support for regulating and managing orchids that can be used in legal trade, as well as for accurately determining the geographical origin of unknown specimens in illegal trade.

2. Results

2.1. Analysis of Amplification Success Rate and Sequence Characteristics

We initially employed five barcode sequences to amplify and sequence 48 samples of *Cymbidium* species. The findings demonstrated a 100% success rate in amplifying the five DNA barcodes (Table 1), along with a 100% success rate in sequencing *matK* and *rbcL* genes. Additionally, the ITS, *psbA-trnH*, and *trnL-F* genes exhibited a sequencing success rate of 95.8%, resulting in obtaining ideal sequences from 46 samples. These results indicate that the amplification and sequencing outcomes of these five barcodes were satisfactory, rendering them suitable for subsequent research on *Cymbidium* using DNA barcoding.

Table 1. Amplification and sequencing of DNA barcoding.

DNA Barcodes	Number of Samples	Number of Successful Amplifications	Amplification Success Rate	Number of Successful Sequences	Sequencing Success Rate
ITS	48	48	100%	46	95.8%
matK	48	48	100%	48	100%
rbcL	48	48	100%	48	100%
psbA-trnH	48	48	100%	46	95.8%
trnL-F	48	48	100%	46	95.8%

Among the single sequences, the analysis of each barcode sequence (Table 2) revealed that the length of sequence alignment was ordered as follows: ITS > *psbA-trnH* > *matK* > *trnL-F* > *rbcL*. The GC content, ranked from highest to lowest, was ITS > *rbcL* > *rnL-F* > *psbA-trnH* > *matK*. Considering the percentage of each site relative to the total length, ITS exhibited the greatest variability, while *rbcL* showed the highest level of conservation. In summary, ITS has the potential to become the DNA barcode of *Cymbidium* species.

Table 2. The sequence features of the DNA barcoding.

Sequence Information	ITS	*matK*	*rbcL*	*trnL-F*	*psbA-trnH*
Comparison length (bp)	885	861	645	775	865
GC content (%)	63.6%	31.1%	41.1%	34.62%	33.36%
Conserved site	672 (75.93%)	784 (91.06%)	627 (97.21%)	721 (93.03%)	732 (84.62%)
Total variation sites	165 (18.64%)	73 (8.48%)	17 (2.64%)	23 (2.97%)	96 (11.10%)
Parsimony-informative site	49	40	10	11	74

Among the combination sequences, the results (Table 3) indicated that the ITS + *matK* + *psbA-trnH* sequence yielded the longest sequence alignment length, and the ITS + *psbA-trnH* sequence resulted in the shortest sequence alignment length. The GC content, when ranked from highest to lowest, was as follows: ITS + *psbA-trnH* > ITS + *matK* > *psbA-trnH* + ITS + *matK* > *matK* + *psbA-trnH*. Considering the percentage of variation sites to the length, the highest proportion was ITS + *psbA-trnH* (17.8 %), followed by ITS + *matK* (17.7 %), with a very minimal difference between them. In terms of the percentage of conserved sites to the length, the proportion of *matK* + *psbA-trnH* was the highest (87.8 %), followed by ITS + *matK* + *psbA-trnH* (83.8 %). Based on the above information, it was concluded that ITS + *psbA-trnH* exhibited the greatest variability, while the *matK* + *psbA-trnH* showed the strongest conservation.

Table 3. DNA combination barcode sequence characteristics.

Sequence Information	ITS + *matK*	*matK* + *psbA-trnH*	ITS + *psbA-trnH*	ITS + *psbA-trnH* + *matK*
Comparison length (bp)	1716	1726	1720	2581
GC content (%)	48.1%	32.5%	50.3%	43.6%
Conserved site	1399 (81.53%)	1516 (87.83%)	1367 (79.48%)	2162 (83.77%)
Total variation sites	303 (17.66%)	169 (9.79%)	306 (17.79%)	368 (14.26%)
Parsimony-informative site	124	114	154	188
Single polymorphic loci	179	54	151	179

2.2. Analysis of Barcoding Gap

The ITS, *matK*, *psbA-trnH*, *rbcL*, and *trnL-F* all produced correct sequencing signals and were used for resolution analyses. The results of genetic distance for the five single-segment barcodes (Table 4) indicated that the average intraspecific genetic distance, from largest to smallest, was ordered as ITS > psbA-trnH > *matK* > *trnL-F* > *rbcL*, and the average interspecific genetic distance followed the same order: ITS > *psbA-trnH* > *matK* > *trnL-F* > *rbcL*. The smallest genetic distance, both interspecific and intraspecific, was observed for *rbcL*. The barcode ITS exhibited larger interspecific and intraspecific genetic distances, and the average interspecific genetic distance for each barcode was greater than the average intraspecific genetic distance. Based on the above information, ITS displayed the highest level of genetic variation, followed by *psbA-trnH*, while *rbcL* showed the lowest level of genetic variation.

Table 4. Comparison of genetic distance differences of single-segment DNA barcodes.

DNA Barcodes	Intraspecific Genetic Distance			Interspecific Genetic Distance		
	Minimum	Maximum	Mean	Minimum	Maximum	Mean
ITS	0	0.09657	0.01759	0	0.14136	0.02890
matk	0	0.02262	0.00335	0	0.03115	0.01208
rbcL	0	0.00782	0.00088	0	0.01731	0.00424
trnL-F	0	0.00980	0.00168	0	0.02004	0.00434
psbA-trnH	0	0.02128	0.00513	0	0.04525	0.01588

Among the combined barcode genetic distance (Table 5), the findings revealed that the order of average intraspecific genetic distance was ITS + *psbA-trnH* > ITS + *matK* > ITS + *matK* + *psbA-trnH* > *matK* + *psbA-trnH*. The order of average interspecific genetic distance was ITS + *psbA-trnH* > ITS + *matK* > ITS + *matK* + *psbA-trnH* > *matK* + *psbA-trnH*. The largest interspecific and intraspecific genetic distance was observed for the ITS + *psbA-trnH*, with the smallest one, *matK* + *psbA-trnH*. The average interspecific genetic distance of barcodes in each combination was greater than the average intraspecific genetic distance. Based on the above information, it is evident that ITS + *psbA-trnH* exhibited a relatively large variation in intraspecific genetic distance, followed by ITS + *matK*.

In this study, we plotted the distributions of interspecific and intraspecific genetic distances for five barcode sequences to evaluate the barcoding gap among different barcodes. The results showed that there is a certain overlap in the distribution of interspecific and intraspecific genetic distances of DNA barcodes in *Cymbidium* species, but the overlap of ITS and *psbA-trnH* barcodes is minimal and skewed towards the extremes compared to other barcodes, with the *psbA-trnH* sequence showing the least overlap (Figure 1). In summary, *psbA-trnH* and ITS sequences are relatively suitable for the identification of *Cymbidium* species.

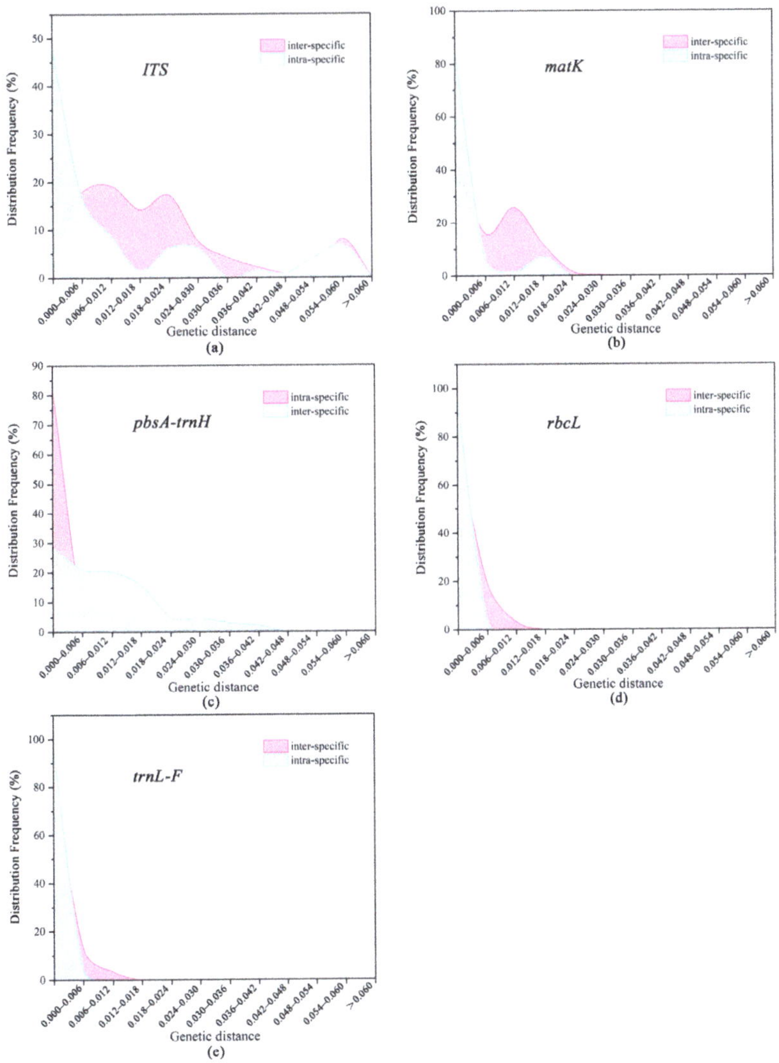

Figure 1. Barcoding gap distribution of five monolithic segments. (**a**) ITS; (**b**) *matK*; (**c**) *psbA-trnH*; (**d**) rbcL; and (**e**) *trnL-F*.

The genetic distance distribution plots of the four combined barcodes reveal that there are certain overlaps in the genetic distance distributions between intraspecific and interspecific barcodes for these combined barcodes (Figure 2). However, the variations in

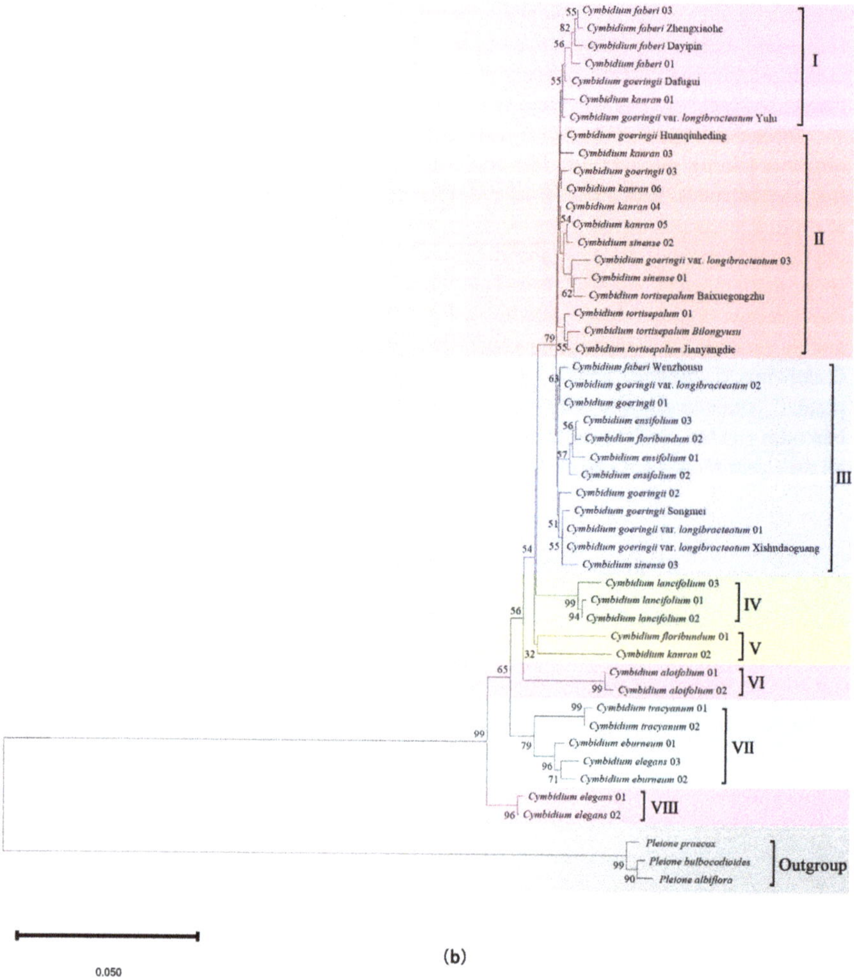

(b)

Figure 3. Phylogenetic tree of some *Cymbidium* species based on *psbA-trnH* and *matK* + *psbA-trnH*. Numbers above the branches indicate bootstrap (BS $\geq$ 50) values. (**a**) *psbA-trnH*, all *Cymbidium* species were classified into five primary clades: I–VI. (**b**) *matK* + *psbA-trnH*, all *Cymbidium* species were classified into five primary clades: I–VIII. The outgroup was represented by three species from *Pleione* D. Don.

The phylogenetic tree, based on ITS + *matK* + *psbA-trnH* of three barcode combinations (Figure 4), reveals that all species within *Cymbidium* were categorized into seven primary clusters. Clade I comprised 11 species, which included one species from sect. *Floribunda* of the subgenus *Cymbidium*, while all the remaining species belong to the sect. *Jensoa* of the subgenus *Jensoa*. All species and varieties of *C. faberi* from the sect. *Jensoa* of the subgenus *Jensoa* were clustered in Clade II. All species of *C. goeringii*, some varieties of *C. tortisepalum*, and the variety of *C. goeringii* known as 'Chunjian', along with its varieties, together form Clade III, which is the most species-rich taxonomic group. The Clade IV was the most complex, with 12 species, including six from the subgenus *Cyperorchis* (three species from sect. *Cyperorchis*, one species from sect. *Eburnea*, and two species from sect. *Iridorchis*), one species from the subgenus *Cymbidium* (*C. floribundum* from sect. *Floribunda*), and three species from sect. *Geocymbidium* of the subgenus *Jensoa*. Clade V was composed of *C. kanran* from the sect. *Jensoa* of the subgenus *Jensoa*.

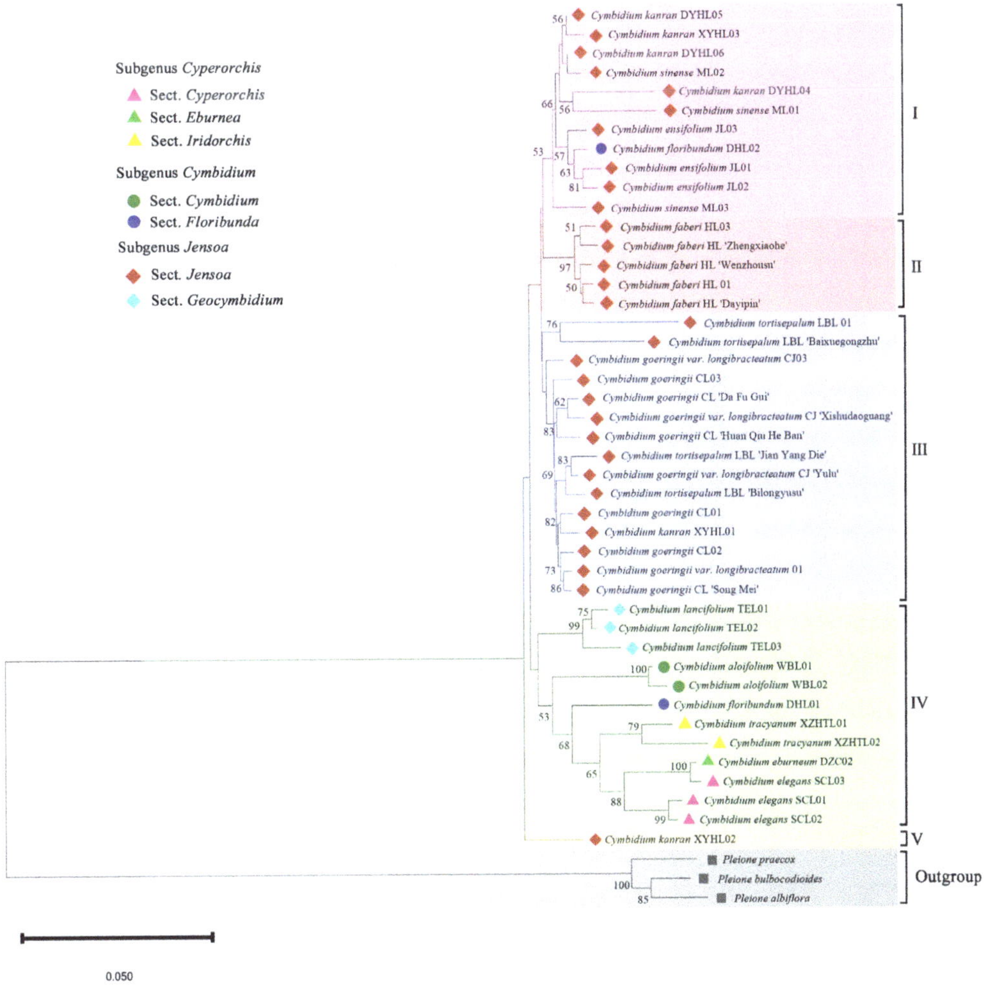

Figure 4. Phylogenetic tree of some *Cymbidium* species based on combined barcodes (ITS + *matK* + *psbA-trnH*). Numbers above the branches indicate bootstrap (BS ≥ 50) values. All *Cymbidium* species were classified into five primary clades: I–V. The outgroup was represented by three species from *Pleione* D. Don.

The phylogenetic analysis showed that unsupported relationships were primarily found within the sect. *Jensoa* of the subgenus *Jensoa* and were often associated with low levels of sequence variation, with the bootstrap value above 50%. The two accessions of the epiphytic orchid (*C. floribundum*) appear in different clades. All epiphytic orchids were clustered in Clade IV, and the terrestrial *C. lancifolium* was also grouped within this clade. Through the analysis of the phylogenetic tree, we have discovered that the *Cymbidium* species with the same life forms exhibit closer affiliation. By integrating barcode sequence characteristics and genetic distance analysis, it was revealed that employing the combined barcode (ITS + *matK* + *psbA-trnH*) technique significantly improves the species identification rate and classification efficiency of *Cymbidium* species. To delve deeper into the taxonomic relationships among *Cymbidium* species, we have also conducted an analysis of the genetic diversity associated with phenotypic traits.

2.4. Genetic Diversity Analysis of Phenotypic Characters in Orchid Germplasm Resources

2.4.1. Diversity Analysis of Quantitative Traits

We analyzed the variation of 11 quantitative traits of *Cymbidium* species and observed that the degree of diversity of different quantitative traits was also different. The coefficient of variation of 11 quantitative traits ranged from 28.2% to 128.9%, demonstrating a considerable degree of trait diversity (Table S2). Among them, the trait with the highest coefficient of variation was the number of flowers in the flower bract, followed by leaf width and the height of scape, indicating that these three traits exhibit high polymorphism. The petal width had the lowest coefficient of variation, followed by sepal length, indicating that these two characters were relatively stable. The average coefficient of variation of other quantitative traits was 36.8%, which was relatively concentrated. These findings reveal a rich genetic variation in phenotypic traits among different varieties of *Cymbidium* species.

The correlation analysis of 11 quantitative traits among orchids revealed that 21 pairs of traits exhibited significant correlations, with all being positive correlations (Table S3). Notably, eight pairs demonstrated significant positive correlations. The height of flower scape, length of leaf, number of flowers, and number of leaves all correlated significantly, indicating a reflection of the plant's robustness. In contrast, leaf width, sepal length, sepal width, and petal width showed negative correlations. The most substantial correlation was found between flower length and labellum length, while the least significant correlation was detected between the number of flowers and sepal length. Significant correlations were also identified between the length and width of the sepal, the length and width of the labellum, and the length of the petal, suggesting a close interrelation between these floral structures.

2.4.2. Diversity Evaluation of Quality Traits

We observed 16 quality traits of *Cymbidium* species phenotypes, including 2 leaf appearance traits, 13 petal appearance traits, and 1 pseudobulb trait. There were 52 variation types among the 16 quality traits, with an average of 3.2 variation types per trait. The Shannon–Wiener index ranged from 0.22 to 2.20 (Table S4), with an average of 1.14, among which the main color variation of the middle sepal was the highest, and the leaf tip pattern variation was the lowest. In the leaf appearance traits of *Cymbidium* species, the tip shape of most varieties is sharp, and the edge of the leaf is slightly serrated. The main color phenotype of the middle sepal was the most abundant, including white, green, yellow, red, purple, and brown, followed by the main color of the middle sepal and the middle petal. Regarding the labellum flaps, the shape and color of the lobes in the labellum lobe of most tested varieties were triangular and pale yellow. In terms of flower spots and stripes, most of the varieties to be tested had streaked sepals without stripes, petals with streaks without stripes, and labellum petals with spots without stripes. Among the characteristics of pseudobulb size, as epiphytic orchids, *C. floribundum*, *C. aloifolium*, *C. elegans*, and *C. tracyanum* have obviously larger pseudobulbs than those of other terrestrial orchids, which aligns with the characteristics of common epiphytic and terrestrial pseudobulbs.

2.4.3. Principal Component Analysis of Phenotypic Traits

After measuring and statistically analyzing 27 quantitative and qualitative traits, we identified eight components with eigenvalues higher than one, which serve as the principal components of the 27 phenotypic traits. These eight principal components cumulatively account for 87.75% of the information in the original traits, with the first and second principal components alone contributing a significant 44.70% (Figure 5). The first principal component is most heavily weighted by the quantity of flowers, closely followed by the height of the scape, indicating that the growth of the scape greatly influences the classification of *Cymbidium* species. In the second principal component, the length of the labellum and petal, and the width of the sepal have substantial weight coefficients, all of which are dimensions of flower size, highlighting the significant role that flower size plays in the classification of *Cymbidium* species. The principal component analysis reveals that these eight components encapsulate the floral traits, encompassing aspects such as the number of flowers, the height of the scape, petals, sepals, labellum, etc. Therefore, the traits of floral phenotype can be used as the main traits for the classification and identification of *Cymbidium* species, while the traits of leaf phenotype and pseudobulb characteristics can serve as supplementary traits for their classification and identification. By describing and analyzing these phenotypic traits as the basis for the classification of *Cymbidium* species, species groups with similar adaptability can be identified, which reflects the genetic and morphological diversity of the species and thus better understands the diversity and evolutionary relationships of *Cymbidium* species.

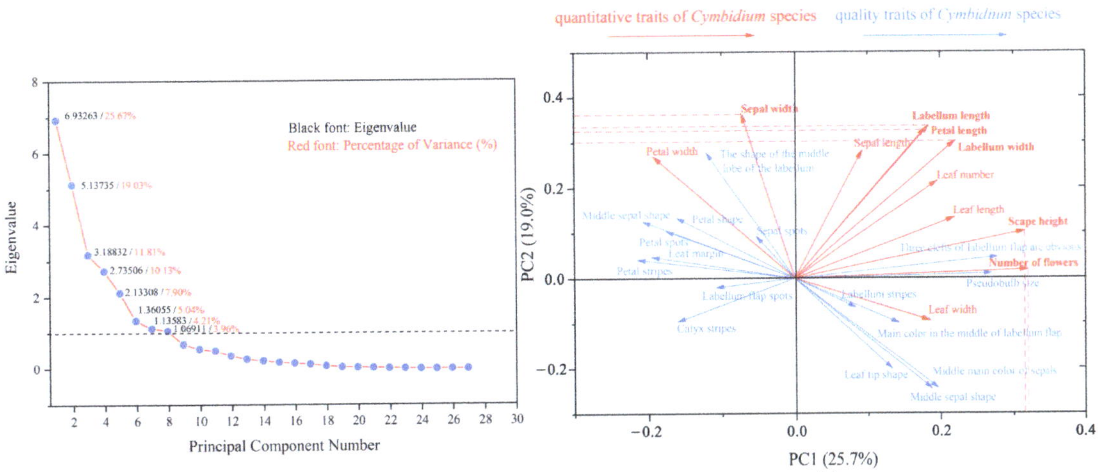

(a) (b)

Figure 5. The principal component analysis of 27 phenotypic trait indicators among *Cymbidium* species. (**a**). The scree plot from principal component analysis. Includes the eigenvalues and contribution rates of each component. (**b**). The loading plot from principal component analysis. Includes the impact of 27 phenotypic traits on the first and second principal components, as well as the contribution rates of the first and second principal components.

2.5. Cluster Analysis of Phenotypic Traits

In order to validate the rationality of trait selection, we conducted in-depth discussions on the correlation between 27 phenotypic traits using R-type clustering analysis. According to the clustering spectrum diagram, when the Euclidean distance was set to 20, we divided the 27 phenotypic traits into 5 groups (Figure 6). Among them, groups A, D, and E contain a relatively large number of phenotypic traits. The five traits of sepal width, petal width, labellum lobe shape, sepal shape, and petal shape in Group A are highly correlated, reflecting the characteristics of flower shape and size. The main color in the middle of the sepals is highly correlated with the main color in the middle of the petals, and the main color in the middle of the labellum is highly correlated with leaf tip traits in Group D. There is also a close correlation between the obvious trichotomy of the labellum, the size of the pseudobulb, and the width of the leaf, and *C. elegans* and *C. tracyanum* can be distinguished from other varieties because of the large pseudobulb and the obvious trichotomy of the labellum. The labellum length, labellum width, petal length, and the sepal length and sepal height of Group E are highly correlated, while the number of leaves, number of flowers, leaf length, and scape height are correlated, reflecting the size of flowers and the abundance of plants. This indicates that among the 27 phenotypic traits of *Cymbidium* species, floral characteristics exhibit higher discriminatory power in distinguishing closely related species within the same genus, thus serving as a pivotal factor for classifying the phylogenetic relationships of *Cymbidium* species using phenotypic traits.

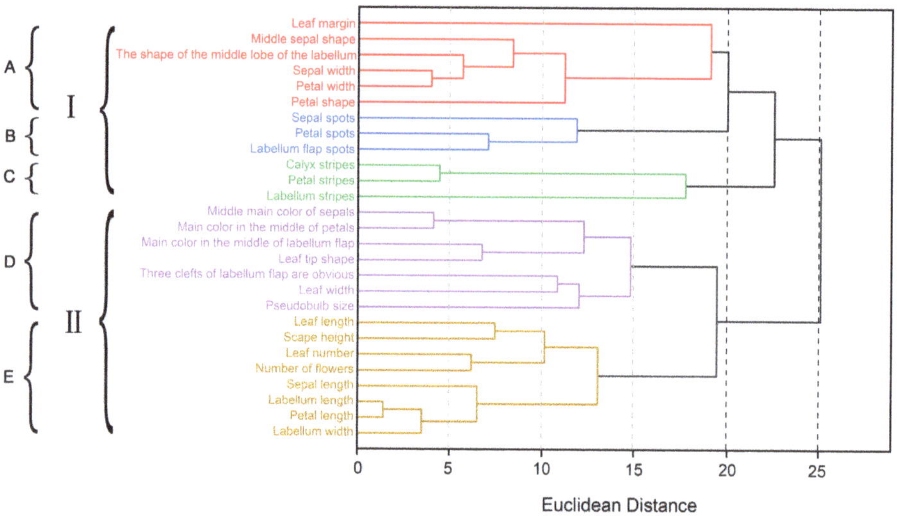

Figure 6. The R-type cluster analysis of 27 phenotypic traits in *Cymbidium* species. When the Euclidean distance was 25 and 20, the 27 phenotypic traits were divided into two groups (I–II) and five groups (A–E).

Based on the analysis of phenotypic traits, we performed dimensionality reduction on 27 phenotypic traits and ultimately simplified them into 21 phenotypic traits, which were subjected to Q-type clustering analysis. The results revealed significant segregation among various *Cymbidium* species, which were divided into seven clusters (Figure 7). *C. faberi*, *C. kanran*, and *C. ensifolium* were predominantly closely related together, characterized by a longer labellum, petals, and a higher number of flowers in the phenotype. At this level, they have a high degree of similarity and are suitable for clustering together. Because of the high similarity of petal stripes and spots and labellum stripes and spots, *C. sinense* and *C. tracyanum* are gathered in one branch. *C. tortisepalum*, *C. goeringii* var. *longibracteatum*, and *C. goeringii* all exhibit narrow leaves and small pseudobulbs in phenotype, with high similarity at the sublevel, making them suitable for clustering together. The leaf length, number of leaves, scape height, number of flowers, three-lobed labellum, and pseudobulb size of *C. elegans* and *C. floribundum* differ greatly from other terrestrial orchids, while the leaf shapes of *C. lancifolium* and *C. aloifolium* are different from other terrestrial orchids, and their phenotypes are significantly different from those of the A, D, and E clusters, so they are all grouped separately.

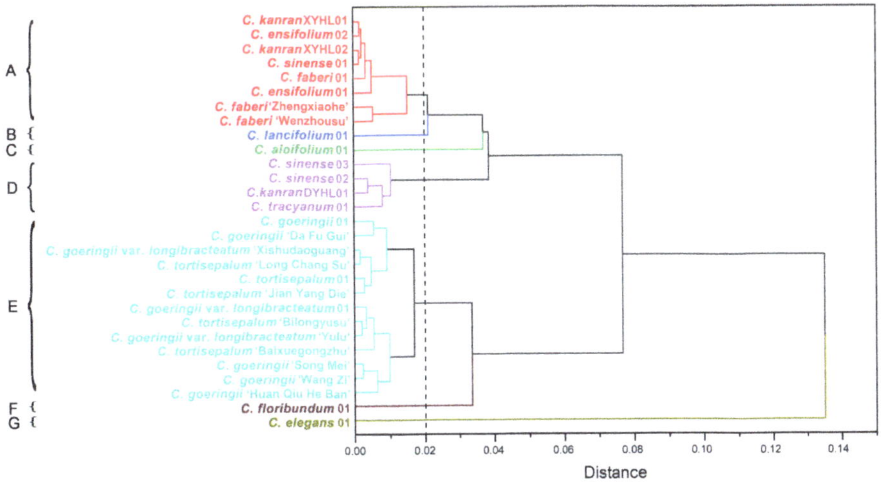

Figure 7. Q-type cluster analysis based on morphological characters depicting the relationship of *Cymbidium* species. All *Cymbidium* species were classified into seven primary clusters: A–G.

2.6. The Support of Molecular Characters for Morphological Features

To further discuss the correlation between the combination of phenotype and DNA barcode identification results, we will also conduct an analysis that compares the phenotypic clustering results with the barcode tree construction results, examining the similarities and differences from the perspective of diverse life forms. According to the analysis of phenotypic traits, Figure 7 reveals that *C. lancifolium*, *C. floribundum*, *C. aloifolium*, and *C. elegans* each form a distinct branch. The phylogenetic tree (Figure 4) constructed by combining DNA barcodes also demonstrates that *C. lancifolium*, *C. floribundum*, *C. aloifolium*, *C. tracyanum*, and *C. elegans* were all clustered separately within individual branches. From the R-type cluster analysis of phenotypic traits (Figure 6), we indicated that floral characteristics are the primary morphological features for orchid classification. The floral diagrams of these plant materials among *Cymbidium* in this study showcased a variety of floral traits, including various flower colors, flower spots, and labellum spot variations. Significant differences in the petal patterns were observed between the four epiphytic orchids (*C. floribundum*, *C. aloifolium*, *C. tracyanum*, and *C. elegans*) and the terrestrial orchids, as well as among these four epiphytic orchids themselves. When classified using DNA barcodes, these four types of epiphytic orchids were divided into one major category, indicating that the formation of different life forms in *Cymbidium* species is closely related to genes, and *Cymbidium* species with significant morphological differences also share close genetic relationships.

Admittedly, there is a significant divergence in floral characteristics between epiphytic and terrestrial orchids (Figure 8). However, *C. lancifolium*, a species of terrestrial plants, exhibits closer affinities with four epiphytic orchids based on both morphological traits and molecular markers. Although phylogenetic trees may have maximum branch support at their nodes, they can still display a notable lack of consistency in phylogenetic signals derived from different genes or regions of the genome, and additional data may not resolve these inconsistencies. Nonetheless, there is substantial potential to understand the basis of phenotypic variation in orchids, from DNA molecular markers to entire genomes, thereby gaining insight into evolutionary changes and their significance within populations.

Figure 8. Flower diagram of phenotypic trait test varieties. The white box contains epiphytic orchids, while the rest are terrestrial orchids. (**A**) *Cymbidium goeringii*; (**B**) *Cymbidium goeringii* 'Song Mei'; (**C**) *Cymbidium goeringii* 'Huan Qiu He Ban' (**D**) *Cymbidium goeringii* 'Da Fu Gui'; (**E**) *Cymbidium goeringii* 'Wang Zi'; (**F**) *Cymbidium faberi*; (**G**) *Cymbidium faberi* 'Wenzhousu'; (**H**) *Cymbidium faberi* 'Zhengxiaohe'; (**I**) *Cymbidium faberi* 'Dayipin'; (**J**) *Cymbidium lancifolium*; (**K**) *Cymbidium tortisepalum*; (**L**) *Cymbidium tortisepalum* 'Baixuegongzhu'; (**M**) *Cymbidium tortisepalum* 'Jian Yang Die'; (**N**) *Cymbidium tortisepalum* 'Bilongyusu'; (**O**) *Cymbidium floribundum*; (**P**) *Cymbidium goeringii* var. *longibracteatum*; (**Q**) *Cymbidium goeringii* var. *longibracteatum* 'Xishudaoguang'; (**R**) *Cymbidium tortisepalum* 'Long Chang Su'; (**S**) *Cymbidium goeringii* var. *longibracteatum* 'Yulu'; (**T**) *Cymbidium aloifolium*; (**U**) *Cymbidium kanran* 01; (**V**) *Cymbidium kanran* 02; (**W**) *Cymbidium kanran* 04; (**X**) *Cymbidium sinense* var. haematodes; (**Y**) *Cymbidium elegans*; (**Z**) *Cymbidium sinense* 'Qihei'; (**a**) *Cymbidium sinense* 'Yangmingjin'; (**b**) *Cymbidium ensifolium* 01; (**c**) *Cymbidium ensifolium* 02; and (**d**) *Cymbidium tracyanum*.

3. Discussion

The species of *Cymbidium* exhibit extensive morphological and genetic variability due to variations in morphology, genetics, and ecological habits among different species, which poses challenges for species identification and classification. The reasons for these differences in *Cymbidium* plants may be due to variations in the genetic material of *Cymbidium* germplasm and the influence of diverse environmental conditions [26–30]. To adapt to local ecological environments, different genetic traits are produced, and morphological characteristics may not accurately reflect the complete genetic information. Thus, this study employs the more stable and precise DNA barcoding technique to assist us in the classification and identification of *Cymbidium* species.

3.1. The Universality and Applicability of Sequences

DNA barcoding technology is an emerging species identification technique that has gained popularity in recent years due to its rapidity, accuracy, and user-friendliness [31]. An ideal DNA barcode should meet three criteria: standardization, extreme simplification, and scalability. It should enable routine and reliable sequencing across diverse sample sets to obtain easily comparable sequences with minimal intraspecific variations but substantial interspecific variations [32]. In this study, we amplified fragments from both nuclear genes (ITS) and chloroplast genes (*matK, rbcL, psbA-trnH, trnL-F*) for barcode analysis. Our findings indicate that there are some differences in PCR amplification and sequencing success rates among the genotypes and loci investigated, which may be related to primer specificity, PCR settings, and the competency of the DNA template. Similar results were also present in previous studies, demonstrating that these factors have a certain impact on the success rates of PCR amplification and sequencing [33]. Sequence feature analysis revealed that ITS exhibited a higher number of variation sites, which was also supported by previous studies on *Uncaria, Rhododendron*, and *Gastrodia elata* [34–36]. Additionally, the chloroplast gene fragment *psbA-trnH* showed a greater number of variation sites and exhibited a more pronounced barcoding gap. However, both the sequences of *trnL-F* and *matK* displayed relatively low variability in this study, while the variability of *rbcL* was predominantly located at the end of candidate chloroplast genomes without a clear barcoding gap. Therefore, based on the characteristics of these five single-fragment DNA barcodes, ITS and *psbA-trnH* have the potential to serve as effective DNA barcodes for the classification and identification of *Cymbidium* species in this study. Moreover, the specificity of DNA barcoding can reveal the high degree of morphological and genetic variability in *Cymbidium* species. These findings are similar to some previous studies, indicating that DNA barcode regions can serve as reliable markers for the identification and classification of *Cymbidium* species and their genotypes [33,37]. To further validate these results and assess the effectiveness of these barcodes across a broader taxonomic range, additional research is needed to discover and explain the potential of DNA barcodes in distinguishing between closely related species.

3.2. Assessment of the Discriminating Power of Single and a Combination of Barcodes

The discriminating power of different DNA barcodes for species identification may vary depending on the taxonomic group studied. Sayed et al. [38] assessed the efficacy of DNA barcodes for species identification and found that *matK* and ITS regions have high universality and sequencing success rates, providing 100% species resolution except for the *trnH-psbA* region. However, in this study, although *matK* and ITS had high sequencing success rates, the *psbA-trnH* sequence, which had a low sequencing success rate, achieved the highest identification success rate and has been successfully applied to identify *Dendro-*

bium and *Dioscorea* [39,40] as well as other plants. Nevertheless, the species identification success rates for the *trnL-F* sequence and *rbcL* sequence were relatively low. Among the five single-fragment DNA barcodes analyzed in this study, the psbA-trnH sequence demonstrated higher specificity and identification success rate, making it more suitable for the classification and identification of *Cymbidium* species. Relying on a single DNA barcode marker may not always provide sufficient variation information for species identification, thereby presenting certain limitations.

Therefore, the utilization of multiple DNA markers for plant identification has been gradually implemented [22], enabling the acquisition of adequate species variation. For instance, the combination of *arpF-atpH* + *pabK-psbL* + *psbA-trnH* as a DNA barcode has been shown to achieve an impressive species identification rate of 98.8% in orchids [41]. When four DNA fragments were employed to identify *Alnus*, it was observed that the resolution of a single fragment was inferior to that of a combination of multiple fragments [42]. Fazekas et al.'s examination of 251 plant individuals demonstrated mere recognition rates of 44% and 45% for *psbK-psbI* and *atpF-atpH*, respectively; however, when *matk* + *psbK-psbI* + *atpF-atpH* were combined as barcodes, the identification success rate climbed to 69% [43]. Our investigation also revealed that the combination of barcode fragments outperformed single fragments in identifying *Cymbidium* species. Among these combinations, ITS + *matK* + *psbA-trnH* exhibited superior resolution with minimal overlap in the "barcoding gap" among other fragment combinations. Similar verification was conducted within the buckwheat genus, where this sequence combination facilitated accurate identification and supported buckwheat's monophyletic grouping [44]. Furthermore, in identifying medicinal orchids, complementary performance was observed among *matK*, ITS and *psbA-trnH* sequences. The ITS + *psbA-trnH* sequence displayed substantial genetic variability with improved "barcoding gap" performance. Nevertheless, its effectiveness in identifying *Cymbidium* species was somewhat limited, possibly attributable to insufficient intraspecific variation or the presence of unusually large intraspecific distances.

3.3. The Phylogeny of the Cymbidium

Incorporating both sequence feature analysis and genetic distance analysis, the phylogenetic tree was constructed based on *psbA-trnH* of the single-fragment barcodes, *matK* + *psbA-trnH* and ITS + *matK* + *psbA-trnH* of the combined barcodes. It was found that the *Cymbidium* species consisted of six major clades, eight major clades, and five major clades, respectively. Based on the analysis of these three phylogenetic trees, the subgenus *Jensoa* does not appear to form a monophyletic group. Among them, the *C. lancifolium* was embedded in the branches of other subgenera in the phylogenetic tree constructed based on ITS + *matK* + *psbA-trnH*, while it formed a separate small branch in the phylogenetic trees constructed based on the *psbA-trnH* barcode and the combined barcode of *matK* + *psbA-trnH*. Although most species of the subgenus *Jensoa* were closely related together in the phylogenetic trees constructed based on the *psbA-trnH* and *matK* + *psbA-trnH*, some varieties of the same species were closely related together with other *Cymbidium* species, such as some varieties of *Cymbidium goeringii* (Figure 3). The existence of these differences may be due to natural hybridization between species [45–48], which led to genetic variation, and may also be due to the lack of plastid information characteristics [22,49], resulting in inconsistent classification results. Which was quite different from the phylogenetic tree by ITS + *matK* + *psbA-trnH*, all samples and varieties of the same species were not separated by the embedding of other species, indicating that the multi-fragment combined barcode enhanced the sequence information characteristics and was more conducive to the classification of the *Cymbidium* species to a certain extent.

During the evolution of orchids, unique characteristics such as deceptive pollination and dust-like, wind-dispersed seeds often lead to gene flow between populations. Deceptive pollination, which involves orchids mimicking sex pheromones and the appearance of specific female insects to sexually lure common male insects as pollinators, allows some orchid species to share pollinators [50]. Therefore, most orchids that use the same pollination strategy and grow in the same habitat share a greater number of insect species than expected. These reproductive strategies make them highly susceptible to interspecific hybridization in nature, resulting in significant genetic heterogeneity between orchid populations [51]. In this study, the analysis of the phylogenetic tree constructed based on the combined barcode revealed that a sample of the epiphytic orchid (*C. floribundum*) was closely related to terrestrial orchids such as *C. ensifolium* in a large clade. It is speculated that this sample of *C. floribundum* may have undergone interspecific hybridization with other *Cymbidium* species. *C. floribundum* primarily relies on insect pollination, and bees are one of the main pollinator groups [52]. Since *C. floribundum* produces a large number of flowers when blooming, it attracts a significant amount of bees as potential pollinators. Given the low specificity of insect pollination, it is likely that during the long cultivation history of *C. floribundum*, cross-pollination with other *Cymbidium* species occurred, leading to genetic changes and variations in the offspring. Additionally, the samples we collected have a relatively longer life history, which may have enabled gene flow between different species via insect pollination in the natural environment. This can lead to genetic variations in *Cymbidium* species without morphological changes, thereby producing classification results that differ from the traditional categorization of *Cymbidium* species.

3.4. Identification and Classification of Cymbidium Species Through the Combination of Morphological Analysis and DNA Barcoding Technology

The traditional morphological analysis [53,54] for classifying orchids has certain limitations and deficiencies, as plant characteristics are finite and varying interpretations of phenotypic traits among individuals can result in inconsistent results. In *Cymbidium* genera, phylogenetic trees based on morphology struggle to achieve precise positioning due to rapid dispersal and hybridization, which can induce swift alterations in morphological traits, thus complicating phylogenetic analyses based on morphology [55]. The integration of DNA barcoding technology with morphological analysis offers a more comprehensive basis for classification, from external characteristics to genetic insights. Previous research on phenotypic traits has found that the species of *C. sinense*, *C. ensifolium*, *C. kanran*, and *C. faberi* were closely related, as were the species of *C. goeringii*, *C. goeringii* var. *longibracteatum*, and *C. tortisepalum* [56]. This supports the results of the morphological analysis in this study. Another study, which utilized both morphological analysis and molecular markers, found that *C. kanran* and *C. sinense* cluster in a group, while all species of *C. goeringii*, *C. goeringii* var. *longibracteatum*, and *C. tortisepalum* cluster together, and *C. faberi* forms a separate branch [56]. Although these research findings [56–58] have many similarities with this study, they only explore the classification relationships of terrestrial orchids within the *Cymbidium* species. However, our study also investigates the taxonomic status and relationships of epiphytic orchids within the *Cymbidium* species, which indicate that the genetic relationships among epiphytic orchids are different from those of terrestrial orchids and that species with the same life form have closer phylogenetic relationships. Furthermore, the difference is that a species of *C. floribundum* was closely related together with two species of *C. ensifolium* on a small branch. These discrepancies between these two research methods have also been observed in the identification of *Amorphophallus* and *Rhododendron* [34,59]. However, discrepancies are noted in the classification relationships among the *Cymbidium* species known as *C. kanran*, *C. floribundum*, and *C. lancifolium*; the taxonomic status of these three species of *Cymbidium* requires further investigation.

It is noteworthy that *C. lancifolium*, which has some phenotypic differences from the typical terrestrial orchids in the *Cymbidium* species, clusters with the majority of terrestrial species within the *Cymbidium* in some research reports, indicating a closer phylogenetic relationship with terrestrial orchids [60,61]. However, other studies [62] on the classification of *Cymbidium* species show that *C. lancifolium* forms a separate cluster, which is consistent with the phenotypic trait clustering results of this study. Although this differs from the phylogenetic tree analysis results within this study, which indicated that *C. lancifolium* grouped with epiphytic orchid species (such as *C. aloifolium*, *C. tracyanum*, *C. elegans*, etc.) in Clade IV (Figure 4), it still indicates a closer phylogenetic relationship with epiphytic orchids. We posit that *C. lancifolium* has a potential that is more inclined towards epiphytism, making it closer to epiphytic orchids. It is speculated that this may be related to the origins and evolution of terrestrial orchids and epiphytic orchids. The origin and evolution of epiphytic orchids are similar to the various origins of ferns, Bromeliaceae, and eudicots [63–66], suggesting that the transformation of epiphytism may rely on certain morphological or genetic prerequisites that are common among members of large taxonomic groups. Given the discrepancies among different research findings [22], there is a need for more accurate molecular identification methods to precisely determine its classification and life form.

Our research findings highlight the practical application of the barcode method in accurately determining the geographical origins of orchid specimens. This capability is of great significance for managing the legal trade in medicinal and ornamental orchids, ensuring that such trade does not adversely affect threatened populations. In addition, by enabling law enforcement agencies to assign geographical origins to unknown specimens extracted in illegal trade, barcode technology has great potential in combating illegal trade and conserving biodiversity. However, we also recognize that, in order to realize its full potential, barcoding should be used in conjunction with other verification methods and implemented within a broader policy and regulatory framework.

3.5. *Cymbidium* Species Diversity

In recent years, research on molecular markers has mainly focused on selecting the most suitable barcode for specific families and genera [18,20]. The use of certain single-gene fragments or combinations currently cannot distinguish all higher plants and is limited to a narrow range, such as family, genus, and species levels [67,68]. Even within a specific range, the results may sometimes be erroneous or contradictory to traditional morphological classification. Moreover, the majority of studies [69–73] on orchid morphology or genetic evolution primarily focus on epiphytic orchids within the entire orchid family or subfamily, with limited analysis conducted on species diversity within *Cymbidium* species and particularly the evolutionary disparities among *Cymbidium* species from different regions.

In this study, the phylogenetic tree constructed using single-fragment barcodes was not as effective in species identification and classification as the phylogenetic tree constructed using combined barcodes. Yang et al. also found that, with the increase in sequences, the phylogenetic resolution and node support values significantly improved, and phylogenetic analyses based on the complete chloroplast genome could overcome the limitations of insufficient DNA sequence sampling [17]. Previous studies, although providing better identification and classification of *Cymbidium* species, did not explore the possible reasons for such classification results. A recent study using the chloroplast genome of *Cymbidium* species speculated that the ancestral life form of *Cymbidium* species was epiphytic and discovered that the northernmost species of the genus had undergone three transitions from epiphytic to terrestrial habits, which seemed to be related to adaptation to the colder northern environment [61]. In this study, a sample of the epiphytic Orchis (*C. floribundum*) collected in Hunan, China, was closely related to some terrestrial Orchis also collected in Hunan in the phylogenetic tree, which is likely due to genetic variation in internal structure for adaptation to local climate change. Climate change is widely recognized as a significant driver of species diversity, and its impact on other plant lineages has been extensively studied [74–76]. In particular, monsoon climates have been shown to shape the evolutionary trajectories of many plant species [77]. Climate change and geographical distribution together influence the life-form changes of *Cymbidium* species and affect their diversity by altering their living environments [61]. In this study, species collected in southern China were predominantly epiphytic, while those collected further north were terrestrial. The phylogenetic tree classification results indicated that species with the same life form had closer phylogenetic relationships. This indicates that the phylogenetic tree constructed using DNA barcoding technology can, to some extent, analyze the geographical origins of species.

The phylogenetic perspective is greatly helpful for understanding the evolutionary development of *Cymbidium* species, but the process of morphological evolution in *Cymbidium* species is highly complex. Amidst the increasing habitat changes caused by humans, many types of orchids are still being discovered and statistically described, which increases the chance of documenting diversity. Therefore, it is imperative to employ genome sequences of *Cymbidium* species in future research to explicate their adaptability to various environments and investigate the evolution of diverse phenotypes and forms. Comparative genomics serves as a potent tool for studying evolution and morphology, while genome assembly will provide invaluable resources for identifying genetic variations associated with ecological traits in *Cymbidium* species and facilitating genomics-assisted breeding.

4. Materials and Method

4.1. Plant Materials

Germplasm resources of the *Cymbidium* species were collected in regions including Hunan, Yunnan, Guangdong, and Fujian, compiling a total of 48 individuals from 30 *Cymbidium* species resources, including species, varieties, and cultivars [7,78]. This study selected 30 species of *Cymbidium.* to serve as taxonomic units for the classification analysis of phenotypic traits. The specimen details are presented in Table S5. To enhance the controllability of the data, 48 individuals from 30 species were used to assess the success rate of PCR amplification and sequencing. Species information is detailed in Table S6. Leaf samples from *Cymbidium* species were collected in batches, and DNA was extracted to evaluate the success rate of PCR amplification and sequencing. All experimental materials are maintained and managed in the flower garden of Hunan Agricultural University, with the requirement that the plants tested be healthy and free from diseases.

4.2. DNA Extraction, Amplification, and Sequencing

The total DNA of *Cymbidium* plants was extracted using the Polysaccharide Polyphenol Plant Genomic DNA Extraction Kit (DP360) from Tiangen Biochemical Technology Co., Ltd. (Beijing, China). The specific steps are described in the instructions. The extracted total DNA was subjected to concentration determination and stored at −20 °C for later use. Through a literature review and repeated amplification tests, the universal primer sequences of ITS, *matK*, *rbcL*, *psbA-trnH*, and *trnL-F* were ultimately selected. The detailed information of the primers and reaction procedures is presented in Table 6, with the primers synthesized by Shanghai Sangon Bioengineering Co., Ltd. (Shanghai, China). PCR amplification products with bright bands, high specificity, and correct fragment size were chosen and submitted to Shanghai Sangon Bioengineering Co., Ltd. for bidirectional sequencing. Because of the varying barcode characteristics, the number of orchid genus sequences obtained from sequencing is unequal (Table S7).

Table 6. PCR primers and reaction procedures of five DNA barcodes.

Barcodes	Primer Name	Primer Sequence (5′–3′)	Reaction Procedure	References
ITS	17SE	ACGAATTCATGGTCCGGTGAAGTGTTCG	95 °C 3 min, 35 cycle (95 °C 15 s, 62 °C 15 s, 72 °C 15 s), 72 °C 5 min	Sun et al., 1994 [79]
	26SE	TAGAATTCCCCGGTTCGCTCGCCGTTAC		Sun et al., 1994
matK	390F	CGATCTATTCATTCAATATTTC	95 °C 3 min, 35 cycle (95 °C 15 s, 46.5 °C 15 s, 72 °C 15 s), 72 °C 5 min	Cuenoud et al., 2002 [80]
	1326R	TCTAGCACACGAAAGTCGAAGT		Cuenoud et al., 2002
psbA-trnH	psbA	GTTATGCATGAACGTAATGCTC	95 °C 3 min, 35 cycle (95 °C 15 s, 55 °C 15 s, 72 °C 15 s), 72 °C 5 min	Sang et al., 1997 [81]
	trnH2	CGCGCATGGTGGATTCACAATCC		Tate, 2002 [82]
rbcL	1F	ATGTCACCACAAACAGAAAC	95 °C 3 min, 35 cycle (95 °C 15 s, 56 °C 15 s, 72 °C 15 s), 72 °C 5 min	Goldman et al., 2001 [83]
	724R	TGCCATGTACCYGCAGTTGC		Goldman et al., 2001
trnL-F	c	CGAAATCGGTAGACGCTACG	95 °C 3 min, 35 cycle (95 °C 15 s, 53 °C 15 s, 72 °C 15 s), 72 °C 5 min	Taberlet et al., 1991 [84]
	f	ATTTGAACTGGTGACACGAG		Taberlet et al., 1991

4.3. Data Analysis

The sequence peak image files obtained from sequencing were corrected and concatenated using the Seqman 7.1 software, and the BLAST (Basic Local Alignment Search Tool) online tool (https://blast.ncbi.nlm.nih.gov/Blast.cgi, accessed on 7 January 2025) was used for alignment verification. With MEGA 11.0 software and a genetic distance model based on the Kimura two-parameter (K2-P) method [85], we calculated the intra- and inter-species genetic distances for each barcode of *Cymbidium* species and compared the sequence differences between intra- and interspecific variations [86–88]. An area chart of the frequency distribution of genetic distance was plotted using Origin 2022, which facilitates a clear observation of whether there is a barcoding gap in each DNA barcode sequence within *Cymbidium* species [87].

Phylogenetic trees provide an intuitive assessment of the species identification ability of DNA barcodes and are among the evaluation criteria for DNA barcodes [89]. In this study, we employed the neighbor-joining method in the MEGA 11.0 software to construct a phylogenetic tree. The K2-P genetic distance model was selected, with the bootstrap test set to 1000 replications [90]. The bootstrap values support rate needs to be greater than or equal to 50% for accurate and reliable identification results.

Phenotypic trait variability and correlation analyses were performed using Excel 2022 and Spss 27.0. A principal component analysis of 27 phenotypic traits was conducted using Origin 2022. Additionally, R-type cluster analysis was employed to perform a correlation analysis on the 27 phenotypic traits, and Q-type cluster analysis based on phenotypic traits was used to conduct a qualitative analysis of the relationships among the 30 species of *Cymbidium*.

4.4. Selection and Determination Methods of Phenotypic Traits

In accordance with the testing guidelines for the specificity, consistency, and stability of orchid species issued by the Ministry of Agriculture of the People's Republic of China (TG/164/3), 27 representative phenotypic traits such as leaf length, sepal length, petal traits, and middle sepal traits were selected, including 11 quantitative traits (Table S8) and 16 qualitative traits (Table S9). During the full-bloom stage of each species, the quantitative traits were measured using tapes and vernier calipers as detailed in Table S8, while qualitative traits were coded and assigned as shown in Table S9. The traits related to color were obtained with reference to the Royal Horticultural Society's standard color card. Ultimately, the average of the actual measured values was calculated, and both the quantitative traits and the assigned qualitative traits were documented and analyzed.

5. Conclusions

Based on the similarities between cluster results and phylogenetic tree results, we verified and selected the combination sequence ITS + *matK* + *psbA-trnH* as a high-quality DNA barcode for *Cymbidium* species. Furthermore, by constructing a phylogenetic tree with combined barcodes, we observed that epiphytic orchid species clustered together on a single clade, and species with similar life forms were closely related. Our study also revealed that phenotypic analysis successfully classified the germplasm resources of *Cymbidium* into seven groups. Floral phenotypic traits were confirmed as the primary characteristics for differentiating both interspecific and intraspecific variations within *Cymbidium* species. The aforementioned findings not only elucidate the application of DNA barcoding techniques for *Cymbidium* species classification but also enhance our comprehension of the taxonomy and geographic distribution of *Cymbidium* species. Simultaneously, they foster deeper contemplation of the taxonomic relationships among diverse life forms of *Cymbidium* species while offering novel ideas and insights for future research on genetic diversity and adaptive evolution within different life forms of *Cymbidium* species.

Supplementary Materials: The following supporting information can be downloaded at https://www.mdpi.com/article/10.3390/plants14040619/s1. Table S1: *Cymbidium* species resolution is based on the method of genetic distance with "best match", "best close match", and a phylogenetic tree of five barcodes and their combination. Table S2: Variation of 11 quantitative traits in *Cymbidium* species. Table S3: Correlation analysis of 11 quantitative traits of *Cymbidium* species. Table S4: Frequency distribution of 16 quality traits of *Cymbidium* species. Table S5: *Cymbidium* species samples and their information used to determine phenotypic traits in this study. All plant samples were maintained and managed in the flower base of Hunan Agricultural University. Table S6: 48 plant samples of *Cymbidium* and information used for DNA extraction in this study. All plant samples were maintained and managed in the flower base of Hunan Agricultural University. Table S7: GenBank accession numbers of the five loci sequences for *Cymbidium* species examined in this study. Table S8: Quantitative traits and test methods of *Cymbidium* spp. Table S9: Quality traits and its assignment of *Cymbidium* spp.

Author Contributions: Conceptualization, Y.P. and F.C.; methodology, H.D.; software, Y.C.; validation, Y.P., X.L. and L.X.; formal analysis, Y.P.; investigation, Y.C.; resources, F.C.; data curation, Y.P. and L.X.; writing—original draft preparation, Y.P.; writing—review and editing, L.X.; visualization, H.D.; supervision, Y.C., X.L. and L.X.; project administration, L.X.; funding acquisition, F.C. All authors have read and agreed to the published version of the manuscript.

Funding: This research was funded by Hunan Provincial Natural Science Foundation of China, grant number No. 2019JJ50232. Supported by the Research Projects on Teaching Reform in Regular Higher Education Institutions in Hunan Province (HNJG-2022-0103), a project funded by the Hunan Provincial Department of Education.

Data Availability Statement: The raw data supporting the conclusions of this article will be made available by the authors upon request.

Acknowledgments: We express our gratitude to all the participants involved in the project and extend our sincere appreciation to the numerous members of the Hunan Mid-Subtropical Quality Plant Breeding and Utilization Engineering Technology Research Center for their invaluable assistance and unwavering support throughout this research endeavor.

Conflicts of Interest: The authors declare no conflicts of interest.

References

1. Chen, X.Q.; Ji, Z.H. *The Complete Book of Chinese Orchids*; China Forestry Publishing House: Beijing, China, 1998; Volume 3, pp. 1–282.
2. Elazazi, E.; Ziems, L.; Mahmood, T. Genotypic Selection Using Quantitative Trait Loci for Better Productivity under High Temperature Stress in Tomato (*Solanum lycopersicum* L.). *Horticulturae* **2024**, *10*, 874. [CrossRef]
3. Yan, H. Morphological and ISSR Molecular Identification of Several *Cymbidium* Species. Master's Thesis, Sichuan Agricultural University, Ya'an, China, 2012.
4. Freudenstein, J.V. Orchid phylogenetics and evolution: History, current status and prospects. *Ann. Bot.* **2024**, mcae202. [CrossRef] [PubMed]
5. Yan, Y.; Cao, B.; Zhang, Y.Q.; Chen, Q.X.; Chen, N.C. Genetic Diversity Analysis of Chinese Orchid Germplasm Resources Based on SRAP Markers. *Chin. J. Trop. Crops* **2020**, *41*, 929–938.
6. Schlechter, R. Die Gattungen *Cymbidium* Sw. und *Cyperorchis* Bl. *Repert. Nov. Specierum Regni Veg.* **1924**, *20*, e96–e110. [CrossRef]
7. Seth, C.J.; Cribb, P.J. A Reassessment of the Sectional Limits in the Genus *Cymbidium* Swartz. In *Orchid Biology, Reviews and Prospectives 3*; Cornell University Press: Ithaca, NY, USA; London, UK, 1984.
8. Puy, D.D.; Cribb, P. *The Genus Cymbidium*; Helm, C., Ed.; The University of Chicago Press: Chicago, IL, USA, 1988; Volume 10.
9. Liu, Z.J.; Chen, X.Q.; Ru, Z.Z. *The Genus Cymbidium in China*; Science Press: Beijing, China, 2006.
10. Hollingsworth, P.M. Refining the DNA Barcode for Land Plants. *Proc. Natl. Acad. Sci. USA* **2011**, *108*, 19451–19452. [CrossRef] [PubMed]
11. Li, D.Z.; Gao, L.M.; Li, H.T.; Wang, H.; Ge, X.J.; Liu, J.Q.; Chen, Z.D.; Zhou, S.L.; Chen, S.L.; Yang, J.B.; et al. Comparative Analysis of a Large Dataset Indicates that Internal Transcribed Spacer (ITS) should be Incorporated into the Core Barcode for Seed Plants. *Proc. Natl. Acad. Sci. USA* **2011**, *108*, 19641–19646. [PubMed]
12. Schindel, D.E.; Miller, S.E. DNA Barcoding a Useful Tool for Taxonomists. *Nature* **2005**, *435*, 17. [CrossRef]
13. Zhang, Z.H.; Jin, Y.Y.; Gao, Y.D.; Zhang, Y.; Ying, Q.C.; Shen, C.J.; Lu, J.J.; Zhan, X.R.; Wang, H.Z.; Feng, S.G. The Complete chloroplast Genomes of Two *Physalis* Species, *Physalis macrophysa* and *P. ixocarpa*: Comparative Genomics, Evolutionary Dynamics and Phylogenetic Relationships. *Agronomy* **2023**, *13*, 135. [CrossRef]
14. Hebert, P.D.N.; Cywinska, A.; Ball, S.L.; Dewaard, J.R. Biological Identifications through DNA Barcodes. *Proc. R. Soc. Lond. B* **2003**, *270*, 313–321. [CrossRef]
15. Ratnasingham, S.; Hebert, P.D.N. Bold: The Barcode of Life Data System. *Mol. Ecol. Notes* **2007**, *7*, 355–364. [CrossRef]
16. Ren, B.Q.; Chen, Z.R. Plant DNA Barcode Technology. *J. Bot.* **2010**, *45*, 1–12.
17. Yang, J.B.; Tang, M.; Li, H.T.; Zhang, Z.R.; Li, D.Z. Complete Chloroplast Genome of the Genus *Cymbidium*: Lights into the Species Identification, Phylogenetic Implications and Population Genetic Analyses. *BMC Evol. Biol.* **2013**, *13*, 84. [CrossRef] [PubMed]
18. Zhang, L.; Huang, Y.W.; Huang, J.L.; Ya, J.D.; Zhe, M.Q.; Zeng, C.X.; Zhang, Z.R.; Zhang, S.B.; Li, D.Z.; Li, H.T.; et al. DNA Barcoding of *Cymbidium* by Genome Skimming: Call for Next-Generation Nuclear Barcodes. *Mol. Ecol. Resour.* **2022**, *23*, 424–439. [CrossRef]
19. Wang, X.M.; Cheng, L.J.; Li, Z.L.; Zhang, Z.C.; Wang, Y.Y. Characterization of the Complete Chloroplast Genome of a Chinese Endangered Species *Cymbidium wenshanense* Y. S. Wu et F. Y. Liu. *Mitochondrial DNA B Resour.* **2023**, *8*, 815–818. [CrossRef] [PubMed]
20. Sharma, S.K.; Dkhar, J.; Kumaria, S.; Tandon, P.; Rao, S.R. Assessment of Phylogenetic Inter-Relationships in the Genus *Cymbidium* (Orchidaceae) Based on Internal Tanscribed Spacer Region of rDNA. *Gene* **2012**, *495*, 10–15. [CrossRef] [PubMed]
21. Choi, S.H.; Kim, M.J.; Lee, J.S.; Ryu, K.H. Genetic Diversity and Phylogenetic Relationships among and within Species of Oriental *Cymbidiums* Based on RAPD Analysis. *Sci. Hortic.* **2006**, *108*, 79–85. [CrossRef]
22. Zhang, G.Q.; Chen, G.Z.; Chen, L.J.; Zhai, J.W.; Huang, J.; Wu, X.Y.; Li, M.H.; Peng, D.H.; Rao, W.H.; Liu, Z.J.; et al. Phylogenetic Incongruence in *Cymbidium* Orchids. *Plant Divers.* **2021**, *43*, 452–461. [CrossRef] [PubMed]
23. Liu, Z.H.; Zeng, X.; Yang, D.; Ren, G.M.; Chu, G.Y.; Yuan, Z.R.; Luo, K.; Xiao, P.G.; Chen, S.L. Identification of Medicinal Vines by ITS2 using Complementary Discrimination Methods. *J. Ethnopharmacol.* **2012**, *141*, 242–249. [CrossRef]
24. Dong, W.P.; Xu, C.; Li, C.H.; Sun, J.H.; Zuo, Y.J.; Shi, S.; Cheng, T.; Guo, J.J.; Zhou, S.L. *Ycf1*, the Most Promising Plastid DNA Barcode of Land Plants. *Sci. Rep.* **2015**, *5*, 8348. [CrossRef]
25. Ning, S.P.; Yan, H.F.; Hao, G.; Ge, X.J. Research Progress on Plant DNA Barcode. *Biodivers. Sci.* **2008**, *5*, 417–425.
26. Zhang, X.S. Evolution and Maintenance of the Environmental Component of the Phenotypic Variance: Benefit of Plastic Traits under Changing Environments. *Am. Nat.* **2005**, *166*, 569–580. [CrossRef]

27. Kolb, A.; Ehrlén, J. Environmental Context Drives Seed Predator-Mediated Selection on a Floral Display Trait. *Evol. Ecol.* **2010**, *24*, 433–445. [CrossRef]
28. Sletvold, N.; Grindeland, J.M.; Ågren, J. Vegetation Context Influences the Strength and Targets of Pollinator-Mediated Selection in a Deceptive Orchid. *Ecology* **2013**, *94*, 1236–1242. [CrossRef]
29. Ning, H.J.; Ao, S.Y.; Fan, Y.Y.; Fu, J.X.; Xu, C.M. Correlation Analysis between the Karyotypes and Phenotypic Traits of Chinese *Cymbidium* Cultivars. *Hortic. Environ. Biotechnol.* **2018**, *59*, 93–103. [CrossRef]
30. Li, W.; Wang, P.; Qi, Q.G.; Zhang, Q.C.; Gao, X.; Lin, M.F.; Cui, Y.T. Phenotypic Diversity and Variation of Lonicera Caerulea Populations in the Changbai Mountain Alongside the Elevation Gradient. *Pol. J. Environ. Stud.* **2021**, *30*, 705–716. [CrossRef]
31. Ren, J.F.; Yang, J.; Li, H.W.; Yun, Y.; Zhang, L. Research Progress and Prospects of DNA Barcoding in Orchids. *Mod. Hortic.* **2015**, *9*, 11–12.
32. Hollingsworth, P.; Graham, S.; Little, D. Choosing and using a Plant DNA Barcode. *PLoS ONE* **2011**, *6*, 236–254. [CrossRef] [PubMed]
33. Mehraj, S.; Parihar, T.J.; Murtaza, D.; Hurrah, A.A.; Wani, I.A.; Lone, F.; Mufti, S.; Zargar, S.M.; Khan, I.; Sheikh, P.A. Newly emerging aquatic macrophytes in Northern Himalayan, Dal Lake of Kashmir Valley identified through DNA Barcoding and Its Antioxidant Profiling. *Ecol. Genet. Genom.* **2023**, *27*, 100162.
34. Cai, H.Y.; Xing, S.T.; He, R.R.; Zhou, B.; Zhao, F. Phylogenetic Analysis of Rhododendron in Mount Taishan Mountain based on ITS2 Sequence and Morphological Characteristics. *Plant Physiol. J.* **2017**, *8*, 1489–1498.
35. Cai, Y.M.; Dai, J.P.; Zheng, Y.X.; Ren, Y.Y.; Chen, H.M.; Feng, T.T.; Gao, X.X.; Zhu, S. DNA Barcode Screening for Molecular Identification of *Uncaria* Plants. *Chin. Tradit. Herb. Drugs* **2022**, *53*, 1828–1837.
36. Wang, D.X. ITS-1 Sequencing and Single Nucleotide Polymorphism Variation Site Analysis of *Gastrodia elata*. *Biotechnology* **2012**, *22*, 48–51.
37. Wu, J.; Li, D.; Boyd, B.; Balan, R.K.; George, S.; Peacock, L.; Pal, C. Comparative Performance of a Multi-Locus Barcoding Approach to enhance Taxonomic Resolution of New Zealand Mosquitoes (Diptera: Culicidae). *Aust. Entomol.* **2023**, *62*, 77–95. [CrossRef]
38. Sayed, H.A.; Mostafa, S.; Haggag, I.M.; Hassan, N.A. DNA Barcoding of *Prunus* Species Collection Conserved in the National Gene Bank of Egypt. *Mol. Biotechnol.* **2023**, *65*, 410–418. [CrossRef]
39. Zhang, J.J. Phylogenetic Analysis of *Dioscorea* Plants Based on *psbA-trnH* Gene and 18S rDNA. Master's Thesis, Inner Mongolia University, Hohhot, China, 2017.
40. Zhou, H.; Ma, S.J.; Yang, P.; Yao, H. Research Progress on Identification of *Dendrobium* Medicinal Plants based on DNA Sequence. *Mod. Tradit. Chin. Med. Mater. Med. World Sci. Technol.* **2015**, *17*, 950–957.
41. Kim, H.M.; Oh, S.H.; Bhandari, G.S.; Kim, C.S.; Park, C.W. DNA Barcoding of Orchidaceae in Korea. *Mol. Ecol. Resour.* **2014**, *14*, 499–507. [CrossRef] [PubMed]
42. Ren, B.Q.; Xiang, X.G.; Chen, Z. Species Identification of *Alnus* (Betulaceae) using nrDNA and cpDNA Genetic Markers. *Mol. Ecol. Resour.* **2010**, *10*, 594–605. [CrossRef]
43. Fazekas, A.; Burgess, K.; Kesanakurti, P.; Graham, S.W.; Newmaster, S.G.; Husband, B.C.; Percy, D.M.; Hajibabaei, M.; Barrett, S.C.H. Multiple Multilocus DNA Barcodes from the Plastid Genome Discriminate Plant Species Equally Well. *PLoS ONE* **2008**, *3*, e2802. [CrossRef]
44. Zheng, Y.D. Molecular Phylogenetic Study of *Fagopyrum* Plants in Southwest China. Master's Thesis, Sichuan Agricultural University, Ya'an, China, 2013.
45. Yu, W.B.; Huang, P.H.; Li, D.Z.; Wang, H. Incongruence between nuclear and chloroplast DNA phylogenies in *Pedicularis* section *Cyathophora* (Orobanchaceae). *PLoS ONE* **2013**, *8*, e74828. [CrossRef]
46. Guo, Y.Y.; Luo, Y.B.; Liu, Z.J.; Wang, X.Q. Reticulate evolution and sea-level fluctuations together drove species diversification of slipper orchids (*Paphiopedilum*) in South-East Asia. *Mol. Ecol.* **2015**, *24*, 2838–2855. [CrossRef]
47. De Ré, F.C.; Robe, L.J.; Wallau, G.L.; Loreto, E.L.S. Inferring the phylogenetic position of the *Drosophila* flavopilosa group: Incongruence within and between mitochondrial and nuclear multilocus datasets. *J. Zool. Syst. Evol. Res.* **2017**, *55*, 208–221. [CrossRef]
48. Kanzi, A.M.; Trollip, C.; Wingfield, M.J.; Barnes, I.; Van der Nest, M.A.; Wingfield, B.D. Phylogenomic incongruence in Ceratocystis: A clue to speciation. *BMC Genom.* **2020**, *21*, 362. [CrossRef]
49. Tang, Y.; Yukawa, T.; Bateman, R.M.; Jiang, H.; Peng, H. Phylogeny and classification of the east Asian *Amitostigma* alliance (Orchidaceae: Orchideae) based on six DNA markers. *BMC Evol. Biol.* **2015**, *15*, 96. [CrossRef] [PubMed]
50. Perkins, J.; Hayashi, T.; Peakall, R.; Flematti, G.R.; Bohman, B. The volatile chemistry of orchid pollination. *Nat. Prod. Rep.* **2023**, *40*, 567–572. [CrossRef] [PubMed]

51. Zhang, G.Q.; Liu, K.W.; Li, Z.; Lohaus, R.; Hsiao, Y.Y.; Niu, S.C.; Wang, J.Y.; Lin, Y.C.; Xu, Q.; Chen, L.J. The *Apostasia* genome and the evolution of orchids. *Nature* **2017**, *549*, 379–383. [CrossRef]
52. Sugahara, M.; Minamoto, T.; Fuchikawa, T.; Michinomae, M.; Shimizu, I. *Apis cerana japonica* discriminates between floral color phases of the oriental orchid, *Cymbidium floribundum*. *Zool. Sci.* **2010**, *27*, 901–906. [CrossRef]
53. Dressler, R.L.; Dodson, C.H. Classification and Phylogeny in the Orchidaceae. *Ann. Mo. Bot. Gard.* **1960**, *47*, 25–68. [CrossRef]
54. Burns-Balogh, P.; Funk, V.A. A phylogenetic Analysis of the Orchidaceae. *Smithson. Contrib. Bot.* **1986**, *61*, 1–79. [CrossRef]
55. Wang, H.C.; Moore, M.J.; Soltis, P.S.; Bell, C.D.; Brockington, S.F.; Alexandre, R.; Davis, C.C.; Latvis, M.; Manchester, S.R.; Soltis, D.E. Rosid Radiation and the Rapid Rise of Angiosperm-Dominated Forests. *Proc. Natl. Acad. Sci. USA* **2009**, *106*, 3853–3858. [CrossRef] [PubMed]
56. Ao, S.Y. *Analysis on Phenotypic Characters and Karyotype of Cymbidium*; Zhejiang A&F University: Hangzhou, China, 2014.
57. Ji, X.B.; Wang, G.D.; Kang, J.C.; Qiao, G.; Wen, X.P. Comparative Analysis of Morphology and RAPD Markers of 12 Orchid Plant Resources in Guizhou Province. *Seed* **2008**, *2*, 56–60.
58. Wei, X.Y.; Liu, H.; Ma, H.; Bie, T.D.; Sun, Y. Analysis of Genetic Diversity and Fingerprint Map Construction of 96 Orchid Germplasm Resources Based on ISSR Markers. *J. Plant Genet. Resour.* **2024**, *25*, 586–599.
59. Yang, Y.Y. Phenotypic Diversity and ITS Marker Analysis of *konjac* Plants. Master's Thesis, Yunnan University, Kunming, China, 2020.
60. Zhe, M.Q.; Zhang, L.; Liu, F.; Huang, Y.W.; Fan, W.S.; Yang, J.B.; Zhu, A.D. Plastid RNA Editing Reduction Accompanied with Genetic Variations in *Cymbidium*, a Genus with Diverse Lifestyle Modes. *Plant Divers.* **2022**, *44*, 316–321. [CrossRef]
61. Chen, H.Y.; Zhang, Z.R.; Yao, X.; Ya, J.D.; Jin, X.H.; Wang, L.; Lu, L.; Li, D.Z.; Yang, J.B.; Yu, W.B. Plastid Phylogenomics Provides New Insights into the Systematics, Diversification, and Biogeography of *Cymbidium* (Orchidaceae). *Plant Divers.* **2024**, *46*, 448–461. [CrossRef] [PubMed]
62. Chen, Z.M.; Gao, L.; Wang, H.Z.; Feng, S.G. Molecular Identification and Phylogenetic Analysis of *Cymbidium* Species (Orchidaceae) Based on the Potential DNA Barcodes *matK*, *rbcL*, *psbA-trnH*, and Internal Transcribed Spacer. *Agronomy* **2024**, *14*, 933. [CrossRef]
63. Crayn, D.M.; Winter, K.; Smith, J.A.C. Multiple Origins of Crassulacean Acid Metabolism and the Epiphytic Habit in the Neotropical Family Bromeliaceae. *Proc. Natl. Acad. Sci. USA* **2004**, *101*, 3703–3708. [CrossRef] [PubMed]
64. Schneider, H.; Schuetipelz, E.; Pryer, K.M.; Cranfill, R.; Magallón, S.; Lupia, R. Ferns Diversified in the Shadow of Angiosperms. *Nature* **2004**, *428*, 553–557. [CrossRef] [PubMed]
65. Schuettpelz, E.; Pryer, K.M. Evidence for a Cenozoic Radiation of Ferns in an Angiosperm–Dominated Canopy. *Proc. Natl. Acad. Sci. USA* **2009**, *106*, 11200–11205. [CrossRef] [PubMed]
66. Zotz, G.; Weigelt, P.; Kessler, M.; Kreft, H.; Taylor, A. EpiList 1.0: A global Checklist of Vascular Epiphytes. *Ecology* **2021**, *102*, e03326. [CrossRef] [PubMed]
67. Raskoti, B.B.; Ale, R. DNA Barcoding of Medicinal Orchids in Asia. *Sci. Rep.* **2021**, *11*, 23651. [CrossRef]
68. Worthy, S.J.; Bucalo, K.; Perry, E.; Reynolds, A.; Cruse-Sanders, J.; Pérez, A.J.; Burgess, K.S. Ability of *rbcL* and *matK* DNA Barcodes to Discriminate between Montane Forest Orchids. *Plant Syst. Evol.* **2022**, *308*, 19. [CrossRef]
69. Givnish, T.J.; Spalink, D.; Ames, M.; Lyon, S.P.; Hunter, S.J.; Zuluaga, A.; Iles, W.J.D.; Clements, M.A.; Arroyo, M.T.K.; Leebens-Mack, J. Orchid phylogenomics and Multiple Drivers of Their Extraordinary Diversification. *Proc. R. Soc. B* **2015**, *282*, 171–180. [CrossRef]
70. Pérez-Escobar, O.A.; Dodsworth, S.; Bogarín, D.; Bellot, S.; Balbuena, J.A.; Schley, R.J.; Kikuchi, I.A.; Morris, S.K.; Epitawalage, N.; Cowan, R. Hundreds of Nuclear and Plastid Loci Yield Novel Insights into Orchid Relationships. *Am. J. Bot.* **2021**, *108*, 1166–1180. [CrossRef]
71. Pérez-Escobar, O.A.; Bogarín, D.; Przelomska, N.A.S.; Ackerman, J.D.; Balbuena, J.A.; Bellot, S.; Bühlmann, R.P.; Cabrera, B.; Cano, J.A.; Charitonidou, M. The Origin and Speciation of Orchids. *New Phytol.* **2024**, *242*, 700–716. [CrossRef]
72. Zhang, W.X.; Zhang, G.Q.; Zeng, P.; Zhang, Y.Q.; Hu, H.; Liu, Z.J.; Cai, J. Genome Sequence of *Apostasia ramifera* Provides Insights into the Adaptive Evolution in Orchids. *BMC Genom.* **2021**, *22*, 536. [CrossRef] [PubMed]
73. Zhang, G.Q.; Xu, Q.; Bian, C.; Tsai, W.C.; Yeh, C.M.; Liu, K.W.; Yoshida, K.; Zhang, L.S.; Chang, S.B.; Chen, F. The *Dendrobium Catenatum* Lindl. Genome Sequence Provides Insights into Polysaccharide Synthase, Floral Development and Adaptive Evolution. *Sci. Rep.* **2016**, *6*, 19029. [CrossRef] [PubMed]
74. Jiang, C.; Tan, K.; Ren, M.X. Effects of monsoon on distribution patterns of tropical plants in Asia. *Chin. J. Plant Ecol.* **2017**, *41*, 1103–1112, (In Chinese with English Abstract).
75. Ding, W.N.; Ree, R.H.; Spicer, R.A.; Xing, Y.W. Ancient orogenic and monsoon-driven assembly of the world's richest temperate alpine flora. *Science* **2020**, *369*, 578–581. [CrossRef]

76. Ye, J.W.; Tian, B.; Li, D.Z. Monsoon intensification in East Asia triggered the evolution of its flora. *Front. Plant Sci.* **2022**, *13*, 1046538. [CrossRef]
77. Ashokan, A.; Xavier, A.; Suksathan, P.; Ardiyani, M.; Leong-Skornicková, J.; Newman, M.; Kress, W.J.; Gowda, V. Himalayan orogeny and monsoon intensification explain species diversification in an endemic ginger (*Hedychium*: Zingiberaceae) from the Indo-Malayan Realm. *Mol. Phylogenet. Evol.* **2022**, *170*, 107440. [CrossRef]
78. Chinese Academy of Sciences. *Flora of China*; Science Press: Beijing, China, 1999; Volume 18, p. 191.
79. Sun, Y.; Skinner, D.; Liang, G.; Hulbert, S.H. Phylogenetic Analysis of Sorghum and Related Taxa using Internal Transcribed Spacers of Nuclear Ribosomal DNA. *Theor. Appl. Genet.* **1994**, *89*, 26–32. [CrossRef] [PubMed]
80. Cuenoud, P.; Savolainen, V.; Chatrou, L.W.; Powell, M.; Grayer, R.; Chase, M.W. Molecular Phylogenetics of Caryophyllales Based on Nuclear 18S rDNA and Plastid *rbcL*, *atpB*, and *matK* DNA Sequences. *Am. J. Bot.* **2002**, *89*, 132–144. [CrossRef]
81. Sang, T.; Crawford, D.J.; Stuessy, T.F. Chloroplast DNA Phylogeny, Reticulate Evolution, and Biogeography of Paeonia (Paeoniaceae). *Am. J. Bot.* **1997**, *84*, 1120–1132. [CrossRef]
82. Simpson, B.B. Paraphyly of Tarasa (Malvaceae) and Diverse Origins of the Polyploid Species. *Syst. Bot.* **2003**, *28*, 723–737.
83. Goldman, D.H.; Freudenstein, J.V.; Kores, P.J.; Molvray, M.; Jarrell, D.C.; Whitten, W.M.; Cameron, K.M.; Jansen, R.K.; Chase, M.W. Phylogenetics of Arethuseae (Orchidaceae) Based on Plastid *matK* and *rbcL* Sequences. *Syst. Bot.* **2001**, *37*, 670–695.
84. Sun, G. Interspecific Polymorphism at Non-Coding Regions of Chloroplast, Mitochondrial DNA and rRNA IGS Region in Elymus Species. *Hereditas* **2002**, *137*, 119–124. [CrossRef] [PubMed]
85. Kumar, S.; Stecher, G.; Tamura, K. MEGA7: Molecular Evolutionary Genetics Analysis Version 7.0 for Bigger Datasets. *Mol. Biol. Evol.* **2016**, *33*, 1870–1874. [CrossRef] [PubMed]
86. Meier, R.; Zhang, G.Y.; Ali, F. The Use of Mean Instead of Smallest Interspecific Distances Exaggerates the Size of the "Barcoding Gap" and Leads to Misidentification. *Syst. Biol.* **2008**, *57*, 809–813. [CrossRef]
87. Chen, S.L.; Yao, H.; Han, J.P.; Liu, C.; Song, J.Y.; Shi, L.C.; Zhu, Y.J.; Ma, X.Y.; Gao, T.; Pang, X.H.; et al. Validation of the ITS2 Region as a Novel DNA Barcode for Identifying Medicinal Plant Species. *PLoS ONE* **2010**, *5*, e8613. [CrossRef] [PubMed]
88. Feng, S.G.; Jiang, M.Y.; Shi, Y.J.; Jiao, K.L.; Shen, C.J.; Lu, J.J.; Ying, Q.C.; Wang, H.Z. Application of the Ribosomal DNA ITS2 Region of *Physalis* (Solanaceae): DNA Barcoding and Phylogenetic Study. *Front. Plant Sci.* **2016**, *7*, 1047. [CrossRef]
89. Carl, R. Woese Institute for Genomic Biology, University of Illinois at Urbana-Champaign. Pattern Analysis of Phylogenetic Trees Could Reveal Connections Between Evolution, Ecology. Available online: https://www.sciencedaily.com/releases/2020/06/200626125018.htm (accessed on 10 February 2025).
90. Somrup, S.; Sangsawang, A.; Mcmillan, N.; Winitchai, S.; Inthoncharoen, J.; Liu, S.K.; Muangmai, N. *Pinctada phuketensis* sp. nov. (Bivalvia, Ostreida, Margaritidae), a New Pearl oyster Species from Phuket, Western Coast of Thailand. *ZooKeys* **2022**, *1119*, 181–195. [CrossRef]

plants

Article

Field Work in Papua New Guinea Documents Seven New Records of a Hemiepiphytic Habit in Ferns

Michael Sundue [1,*] and Heveakore Maraia [2,3]

[1] Royal Botanic Garden Edinburgh, 20a Inverleith Row, Edinburgh EH3 5LR, UK
[2] Biology Center of the Czech Academy of Sciences, Institute of Entomology, Branišovská 1160/31, 370 05 České Budějovice, Czech Republic
[3] Department of Ecosystem Biology, Faculty of Science, University of South Bohemia, Branišovská 1645/31a, 370 05 České Budějovice, Czech Republic
* Correspondence: msundue@rbge.org.uk

Abstract: Hemiepiphytes have captured the attention of biologists since they seemingly hold clues to the evolution of epiphytes themselves. Hemiepiphytes are known to occur sporadically in the leptosporangiate ferns, but our understanding of their evolution remains limited by the relatively small number of detailed observations. This study adds to our knowledge by documenting seven species previously assumed to be holoepiphytes. This finding was based on fieldwork conducted in the Baining Mountains of Papua New Guinea that resulted in 319 collections representing 206 species. Approximately 3% of these species were hemiepiphytes: *Asplenium acrobryum*, *A. amboinense*, *A. scandens*, *A. scolpendropsis*, *Crepidomanes aphlebioides*, *Leptochilus macrophyllus*, and *Sphaerostephanos scandens*. All started growth as low-trunk epiphytes, and later, as larger climbing plants, exhibited strongly dimorphic roots consisting of short clasping ones that affixed the rhizome to the trunks and long feeding roots that entered the soil. Most of the seven hemiepiphyte species that we found exhibited distichous phyllotaxy and dorsiventrally flattened rhizomes, suggesting morphological convergence associated with this habit in four families. These new records suggest that large hemiepiphytic clades occur in *Asplenium* and *Leptochilus*. Our observations expand the geographic and taxonomic breadth of hemiepiphytic ferns, provide a baseline estimate of their diversity within a tropical flora, and offer morphological and phylogenetic clues to uncover additional records.

Keywords: epiphyte; morphological convergence; plant habit; New Britain

1. Introduction

Most ferns species are terrestrial understory perennial herbs, but a variety of other habits are known, including lithophytic, arborescent, vines, scandent disturbance colonizers, and epiphytes [1,2]. Epiphytes are particularly well developed in ferns; ferns comprise 3% of the world's vascular flora, but nearly 10% of all epiphytes [3]. Hemiepiphytes also occur in the ferns. They begin life as low-trunk epiphytes [4] but later connect to the ground via roots [5,6]. The recognition of hemiepiphytes in ferns dates back at least 20 years, probably longer, but many early authors did not apply the term as it is currently understood, and those records need to be reassessed [7–13]. Nitta and Epps [14] established a new wave of publications in their study of *Vandenboschia collariata* (Bosch) Ebihara & K. Iwats. by explicitly documenting its life history starting as gametophytes establishing on trunk bases and later connecting to the soil via roots. This inspired additional studies, and within the last fifteen years, about 10 new examples have been documented. In many of these cases, a holoepiphytic (on trees, disconnected from the ground) habit had been previously assumed, and the hemiepiphytic habit was only determined through careful field study [5,15–17]. In others, the habit was noted in the description of a new species [18–21]. These new records have captured the attention of biologists since they seemingly hold clues to the evolution of epiphytes themselves, with recent investigations asking whether hemiepiphytes have evolved from terrestrial, epiphytic, or epipetric ancestors [22–25]. To date, all three of these evolutionary histories have been supported. To further explore this topic, we searched for new hemiephytic records in Papua New Guinea, an understudied biodiversity hotspot [26].

2. Results

We found seven species of hemiepiphytes: *Asplenium acrobyrum*, *A. ambioinsene*, *A. scandens*, *A. scolopendropsis*, *Crepidomanes aphlebioides*, *Leptochilus macrophyllus*, and *Sphaerostephanos scandens*. In each species we observed the second and third criteria (see Section 4) used to diagnose them as hemiepiphytes; namely, that they make root contact with the soil, and the rhizomes are only found upon trees—never on the forest floor (Table 1).

Table 1. Habit criteria scored for the seven species of hemiepiphytes found in this study. Check marks indicate that the observation was confirmed.

Species	Plants Are Low-Trunk Epiphytes	Plants Never Seen Growing Terrestrially	Gametophytes or Young Plants Observed Upon Phorophyte	Roots Connected to Ground
Asplenium acrobryum	✓	✓	✓	✓
Asplenium amboinense	✓	✓		✓
Asplenium scandens	✓	✓		✓
Asplenium scolopendropsis	✓	✓	✓	✓
Crepidomanes aphlebioides	✓	✓		✓
Leptochilus macrophyllus	✓	✓		✓
Sphaerostephanos scandens	✓	✓		✓

In both *Asplenium amboinsense* and *A. scolopendropsis*, we also confirmed that gametophytes established upon the phorophyte (criterion number one). We observed root dimorphism in each of these species, which is one of the morphological traits we expect to observe in hemiepiphytes. All our hemiepiphytic species except *A. scandens* had distichous phyllotaxy (Table 2). Dorsiventrally flattened rhizomes were observed in *A. amboinense*, *A. scandens*, and *Leptochilus macrophyllus*; the rhizomes were terete or nearly so in *Asplenium scolopendropsis, Crepidomanes aphlebioides*, and *Sphaerostephanos scandens*. While all ferns exhibit, to some degree, heteroblastic leaf sequences from small to large leaves, we observed an abrupt transition from small to large in *A. scolopendropsis* after roots had contacted the soil. *Crepidomanes aphlebioides* exhibited leaf dimorphy, but we are uncertain of the relation between leaf dimorphy and soil contact of the roots in that species.

Table 2. Observed morphological characters of the hemiepiphytic habit. Check marks indicate that the observation was confirmed, and "no" indicates it was confirmed to not be the case.

Species	Gametophytes Non-Cordiform	Rhizomes Dorsiventrally Flattened	Roots Dimorphic with "Clasping" and "Feeder" Roots	Leaves with a Developmental Series That Changes upon Contact with the Ground	Phyllotaxis Distichous or Leaves All Emerging from One Side
Asplenium acrobryum		✓	✓		✓
Asplenium amboinense		✓	✓		✓
Asplenium scandens		✓	✓		no
Asplenium scolopendropsis	✓	no	✓	✓	✓
Crepidomanes aphlebioides		no	✓		✓
Leptochilus macrophyllus		✓	✓		✓
Sphaerostephanos scandens		no	✓		✓

Our expedition to the Baining Mountains resulted in 319 collections representing 206 species of ferns and lycophytes. Of these species, seven are hemiepiphytes; thus, this life form comprises at least 3% of the fern and lycophyte species diversity in the Baining Mountains. In contrast, 43 species (~21%) were interpreted as holoepiphytes. The remaining 156 (76%) were terrestrial with a variety of habits, including climbers, scandent species, and tree ferns.

2.1. Enumeration of Species

2.1.1. *Asplenium acrobryum* Christ. (Aspleniaceae)

Observations—*Asplenium acrobryum* (Figure 1A) is a short-creeping low-trunk hemiepiphyte with large simple arching leaves (Figure 1B) that spread away from the phorophyte. The rhizome bears dorsal leaves (Figure 1D) and lateral roots (Figure 1D). The roots are dimorphic, and both types of roots, clasping and feeding (Figure 1E), are abundant. The long feeder roots branched profusely upon contacting the soil (Figure 1F), but not before then. The leaves also bear subapical proliferous buds (Figure 1C).

Figure 1. *Asplenium acrobryum*. (**A**) Sporophyte. (**B**) Sori. (**C**) Subapical proliferous bud. (**D**) Rhizome with dense clasping roots. (**E**) Descending feeder roots on the trunk base. (**F**) Feeder roots removed from soil at contact point. (**A–F**) *Sundue & Maraia 4250* (VT).

Vouchers—East New Britain Province: Wild Dog Camp, former site of SINIVIT Wild Dog Mine, Bootsiqui trail, −4.62616 152.04406, 1012 m, *Sundue & Maraia 4250* (BISH, LAE, VT). Madang Province: Madang, Bundi, −5.759149 145.235891, 1800 m *Sundue 3773* (LAE, VT, UC). Madang Province: Madang, Bundi, −5.759149 145.235891, 1800 m *Sundue 3775* (LAE, VT, UC).

2.1.2. *Asplenium amboinense* Willd. (Aspleniaceae)

Observations—*Asplenium amboinense* (Figure 2A) juvenile plants (Figure 2B) were observed establishing as low-trunk epiphytes. Rhizomes are strongly dorsiventral with dorsal leaves and lateral roots (Figure 2D,E). The roots are conspicuously dimorphic, with short clasping roots (Figure 2E), and elongate feeder roots (Figure 2D). The simple and entire leaves spread from the phorophyte (Figure 2B,C) and bear proliferous buds at the lamina apices (Figure 2A).

Figure 2. *Asplenium amboinense*. (**A–C**) Sporophyte growing as a low-trunk epiphyte. (**B**) Young sporophyte establishing upon the phorophyte. (**D**) Sporophyte with feeder roots (black arrow). (**E**) Dorsiventral rhizome showing leaves departing from the upper surface and clasping roots (white arrow). (**F**) Mature leaf with sori. (**A–F**) *James & Sundue 1519*.

Vouchers—Northern Province: Akupe Camp, Kuriae River, Sibium Mountains, Umate Village., −9.285825 148.27266, 740 m, *James & Sundue 1519* (BISH, LAE, VT). Madang, Bundi; −5.758982, 145.186093, 2200 m, *Sundue 3825* (LAE, VT, UC).

2.1.3. *Asplenium scandens* J. Sm. (Aspleniaceae)

Observations—*Asplenium scandens* is a long-creeping low-trunk hemiepiphyte with divided leaves (Figure 3A,F) that spread away from the phorophyte (Figure 3B). The leaves are held in distichous arrangement, but close observation indicates that the phyllotaxis develops as helical. The rhizome is prominently dorsiventral (Figure 3D), with dorsal leaves and lateral roots (Figure 3D). The roots are sparsely produced, but strongly dimorphic with both clasping and feeding roots (Figure 3E).

Figure 3. *Asplenium scandens.* (**A**) Habit of sporophyte growing as low-trunk epiphyte. (**B**) Creeping rhizome with leaves directed toward the ventral side. (**C**) Ventral surface of rhizome showing short clasping roots and elongate feeder roots (arrow). (**D**) Apex of rhizome showing phyllotaxis. The light-green longitudinal lines on the leaves are aerophores. (**E**) Rhizome cross-section showing dorsiventral compression. (**F**) Detail of sori. (**A–F**) *Sundue & Maraia 4333* (VT).

Vouchers—East New Britain Province: Wild Dog Camp, former site of SINIVIT Wild Dog Mine, slopes above stream, −4.62616 152.04406, 1012 m, *Sundue & Maraia 4333* (BISH, LAE, VT). Madang. Bundi, 5°45′32.3958″S, 145°14′9.2076″E, 1800 m, *Sundue 3786* (LAE, VT, UC)

2.1.4. *Asplenium scolopendropsis* F. Muell. (Aspleniaceae)

Observations—*Asplenium scolopendropsis* gametophytes and juvenile plants (Figure 4A–D) were observed establishing on the lower portions of trunks. Rhizomes were long-creeping, with conspicuously distichous phyllotaxis (Figure 4E–H). The roots were dimorphic, with short clasping roots, and elongate feeder roots that connect to the soil. Leaf development exhibited strong heteroblasty, with the smallest (earliest produced) leaves simple and entire (Figure 4A), the next set of leaves deeply pinnatifid (Figure 4B–E), and the largest leaves proximally divided and distally entire (Figure 4G), or sometimes irregularly serrate (Figure 4H). Our observations indicate that the fully entire leaves characteristic of the mature plants do not develop until after roots contact the soil (Figure 4I).

Figure 4. *Asplenium scolopendropsis*. (**A**) Juvenile sporophyte developing from gametophyte with first leaf. (**B–D**) Sporophyte showing early leaf transition series. (**E**) Small sporophyte climbing phorophyte showing distichous leaf arrangement and transition from simple to divided leaves. (**F–H**) Mature sporophyte showing transition from divided to simple leaves. (**I**) Mature sporophyte. (**A–I**) *James & Sundue 1503* (VT).

Vouchers—Northern Province, Akupe Camp, Kuriae River, Sibium Mountains, Umate Village, −9.285825 148.27266, 740 m, *James & Sundue 1503* (BISH, LAE, VT); East New Britain Province: Wild Dog Camp, former site of SINIVIT Wild Dog Mine, Collected in ravine below Wild Dog Camp, −4.62616 152.04406, 1012 m, *Sundue & Maraia 4409* (BISH, LAE, VT).

2.1.5. *Crepidomanes aphlebioides* (Christ) I. M. Turner (Hymenophyllaceae)

Observations—*Crepidomanes aphlebioides* was seen growing as a short-creeping, low-trunk hemiepiphyte growing on both large (Figure 5A,B) and small (Figure 5C) trees. Plants were seen to have root dimorphy, and the feeder roots were observed to connect with the ground (Figure 5A arrows). *Crepidomanes aphlebioides* has dimorphic fronds, with short sessile (epetiolate) fronds, and larger petiolate fronds. The short fronds are highly dissected, with elongate filiform segments that spread in various directions (Figure 5C,E). In some cases, the short fronds cover the rhizome (Figure 5B). The larger fronds are planar and spread away from the rhizome (Figure 5D). Whether the development of these different frond types is related to the transition from holoepiphyte to hemiepiphyte was not determined by us.

Figure 5. *Crepidomanes aphlebioides*. (**A**) Rhizome with elongate feeder roots growing along phorophyte (arrows). (**B**). Rhizome covered by aphlebieae (reduced leaves). (**C**) Ventral surface of rhizome with mass of roots. (**D**) Mature sporophyte growing as low-trunk epiphyte. (**E**) Rhizome with aphlebieae and clasping roots (arrow); apex (not seen) toward the right. (**F**) Detail of frond. (**G**) Detail of frond with receptacle. (**A–G**) *Sundue & Maraia 4214* (VT).

Vouchers—East New Britain: Wild Dog Camp, former site of SINIVIT Wild Dog Mine, Omrock trail, −4.62616 152.04406, 1012 m, *Sundue & Maraia 4214* (BISH, LAE, VT).

2.1.6. *Leptochilus macrophyllus* (Blume) Noot.

Observations—*Leptochilus macrophyllus* occurred as a low-trunk hemiepiphyte (Figure 6A,B) The rhizome is long creeping and dorsiventrally compressed with dorsal-lateral leaves and lateral roots (Figure 6C,D). The rhizomes also bear lateral branch buds (Figure 6C), and some older populations seemed to be spreading by branching (Figure 6A). The roots were dimorphic with short clasping roots adhering to the phorophyte, and elongate feeder roots that connected with the soil. The simple entire leaves were arching and generally spreading away from the phorophyte. Besides living trees, we also observed this species on a fallen dead branch (Figure 6A); whether the *L. macrophyllus* established on a living or dead tree was unclear to us.

Figure 6. *Leptochilus macrophyllus*. (**A**,**B**) mature sporophyte growing as low-trunk epiphyte. (**C**) Ventral surface of dorsiventrally compressed rhizome with lateral clasping roots. (**D**) Dorsal surface of rhizome with distichous phyllotaxis. (**E**) Mature sori. (**A–E**) *Sundue & Maraia* 4398 (VT).

Vouchers—East New Britain: Wild Dog Camp, former site of SINIVIT Wild Dog Mine, Walk to Camp III at Regess, −4.67986 152.01701, 1355 m, *Sundue & Maraia 4398* (BISH, LAE, VT).

2.1.7. *Sphaerostephanos scandens* Holttum (Thelypteridaceae)

Observations—*Sphaerostephanos scandens* occurred as a low-trunk epiphyte on the root mantle of tree fern trunks (Figure 7A,B). The long-creeping rhizomes were radially symmetrical (Figure 7C), with short clasping roots and many elongate feeder roots (Figure 7D) connecting to the ground. The leaves emerged in spiral phyllotaxis and spread away from the phorophyte (Figure 7A,B,E). We found only a single population, all of which grew on tree fern trunks.

Figure 7. *Sphaerostephanos scandens.* (**A**,**B**) Sporophyte growing as a low-canopy epiphyte upon a tree fern trunk. (**C**) Rhizome apex showing radially symmetrical growth. The light-colored longitudinal lines are aerophores. (**D**) Rhizome with petioles departing petioles and elongate feeder roots. (**E**) Sporophyte removed from the phorophyte. (**F**) Abaxial lamina surface with mature sori (arrow). (**A**–**F**) *Sundue & Maraia 4420* (VT).

Vouchers—East New Britain: Wild Dog Camp, former site of SINIVIT Wild Dog Mine, Walk to Camp III at Regess, −4.67986 152.01701, 1355 m, *Sundue & Maraia 4420* (BISH, LAE, VT).

3. Discussion

Our results for the Baining Mountains indicate that at least 3% of the New Britain fern flora are hemiepiphytic. This percentage is small compared to the nearly 21% of species that were holoepiphytes, but an important finding because they represent a very different ecological niche. These new records occur in taxonomically disparate groups of eupolypod ferns—Aspleniaceae, Hymenophyllaceae, Polypodiaceae, and Thelypteridaceae; these families are not closely related to each other but show signs of morphological convergence to this habit. All species were found to have dimorphic roots (i.e., roots differentiated as either feeding or clasping roots), and all but *A. scandens* were found to have a distichous phyllotaxy (Table 2). Three of the seven species exhibited dorsiventrally flattened rhizomes, and two of them exhibited dramatic changes in leaf morphology throughout the leaf developmental series. These results reinforce the hemiepiphytic syndrome described by Testo and Sundue [5] and suggest that these traits can guide the discovery of other hemiepiphytic ferns.

Our results also reveal that both *Asplenium* and *Leptochilus* have clades comprised primarily of hemiepiphytic species. In *Asplenium*, the three species reported here were found to be closely related by Xu et al. [27]. They found a clade comprising *A. amboinsense, A. marattioides, A. scandens*, and *A. scolpendropsis*, which together were sister to other simple-leaved species, their "Neottopteris" clade. *Asplenium marattioides* has not been reported as a hemiepiphyte, but examination of *A. marattioides* herbarium specimens online (e.g., *Braithwaite 4528*, L; *Fawcett 687*, MICH; *Game 95/295*, VT) show traits characteristic of hemiepiphytes including a dorsiventral rhizome, root dimorphy, and strong morphological differences throughout the leaf developmental series. It should be targeted for future study. Species similar to *A. scolopendropsis* such as *A. schizocarpum*, and *A. translucens* should also be targeted. Xu et al. [27] also reported *A. simplicifrons* as belonging to this clade, but it is epipetric and holoepiphytic. However, the sample Xu et al. [27] used was reported to be from Papua New Guinea, where *A. simplicifrons* does not occur. Given that it was nested within two samples of *A. amboinense* in their results, it seems likely that it is a misidentified sample of that species. Ohlsen et al. [28] sampled *A. simplicifrons* from near the type locality and resolved it in an unrelated clade.

The recognition of a hemiepiphytic clade in *Asplenium* provides an opportunity to infer the evolution of habit in these clades. The hemiepiphytic clade is sister to the *Asplenium nidus* group (bird's nest ferns) *sensu lato* in Xu et al. [27]. That clade is large and variable but is mostly epiphytic. Sister to both of these are the *A. scolopendrioides* clade and then the *A. ceterach* clade, which are terrestrial or epipetric. Although there is a great deal of disparity between these clades, the topology does indicate a transition from an ancestrally terrestrial/epipetric habit diverging into hemiepiphytic and holoepiphytic sister groups.

Our results also reveal phylogenetic patterns in *Leptochilus*. Three other species have been reported as hemiepiphytes, namely, *L. brevipes*, *L. ornithopus*, and *L. scandens*. Together, these reports suggests that hemiepiphytes might be widespread in the genus but overlooked. In addition, other evidence suggests that *Leptochilus* may often be facultatively hemiepiphytic. Yu et al. [21] reported that *L. brevipes* grows both hemiepiphytically and epipetrically, and Chen et al. [25] reported 15 species of *Leptochilus* as growing terrestrially in their phylogenetic reconstruction of habit, but several of these species are also reported as epiphytes [29]. Chen et al. [25] did not distinguish hemiepiphytic from terrestrial habits and no justification or evidence in support of the habits was provided. The discrepancy in reported habits suggest that some of those species may in fact be facultatively hemiepiphytic.

Crepidomanes aphlebioides was previously reported as hemiepiphytic by Hennequin et al. [8] but without documentation. Our results confirm that the species establishes on trees and then connects to the soil via elongate feeder roots. However, collection notes would indicate that at the southern edge of its range in Queensland, Australia, it grows epipetrically (*Gray 8183*, MELU; *Jones 18611*, MELU; *Ohlsen & Field s.n.* MELU). This raises the question of how and when habits shift. In their phylogenetic results, Hennequin et al. [8] resolved *C. aphlebioides* as sister to a clade of terrestrial species including *C. intermedium* (Bosch) Ebihara & K. Iwats., *C. grande* (Copel.) Ebihara & K. Iwats, and *C. thysanostomum* (Makino) Ebihara & K. Iwats. Other closely related clades comprised epiphytic species, and *Vandenboschia*, which includes several hemiepiphytes [7,8,14]. The phylogenetically clustered distribution of terrestrial and hemiephytic species in their results implies a frequent shift between these two habits evolving from an epiphytic ancestral habit.

When he described *Sphaerostephanos scandens*, Holttum [30] reported it as an epiphyte, but he may not have ever observed the species himself, and he may not have considered the distinction between holoepiphytic and hemiepiphytic habits in ferns. He did, however, describe it as climbing, which is consistent with our observations. Our interpretation that this species is a hemiepiphyte is noteworthy given that the other estimated 1200 species of Thelypteridaceae are reportedly found on soils, rocks, or in wetlands. This would suggest that *S. scandens* most likely evolved from terrestrial ancestors.

Holttum suggested that *S. scandens* is related to *S. invisus* (Forst. f.) Holttum and *S. mundas* (Rosenst.) Holttum (as *Dryopteris farinosa* Brause) and "a few other species in New Guinea which have climbing rhizomes", but without further details. In their phylogenomic analysis, Fawcett et al. [31] supported the close relationship of *S. scandens* and a plant identified as *S.* aff. *mundas* but resolved *S. invisus* within the distantly related *Strophocaulon*. Whether a clade of hemiepiphytes remains to be discovered within *Sphaerostephanos* requires further investigation.

4. Materials and Methods

Field work was primarily conducted during a 2016 expedition to the Baining Mountains in East New Britain, but also draws from observations made in two other expeditions, a 2013 expedition to the Sibium Mountains in Northern (Oro) Province, and a 2014 expedition to Mt. Wilhelm in Chimbu and Madang provinces. Observations were made from one or more populations in the field and further augmented with herbarium study. When possible, all life stages were photographed in situ.

Documenting a hemiepiphytic habit is undertaken based on three distinct established criteria. The first is that gametophytes establish epiphytically on trees. Gametophyte identification can be confirmed by finding thalli persisting on juvenile sporophytes. This is conducted by searching tree trunks for successively younger sporophytes during which a developmental series of leaf morphology is discovered, allowing for the identification of the youngest sporelings. These youngest plants often retain gametophyte thali. The second criterion is that sporophytes must be observed to contact the soil via roots. This is conducted by carefully removing roots from bark of the phorophyte (host tree) and following them to the forest floor. The roots of holoepiphytes, in contrast, have roots that do not contact the soil. The third criterion required is that sporophytes never establish on the ground. This observation serves to distinguish obligate hemiepiphytes from facultative hemiepiphytes. Previous studies emphasized the need to distinguish primary hemiepiphytes from secondary hemiepiphytes; however, we agree with Zotz et al. [13] that secondary hemiepiphytes have never clearly been demonstrated for ferns or any other plant group. Therefore, "secondary hemiepiphytes" are not further considered here.

The gold standard for establishing a species as a hemiepiphyte is the positive observation of all three criteria. However, observing only the second and third criteria is sufficient in many cases. Given that sporophytes can only establish where a gametophyte had been present, the presence of a sporophyte implies where the spore germinated, and the gametophyte began growth. We believe this is a safe assumption to make for many small to medium-sized plants. However, we do not recommend making this assumption in the case of plants that have been established for long periods of time, with elongate and complicated rhizome systems such as those observed by Gay [32] for *Mickelia guianensis* (Aublet) R. C. Moran, Sundue & Labiak, or which may be climbing or scandent, as they could easily mislead observers.

Beyond these criteria, we also looked for morphological features that are expected to occur in hemiepiphytes, as described in Testo & Sundue [33]; specifically, non-cordiform (i.e., filamentous, ribbon, or strap-shaped) gametophytes, dorsiventrally flattened rhizomes, a distinctive leaf developmental series that changes with root–soil contact, and root dimorphism where the plants have distinctly short "clasping" roots that serve to attach the rhizome to the phorophyte and elongate "feeder" that descend to the ground.

Observations were made concurrently with a biological inventory of the ferns in the Sibium Mountains. Collections were opportunistic and not designed with a statistical methodology; nonetheless, we report the total number of collections and the total number of confirmed hemiepiphytes to provide a baseline quantification of the percentage of this habit in a tropical fern flora.

Specimens were identified through direct comparison of collections at BM, E, GH, K, UC, and VT, or through comparison of images available through www.pteridoportal.org. We also relied upon relevant taxonomic literature [33–40].

5. Conclusions

This study advances our understanding of fern ecology by expanding the taxonomic breadth and number of ferns known to grow hemiepiphytically. We found that approximately 3% of fern species in the Baining Mountains were hemiepiphytes. Our records in *Asplenium* and *Leptochilus* indicate a strong phylogenetic pattern in those genera. Additionally, our findings of hemiepiphytism in *Sphaerostephanos* (Thelypteridaceae) and *Crepidomanes* (Hymenophyllaceae) indicate that other species in those genera should be examined for further instances of hemiepiphytism.

Author Contributions: Conceptualization and methodology, M.S. and H.M.; investigation, resources, and data curation, H.M. and M.S.; writing—original draft preparation, M.S.; review and editing, H.M. All authors have read and agreed to the published version of the manuscript.

Funding: This research was funded by the United States National Science Foundation, NSF DEB-2045325.

Data Availability Statement: Original photographs for all collections are freely available via www.fernsoftheworld.com.

Acknowledgments: We would like to thank Allen Allison of the Bernice P. Bishop Museum for including us in the Baining Mountain expedition, and MS on the Sibium Mountains expedition. Michael Kessler from the University of Zurich is thanked for including MS on the Mt. Wilhelm expedition. We acknowledge Thomas Magun and Bernard Sule of PNG Forest Research Institute, LAE herbarium, for assisting us in mounting and cataloguing our field-preserved materials. Cheng-Wei Chen (TAIF), Susan Fawcett (UC), and Alan Smith (UC) assisted with some specimen identifications. Robbin Moran, Sabine Hennequin, and two anonymous reviewers provided excellent feedback that helped improve the manuscript.

Conflicts of Interest: The authors declare no conflicts of interest.

References

1. Tryon, R.M.; Tryon, A.F. *Ferns and Allied Plants: With Special Reference to Tropical America*; Springer Science & Business Media: New York, NY, USA, 2012.
2. Shang, H.; Sundue, M.; Wei, R.; Wei, X.P.; Luo, J.J.; Liu, L.; Schwartsburd, P.B.; Yan, Y.H.; Zhang, X.C. Hiya: A new genus segregated from *Hypolepis* in the fern family Dennstaedtiaceae, based on phylogenetic evidence and character evolution. *Mol. Phylogenetics Evol.* **2018**, *127*, 449–458. [CrossRef] [PubMed]
3. Sundue, M.A.; Testo, W.L.; Ranker, T.A. Morphological innovation, ecological opportunity, and the radiation of a major vascular epiphyte lineage. *Evolution* **2015**, *69*, 2482–2495. [CrossRef] [PubMed]
4. Watts, J.L.; Moran, R.C.; Watkins, J.E., Jr. *Hymenasplenium volubile*: Documentation of its gametophytes and the first record of a hemiepiphyte in the Aspleniaceae. *Ann. Bot.* **2019**, *124*, 829–835. [CrossRef] [PubMed]
5. Testo, W.; Sundue, M. Primary hemiepiphytism in *Colysis ampla* (Polypodiaceae) provides new insight into the evolution of growth habit in ferns. *Int. J. Plant Sci.* **2014**, *175*, 526–536. [CrossRef]
6. Benzing, D.H. Epiphytism: A preliminary overview. In *Vascular Epiphytes: General Biology and Related Biota*; Ashton, P.S., Hubbell, S.P., Janzen, D.H., Marshall, A.G., Raven, P.H., Tomlinson, P.B., Eds.; Cambridge University Press: Cambridge, UK, 1990; pp. 1–42.
7. Dubuisson, J.Y.; Hennequin, S.; Rakotondrainibe, F.; Schneider, H. Ecological diversity and adaptive tendencies in the tropical fern *Trichomanes* L. (Hymenophyllaceae) with special reference to climbing and epiphytic habits. *Bot. J. Linn. Soc.* **2003**, *142*, 41–63. [CrossRef]
8. Hennequin, S.; Schuettpelz, E.; Pryer, K.M.; Ebihara, A.; Dubuisson, J.Y. Divergence times and the evolution of epiphytism in filmy ferns (Hymenophyllaceae) revisited. *Int. J. Plant Sci.* **2008**, *169*, 1278–1287. [CrossRef]
9. Kreier, H.P.; Zhang, X.C.; Muth, H.; Schneider, H. The microsoroid ferns: Inferring the relationships of a highly diverse lineage of Paleotropical epiphytic ferns (Polypodiaceae, Polypodiopsida). *Mol. Phylogenetics Evol.* **2008**, *48*, 1155–1167. [CrossRef]
10. Tsutsumi, C.; Kato, M. Rhizome morphology and patterns of leaf production of secondary hemiepiphytic *Oleandra pistillaris* (Oleandraceae). *Bull. Natl. Mus. Nat. Sci. Ser. B* **2010**, *36*, 21–25.
11. Canestraro, B.K.; Moran, R.C.; Watkins, J.E., Jr. Reproductive and physiological ecology of climbing and terrestrial *Polybotrya* (Dryopteridaceae) at the La Selva Biological Station, Costa Rica. *Int. J. Plant Sci.* **2014**, *175*, 432–441. [CrossRef]

12. Zotz, G. 'Hemiepiphyte': A confusing term and its history. *Ann. Bot.* **2013**, *111*, 1015–1020. [CrossRef]
13. Zotz, G.; Almeda, F.; Bautista-Bello, A.P.; Eskov, A.; Giraldo-Cañas, D.; Hammel, B.; Harrison, R.; Koester, N.; Krömer, T.; Lowry, P.P., II; et al. Hemiepiphytes revisited. *Perspect. Plant Ecol. Evol. Syst.* **2021**, *51*, 125620. [CrossRef]
14. Nitta, J.H.; Epps, M.J. Hemi-epiphytism in *Vandenboschia collariata* (Hymenophyllaceae). *Brittonia* **2009**, *61*, 392–397. [CrossRef]
15. Lagomarsino, L.P.; Grusz, A.L.; Moran, R.C. Primary hemiepiphytism and gametophyte morphology in *Elaphoglossum amygdalifolium* (Dryopteridaceae). *Brittonia* **2012**, *64*, 226–235. [CrossRef]
16. Fawcett, S.; Sundue, M. Evidence of primary hemiepiphytism in *Pleopeltis bradeorum* (Polypodiaceae). *Brittonia* **2016**, *68*, 187–194. [CrossRef]
17. Labiak, P.H.; Matos, F.B.; Rouhan, G.; Hanks, J.G.; Moran, R.C. Notes on the Taxonomy and Growth Habits of Three Species of *Campyloneurum* (Polypodiaceae) from Southeastern Brazil. *Am. Fern J.* **2017**, *107*, 1–20. [CrossRef]
18. Adjie, B.; Kurniawan, A.; Sahashi, N.; Watano, Y. *Dicksonia timorense* (Diksoniaceae), a hemi-epiphytic new species of tree fern endemic on Timor Island, Indonesia. *Reinwardtia* **2012**, *13*, 357–362.
19. Fujiwara, T.; Quang, B.H.; Tagane, S.; Murakami, N.; Oguri, E. *Leptochilus ornithopus* (Polypodiaceae), a new hemiepiphytic fern species from central highlands of Vietnam. *Phytotaxa* **2023**, *584*, 149–160. [CrossRef]
20. Wei, H.J.; Huang, Y.; Chen, B. *Leptochilus scandens* (Polypodiaceae), a new microsoroid fern species from Guangdong, China. *Phytotaxa* **2023**, *618*, 79–85. [CrossRef]
21. Yu, Z.-Y.; Liang, Z.-L.; Hu, Y.-P.; Liang, Z.-L.; He, H.-L.; Li, J.-H.; Zhang, L. *Leptochilus brevipes* (Polypodiaceae), a new fern species from southeastern Yunnan, China based on morphological and molecular evidence. *Phytotaxa* **2024**, *634*, 143–152. [CrossRef]
22. Moran, R.C.; Labiak, P.H.; Sundue, M. Phylogeny and character evolution of the bolbitidoid ferns (Dryopteridaceae). *Int. J. Plant Sci.* **2010**, *171*, 547–559. [CrossRef]
23. Watkins, J.E., Jr.; Cardelús, C.L. Ferns in an angiosperm world: Cretaceous radiation into the epiphytic niche and diversification on the forest floor. *Int. J. Plant Sci.* **2012**, *173*, 695–710. [CrossRef]
24. Testo, W.; Sundue, M. A 4000-species dataset provides new insight into the evolution of ferns. *Mol. Phylogenetics Evol.* **2016**, *105*, 200–211. [CrossRef] [PubMed]
25. Chen, C.C.; Hyvönen, J.; Schneider, H. Re-terrestrialization in the phylogeny of epiphytic plant lineages: Microsoroid ferns as a case study. *J. Syst. Evol.* **2023**, *61*, 613–626. [CrossRef]
26. Cámara-Leret, R.; Frodin, D.G.; Adema, F.; Anderson, C.; Appelhans, M.S.; Argent, G.; Arias Guerrero, S.; Ashton, P.; Baker, W.J.; Barfod, A.S.; et al. New Guinea has the world's richest island flora. *Nature* **2020**, *584*, 579–583. [CrossRef]
27. Xu, K.W.; Zhang, L.; Rothfels, C.J.; Smith, A.R.; Viane, R.; Lorence, D.; Wood, K.R.; Chen, C.W.; Knapp, R.; Zhou, L.; et al. A global plastid phylogeny of the fern genus *Asplenium* (Aspleniaceae). *Cladistics* **2020**, *36*, 22–71. [CrossRef] [PubMed]
28. Ohlsen, D.J.; Perrie, L.R.; Shepherd, L.D.; Brownsey, P.J.; Bayly, M.J. Phylogeny of the fern family Aspleniaceae in Australasia and the south-western Pacific. *Aust. Syst. Bot.* **2015**, *27*, 355–371. [CrossRef]
29. Zhang, X.-C.; Nooteboom, H.P. Leptochilus. In *Flora of China*; Wu, Z.-Y., Raven, P.H., Hong, D.-Y., Eds.; Science Press: Beijing, China; Missouri Botanical Garden Press: St. Louis, MO, USA, 2013; Volume 2–3, pp. 833–838.
30. Holttum, R.E. The family Thelypteridaceae in the Pacific and Australasia. *Allertonia* **1977**, *1*, 169–234.
31. Fawcett, S.; Smith, A.R.; Sundue, M.; Burleigh, J.G.; Sessa, E.B.; Kuo, L.Y.; Chen, C.W.; Testo, W.L.; Kessler, M.; Barrington, D.S. A global phylogenomic study of the Thelypteridaceae. *Syst. Bot.* **2021**, *46*, 891–915. [CrossRef]
32. Gay, H. The architecture of a dimorphic clonal fern, *Lomagramma guianensis* (Aublet) Ching (Dryopteridaceae). *Bot. J. Linn. Soc.* **1993**, *111*, 343–358. [CrossRef]
33. Chen, C.W.; Perrie, L.; Glenny, D.; Chiou, W.L. *Sol Amazing: Lycophytes & Ferns of the Solomon Islands*; National Museum of Natural Science: Taichung City, Taiwan, 2017.
34. Chen, C.W.; Perrie, L.; Glenny, D.; Chiou, W.L.; Fawcett, S.; Smith, A.R.; Parris, B.S.; Ebihara, A.; Ohlsen, D.; Lehtonen, S.; et al. An annotated checklist of lycophytes and ferns of the Solomon Islands. *Fern Gaz.* **2022**, *21*, 292–497.
35. Johns, R.J.; Edwards, P.J.; Utteridge, T.M.A.; Hopkins, H.C.F. *A Guide to the Alpine and Subalpine Flora of Mount Jaya*; Royal Botanic Gardens, Kew: Richmond, UK, 2006.
36. Piggott, A.G. *Ferns of Malaysia in Colour*; Tropical Press: Kuala Lumpur, Malaysia, 1988.
37. Kato, M. *Illustrated Flora of Ferns & Fern Allies of South Pacific Islands*; National Museum of Nature and Science (Tokyo): Tokyo, Japan; Tokai University Press: Shizuoka, Jpan, 2008.
38. Holttum, R.E. *A Revised Flora of Malaya Volume II Ferns of Malaya*; US Government Printing Office: Washington, DC, USA, 1968.
39. Nooteboom, H.P. *Flora Malesiana. Series II, Ferns and Fern Allies*; National Herbarium of The Netherlands: Leiden, The Netherlands, 2012; Volume 4.
40. Holttum, R.E. *Flora Malesiana, Series II: Pteridophyta, Volume 2, Part 5, Thelypteridaceae*; American Fern Society: St. Louis, MO, USA, 1983.

plants

Article

Exploring Co-Occurrence Patterns to Understand Epiphyte–Liana Interactions

Sergio J. Ceballos [1,2,*], Ezequiel Aráoz [1,2] and Tobías Nicolás Rojas [1,3]

1 Instituto de Ecología Regional (IER), Universidad Nacional de Tucumán (UNT)-Consejo Nacional de Investigaciones Científicas y Técnicas (CONICET), Yerba Buena 4107, Tucumán, Argentina; ezequielaraoz@gmail.com (E.A.); tobiasnrojas@gmail.com (T.N.R.)

2 Facultad de Ciencias Naturales e Instituto Miguel Lillo, Universidad Nacional de Tucumán, San Miguel de Tucumán 4000, Tucumán, Argentina

3 Mueller Laboratory, Biology Department, The Pennsylvania State University, University Park, PA 16802, USA

* Correspondence: serceballos@gmail.com

Abstract: Although epiphytes and lianas share the same habitat, most research has treated these two groups independently. This study aimed to evaluate the co-occurrence of vascular epiphytes and lianas in the subtropical montane forests of northwestern Argentina. We recorded epiphyte cover and liana basal area on trees $\geq$ 10-cm-dbh in 120 20 $\times$ 20 m plots in the Sierra de San Javier (Tucumán, Argentina). Of the 2111 trees sampled, 727 (34%) hosted lianas, and 1095 (52%) hosted epiphytes. Both plant groups were found together on 20% of the sampled trees. The species richness of lianas and epiphytes, along with liana basal area and epiphyte cover, increased with tree diameter and reached higher values in mature forests compared to successional forests. Both groups colonized the same canopy tree species with larger diameters, whereas smaller trees were typically colonized by either lianas or epiphytes, but not both. Epiphyte species were more likely to co-occur with liana species with specialized climbing mechanisms. Tree size and forest type (mature vs. successional) emerged as key factors influencing the co-occurrence of lianas and epiphytes in these forests. This study establishes a basis for future research into the interactions between lianas and epiphytes, seeking to determine whether they co-occur in the same habitats.

Keywords: canopy ecology; competition; species interactions; trees; Yungas

1. Introduction

Trees have outstanding ecological value in supporting the biodiversity of the forest canopy [1,2]. In particular, for lianas and epiphytes, trees play a fundamental role, by providing support and a diversity of microhabitats [3,4]. Lianas climb tree trunks in search of light, while epiphytes germinate and root on trees without parasitizing them [2,5–7]. Although both plant groups rely on trees for structural support, our understanding of how lianas and epiphytes interact remains limited. Despite sharing the same trees, most research has examined these plant groups separately, focusing on liana–tree or epiphyte–tree relationships [2,8,9]. Consequently, little is known about whether lianas and epiphytes compete against or facilitate each other on shared trees.

Lianas and epiphytes exhibit temporal segregation during forest succession. In tropical forests, the abundance and richness of these two plant groups increase at different stages of succession in a predictable pattern [4,10]. Specifically, in forests regenerating from abandoned agricultural fields, pastures, or logged areas, lianas colonize rapidly during the first 30 years of forest recovery [10–13]. Lianas favor the conditions typical of early successional forests, such as high light availability, low canopy height, and the abundance of small-diameter trees suitable for climbing [10,11]. As forests mature, liana abundance decreases, while epiphyte abundance increases [14–16]. Epiphytes colonize more slowly during succession, becoming more prevalent in older forests where large-diameter trees are more common [4,17,18]. These larger trees provide more surface, microhabitats, and longer time exposure for epiphyte colonization [9,17,19].

In addition to temporal segregation, lianas and epiphytes may also segregate spatially due to differences in their microhabitat niches [20,21]. For instance, hemiepiphytes have been observed to thrive better in moist sites compared to lianas [21–23]. Conversely, lianas often form dense clumps around trees at the edges of dry forests, producing large amounts of foliage above their hosts and leaving limited space for epiphytic plants [20]. Interactions with other organisms could also contribute to the segregation of lianas and epiphytes. For example, in a Bornean tropical rainforest, a species of ant that forms symbiotic relationships with two epiphytic ferns actively prunes and removes lianas [24]. However, research remains limited, and general mechanisms explaining the interactions between lianas and epiphytes beyond their temporal and spatial segregation are yet to be established.

As in other regions, lianas and epiphytes in the subtropical montane forests of Argentina have been studied separately [25,26]. Research in this area has demonstrated that both lianas and epiphytes are significant components of biodiversity, particularly abundant in mature forests [9,27]. A study performed in these forests showed that large and well-lit trees were more likely to support lianas and hosted a greater liana abundance than small and shaded trees [27]. Similarly, another study revealed that larger-diameter trees supported more vascular epiphyte species than smaller-diameter trees in the same area [9]. Additionally, the interactions between these plant groups and host tree characteristics have been analyzed, but only in mature forests [9,27]. Given that both groups are abundant in similar habitats, these subtropical forests provide an excellent setting for investigating their interactions.

To better understand the interactions between lianas and epiphytes, it is important to determine whether they share the same host tree species and coexist on the same trees within the same forests. In this study, we combined two datasets of lianas and epiphytes collected from the same mature and successional forest plots as an initial step to investigate their interactions. The objective of this study was to assess the co-occurrence of vascular epiphytes and lianas—plants that depend on trees for support—in the subtropical montane forests of northwestern Argentina. Specifically, we aimed to carry out the following actions: 1. Compare the occurrence of lianas and epiphytes at three ecological scales (i.e., tree zone, tree, and forest patch). 2. Test whether forest type (i.e., successional vs. mature) or tree size explains the species richness of lianas and epiphytes, liana basal area, and epiphyte cover. 3. Explore co-occurrence patterns among liana, epiphyte, and tree species using interaction networks. 4. Analyze the spatial association of trees hosting lianas and epiphytes. We expected the following outcomes: 1. Lianas and epiphytes will coexist on the same trees and in the same areas of each tree, though their occurrence will vary across different forest patches. 2. Since both lianas and epiphytes require space on trees to grow, the richness and abundance of these groups will increase on large-diameter trees in mature forests. 3. Lianas and epiphytes will interact with the same set of tree species. 4. Trees hosting lianas will be spatially associated with those hosting epiphytes.

2. Materials and Methods

2.1. Study Area

We conducted this study in Parque Sierra de San Javier (26°46′3.5″ S, 65°29′22.5″ W), a 14,000 ha protected area located between 600 and 1800 m asl in Tucumán, northwest Argentina (Figure 1). The area lies 13 km west of Gran San Miguel de Tucumán, a city of approximately one million inhabitants, whose influence has shaped the land cover and forest dynamics of Sierra de San Javier over the past centuries [28]. Historically, the main activities in the Sierra included agriculture and livestock grazing, which replaced previously forested areas. However, the establishment of the protected area in 1973, along with rural outmigration and reduced land use intensity, has transformed the landscape into a mosaic of shrublands and successional forests [28,29].

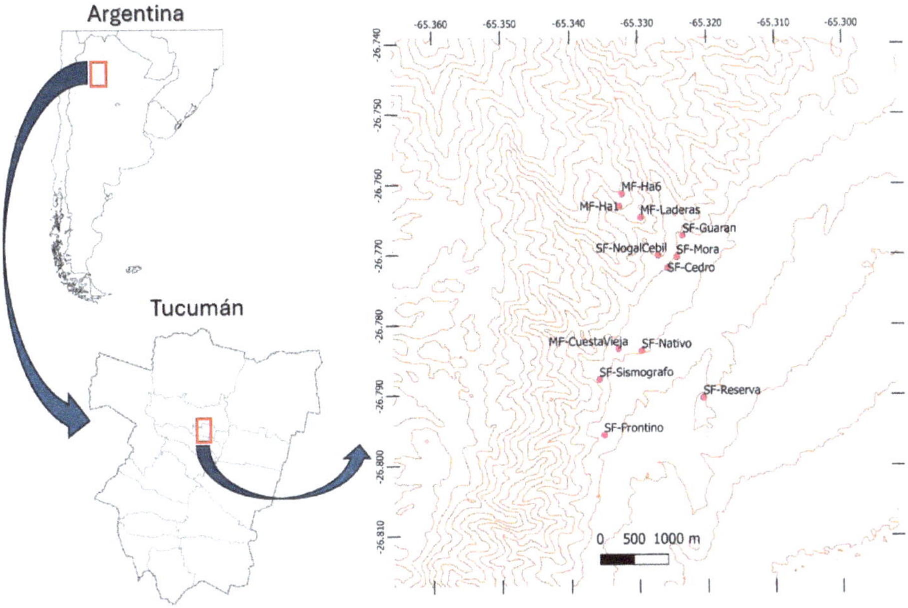

Figure 1. Distribution of the 12 forest patches in the Sierra de San Javier, Tucumán, Argentina. SF and MF are successional and mature forests, respectively. Red boxes indicate the detailed areas shown in the subsequent sub-figure.

The study area has a subtropical climate characterized by a pronounced dry season from April to October and occasional winter frosts [30]. Mean annual precipitation ranges from 1300 to 1500 mm, and the mean annual temperature is 18 °C [31].

The vegetation corresponds to subtropical montane forests, representing the southernmost extension of Neotropical Andean forests [32]. This study focused on low mountain forests (700–1000 m asl), including mature and successional forests. Mature forests are dominated by tree species such as *Blepharocalyx salicifolius*, *Ocotea porphyria*, and *Pisonia zapallo*, while the subcanopy is primarily composed of *Eugenia uniflora*, *Piper tucumanum*, and *Allophylus edulis* [27,33]. Successional forests, which have developed mainly on abandoned croplands, contain native tree species such as *Tecoma stans*, *Heliocarpus popayanensis*, *Parapiptadenia excelsa* and *Tipuana tipu*, along with exotic tree species such as *Ligustrum lucidum* and *Morus alba* [29].

In the Sierra de San Javier, both lianas and epiphytes are abundant and diverse in the Yungas forests [34,35]. They occur on approximately 62–65% of trees with a diameter greater than 10 cm [25–27]. In the area, mature forests provide optimal habitats for lianas and epiphytes due to the presence of treefall gaps and large-diameter trees [34,36]. However, these plants are also present in successional forests that have regenerated from abandoned lands that were once used for farming and cattle grazing. The most abundant liana species are *Cissus striata*, *Celtis iguanaea*, and *Chamissoa altissima* [37]. The most common epiphyte species are *Pleopeltis tweediana*, *Microgramma squamulosa*, and *Aechmea distichantha* [34].

2.2. Data Collection

For this study, we considered the following definitions for lianas and epiphytes. Lianas are climbing plants with true wood or persistent, fibrous stems, which germinate on the ground and rely on external physical support to ascend to the canopy [38]. Epiphytes are plants that germinate and root non-parasitically on other plants [2]. The study area lacks hemiepiphytes and nomadic vines, so there was no ambiguity in distinguishing between lianas and epiphytes.

Lianas, vascular epiphytes, and trees were surveyed in 12 forest patches located at elevations between 700 and 1000 m above sea level in Parque Sierra de San Javier (Figure 1). Our study included eight successional forest patches, along with four mature forests that did not show recent signs of human disturbance [29,34,35].

In every forest patch we sampled ten 20 × 20 m plots. Most plots were spaced 20 m apart and located more than 10 m from the forest edge. An exception was one forest patch ("Cedro", Figure 1), which included nine contiguous plots (0.36 ha) monitored every 5 years by the Red Subtropical de Parcelas Permanentes (RedSPP) (https://ier.conicet.gov.ar/red-subtropical-de-parcelas-permanentes-redspp/ (accessed on 28 November 2024)) and one additional plot located 20 m away. Within each plot, we identified and measured the diameter at breast height (dbh) of all trees $\geq$ 10 cm. In the Cedro forest patch, we also mapped the location of each tree using an x–y coordinate system to perform spatial analyses in the nine contiguous plots.

In each plot, we identified and measured the diameter of lianas supported by trees, focusing on woody climbers with a diameter of $\geq$1 cm at 130 cm from the main rooting point [38]. Additionally, we identified vascular epiphyte species and estimated their total cover per tree using binoculars from the ground. We acknowledge that ground-based observations may underestimate the richness and frequency of epiphytes [39]. However, we selected this method to sample a large number of trees in an area with relatively low epiphyte richness [26]. Cover was estimated using a modified Braun-Blanquet scale, considering the percentage of the tree surface covered by epiphytes (1 = 1 to 5%, 2 = 6 to 25%, 3 = 26 to 50%, 4 = 51 to 75%, and 5 = 76 to 100%) [40]. Additionally, we recorded the number of occurrences of lianas and epiphytes on the crowns and trunks of the trees.

2.3. Data Analysis

We compared the occurrence of lianas and epiphytes across three ecological scales: tree zone, tree, and forest (objective 1). At the tree zone scale, we compared the frequency of occurrence of lianas and epiphytes between the trunk and crown of trees using generalized linear models (GLMs) with a negative binomial distribution. At the tree scale, we used Chi-square tests to analyze the associations between the presence and absence of lianas and epiphytes in trees of four different size classes. At the forest scale, we compared the number of trees colonized by lianas or epiphytes across 12 forest patches using generalized linear models (GLMs) with a negative binomial distribution.

To analyze the effects of forest type (i.e., successional or mature) and tree size (dbh) on the richness of lianas and epiphytes, liana basal area, and epiphyte cover, we performed four Bayesian mixed-effects linear regressions (objective 2). In these models, forest type and tree size were the explanatory variables, and tree species was set as a random variable. This way, we developed a model for each response variable (i.e., liana basal area, liana richness, epiphyte richness, and epiphyte cover). Liana abundance posteriors were estimated using a log-normal hurdle distribution, due to the high number of zeros in the sampled trees. For the abundance of epiphytes, we used cumulative distribution, since the response variable is ordinal. For the richness of lianas and epiphytes, we used a hurdle Poisson distribution. In all cases, we ran four Markov chains of 8000 iterations each. The priors were weakly informative in each of the four models (Supplementary Materials S1). All models converged successfully and reached acceptable Rhat and effective sample size (ESS) values (Rhat < 1.01 and ESS > 1000; Supplementary Materials S1).

We explored the relationships between liana and tree species, epiphytes and trees, and lianas and epiphytes using interaction networks (objective 3). To describe the frequency of interactions between lianas and trees, we considered the total basal area of liana species associated with each tree species. For the epiphyte–tree interaction network, we described the frequency of interactions based on the number of zones [41] occupied by each epiphyte species on each tree species. For the liana–epiphyte interaction network, we only included trees where both lianas and epiphytes occurred together. All interaction networks were generated using the bipartite library in R [42]. In addition, we compared the number of epiphyte co-occurrences with lianas that use specialized climbing mechanisms versus those that merely lean on tree branches, using a generalized linear model with a negative binomial distribution.

We conducted a point pattern analysis to investigate the spatial association of trees hosting lianas or epiphytes, at different distances (objective 4). This analysis was carried out within the 0.36 ha permanent plot ("Cedro"), where the trees were mapped using an x- and y-coordinate system. The input for our analyses included tree locations, as well as the presence of lianas and epiphytes, as categorical marks. To assess these spatial patterns, we employed the K function, which examines the frequency of additional points (e.g., a tree) at increasing distances (radii) from each focal point. To reduce the effect of autocorrelation due to cumulative counts of additional points, we only considered points within two-meter-wide rings. To do so, the K value of each radius was recalculated by subtracting the K value corresponding to a two-meter-lower radius. We analyzed the spatial pattern of marked points (e.g., trees with lianas or epiphytes) and the frequency of crossed interactions (i.e., we evaluated potential interactions between lianas and epiphytes at different distances). We contrasted the observed pattern with a null model using Monte Carlo permutations. Specifically, we conducted 10,000 simulations to establish confidence intervals defined by maximum and minimum values at a significance level of $\alpha = 0.05$. If the observed pattern falls within the confidence intervals, it suggests that the distribution of trees with lianas and epiphytes is consistent with the noninteraction hypothesis. Conversely, if the observed pattern deviates from these intervals, it indicates that the probability of finding a tree with epiphytes or lianas at different distances either increases or decreases. In the permutations, we considered the influence of tree size on the presence of lianas and epiphytes by adjusting binomial models, and used the predicted probabilities to reshuffle marks. Spatial analyses were performed using the spatstat package in R [43,44].

3. Results

3.1. Liana and Epiphyte Occurrences

In this study, we surveyed 2111 trees, of which 727 (34%) were hosts of lianas and 1095 (52%) of epiphytes. Of the total number of trees with lianas, only 44 trees (6%) had exclusively lianas and no epiphytes. In contrast, of the total number of trees with epiphytes, 62% hosted exclusively epiphytes and no lianas. Each tree had a median of one liana species (range: 1–5) and a mean basal area of 0.0025 square meters of lianas (range: 0.0001–0.0295). The trees had a median of two epiphyte species (range: 1–9) and a median epiphyte cover of 1 (range: 1–5).

At the tree zone scale, there were no significant differences between liana and epiphyte occurrences in the trunk (estimate = 0.06, z = 0.24, p = 0.81) and crown of trees (estimate = −0.31, z = −1.17, p = 0.24). The crown had a significantly higher occurrence of epiphytes compared to the trunk (estimate = −0.55, z = −2.72, p = 0.006; Figure 2). In contrast, there was no significant difference between the trunk and crown for liana occurrence (estimate = −0.18, z = −0.56, p = 0.57; Figure 2).

At the tree scale, the presence and absence of lianas and epiphytes were not significantly associated across four tree diameter classes (Table 1). Lianas and epiphytes shared few small trees (i.e., dbh < 13.15 cm), and almost half of the trees larger than 33.76 cm in dbh (Table 1). However, in both cases, there was no deviation from the expected proportions of lianas and epiphytes colonizing the trees.

Table 1. Number of trees hosting lianas and epiphytes per dbh class.

Tree dbh Quartiles	No. of Trees	No. of Trees Hosting Lianas	No. of Trees Hosting Epiphytes	No. of Trees Shared Between Lianas and Epiphytes	Chi-Square Test (p-Value)
1 (<13.15 cm)	528	123	130	27 (11.9%)	0.44 (0.51)
2 (>13.15–<19.75)	535	156	223	62 (19.6%)	0.24 (0.62)
3 (>19.75–<33.76)	521	178	319	111 (28.8%)	0.08 (0.77)
4 (>33.76)	527	270	423	215 (45%)	0.07 (0.79)

On average, 20% of the trees hosted both lianas and epiphytes in most forest patches (Table 2). In certain forest patches, there was a significant prevalence of trees hosting epiphytes compared to the number of trees with lianas (e.g., Reserva, NogalCebil, Frontino, and CuestaVieja; Tables 2 and 3). In contrast, in one of the forest patches (i.e., Mora), there were twice as many trees colonized by lianas as by epiphytes (Tables 2 and 3).

Table 2. Number of trees hosting lianas and epiphytes per forest patch.

Forest Patch Label	Forest Type	No. of Trees	No. of Trees Hosting Lianas	No. of Trees Hosting Epiphytes	No. of Trees Shared Between Lianas and Epiphytes
Guarán	Successional	197	91	110	60 (42.6%)
Nativo	Successional	182	33	49	17 (26.2%)
Reserva	Successional	226	7	120	4 (3.3%)

Table 2. *Cont.*

Forest Patch Label	Forest Type	No. of Trees	No. of Trees Hosting Lianas	No. of Trees Hosting Epiphytes	No. of Trees Shared Between Lianas and Epiphytes
Mora	Successional	141	82	43	32 (34.4%)
NogalCebil	Successional	146	24	95	21 (21.4%)
Cedro	Successional	147	73	55	37 (40.7%)
Sismógrafo	Successional	173	53	61	19 (20%)
Frontino	Successional	168	32	134	27 (19.4%)
CuestaVieja	Mature	190	63	147	47 (28.8%)
Laderas	Mature	135	61	90	52 (52.5%)
Ha1	Mature	210	105	102	54 (35.3%)
Ha6	Mature	196	103	86	45 (31.3%)

Table 3. Results of generalized linear models comparing the number of trees colonized by lianas or epiphytes per forest patch. * indicate that differences are statistically significant.

Forest Patch Label	Estimate	z Value	*p* Value
Guarán	−0.18	−0.85	0.39
Nativo	−0.39	−1.10	0.27
Reserva	−2.84	−7.31	0.001 *
Mora	0.64	2.88	0.003 *
NogalCebil	−1.37	−5.03	0.001 *
Cedro	0.28	1.02	0.31
Sismógrafo	−0.14	−0.58	0.55
Frontino	−1.43	−7.27	0.001 *
CuestaVieja	−0.84	−4.15	0.001 *
Laderas	−0.38	−2.34	0.09
Ha1	0.01	0.13	0.89
Ha6	0.18	1.20	0.22

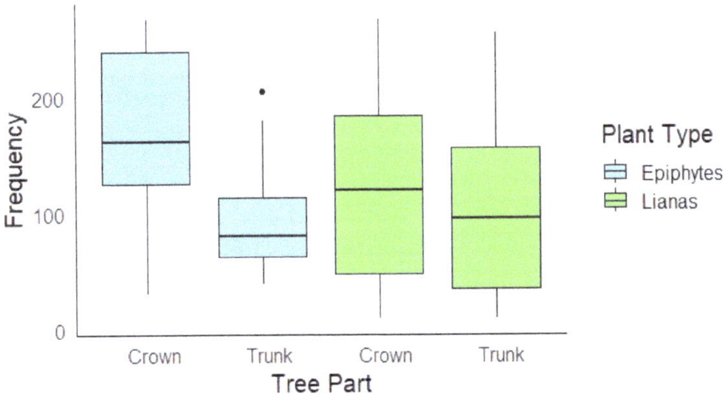

Figure 2. Number of occurrences of lianas and epiphytes on the crown and trunk of trees. The isolated point represents an outlier, a value beyond 1.5 times the interquartile range from the first and third quartiles.

3.2. The Influence of Forest Type and Tree Size on Lianas and Epiphytes

The type of forest influenced liana basal area, epiphyte cover, and the species richness of both lianas and epiphytes in a similar way. Trees in mature forests tended to host more species of lianas and epiphytes compared to those in successional forests (Figure 3a,b). Similarly, tree size was positively associated with liana and epiphyte species richness, liana basal area, and epiphyte cover (Figure 3c,d). Details on the estimations are provided in Supplementary Materials S1.

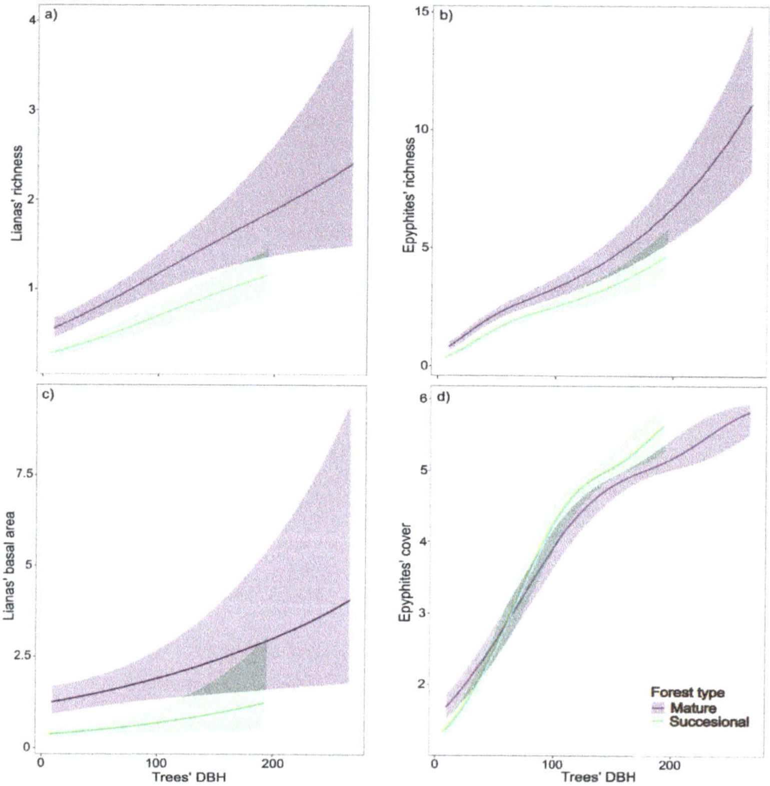

Figure 3. Comparison between mature and successional forests with respect to the association between trees´ dbh and (**a**) the species richness of lianas, (**b**) species richness of epyphites, (**c**) basal area of lianas and (**d**) epyphites' cover.

3.3. Interaction Networks Between Lianas, Epiphytes, and Trees

In general, lianas and epiphytes interacted with the same tree species (Figures 4 and 5). However, there were differences in the frequency of interactions that trees had with liana and epiphyte species. The liana species that showed the highest frequency of interaction with tree species were *Celtis iguanaea*, *Cissus striata*, *Quechualia fulta*, and *Dolichandra unguis-cati*. Liana species interacted most frequently with canopy tree species such as *Ocotea porphyria* and *Terminalia triflora*, as well as with species in the middle and lower strata, such as *Myrcianthes pungens*, *Piper tucumanum*, *Eugenia uniflora*, and *Morus alba* (Figure 4). In contrast, epiphytes interacted almost exclusively with canopy trees such as *Ocotea porphyria*, *Terminalia triflora*, and *Parapiptadenia excelsa*. *Ocotea porphyria* was the tree species that had the most interactions with epiphyte species, including *Pleopeltis tweediana*,

Peperomia tetraphylla, *Microgramma squamulosa*, *Aechmea distichantha*, and *Peperomia theodori* (Figure 5). Among the epiphyte–liana interaction network, epiphyte species co-occurred most frequently with *Cissus striata* and *Dolichandra unguis-cati* (Figure 6). There were significantly more epiphyte occurrences on trees with liana species that use specialized climbing mechanisms than on trees with lianas that merely lean on tree branches (estimate = 1.21, z = 2.74, $p = 0.006$).

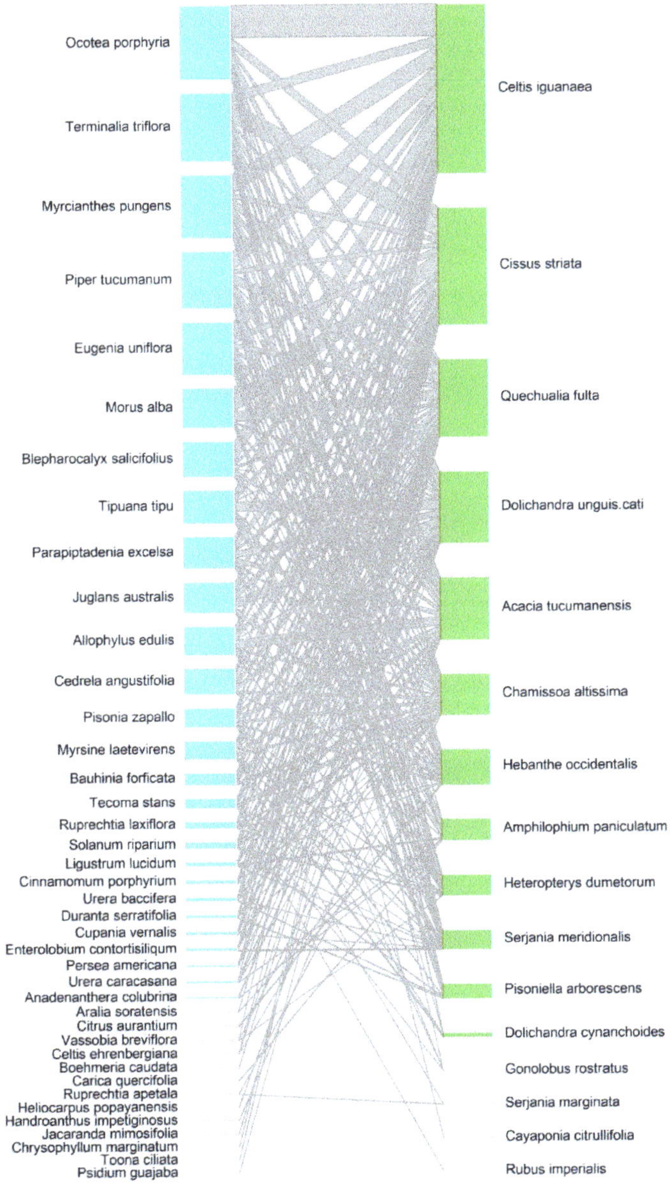

Figure 4. Interaction network between species of trees (**left**) and lianas (**right**). The size of the boxes represents the frequency of interactions between species.

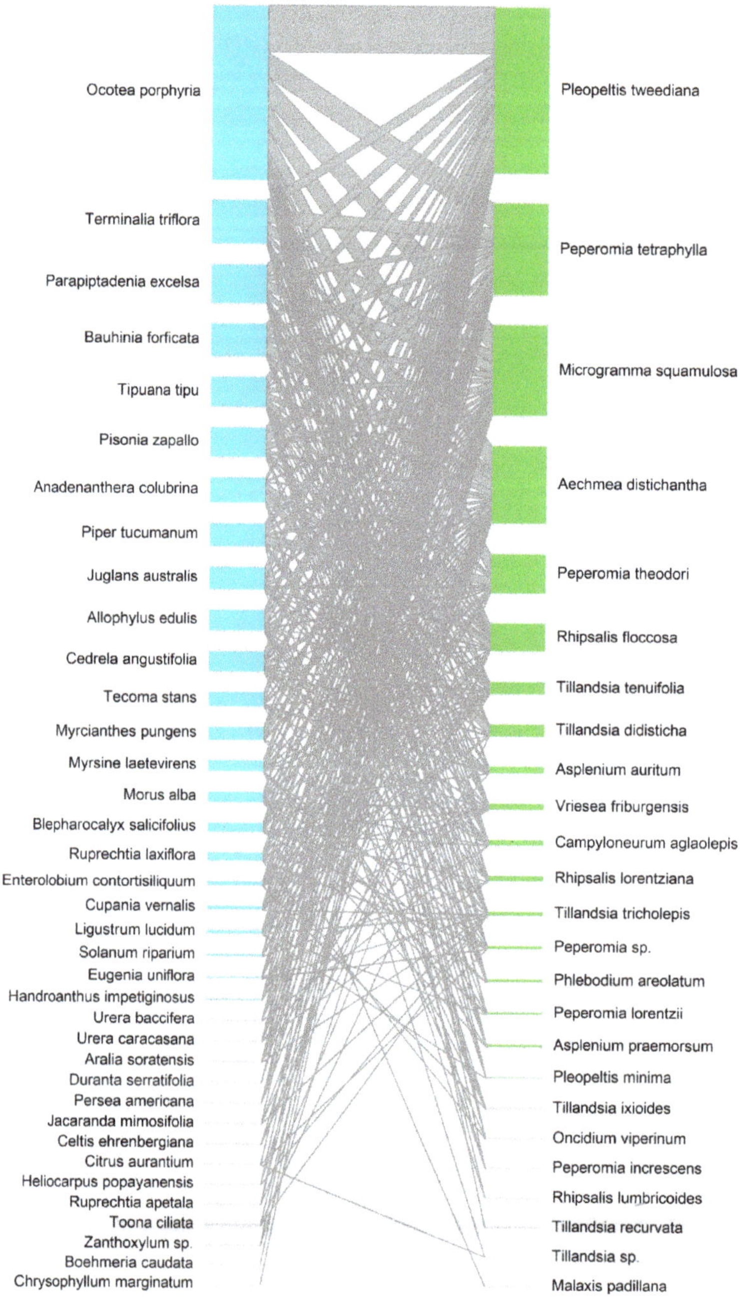

Figure 5. Interaction network between species of trees (**left**) and epiphytes (**right**). The size of the boxes represents the frequency of interactions between species.

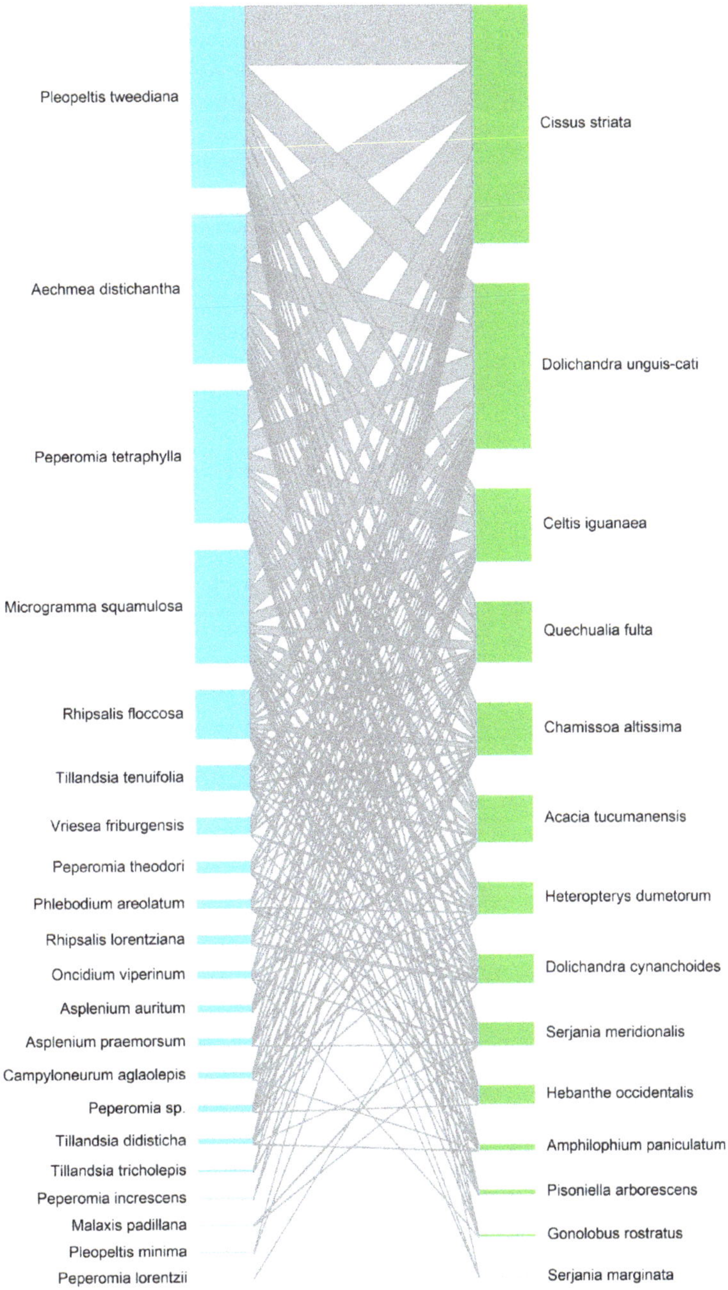

Figure 6. Interaction networks between species of epiphytes (**left**) and lianas (**right**). The size of the boxes represents the frequency of interactions between species.

3.4. Spatial Patterns of Trees Hosting Lianas and Epiphytes

The contiguous 0.36 ha plot contained 120 trees; 66 trees hosted at least one liana and 39 hosted at least one epiphyte. Point pattern analysis revealed that the likelihood of encountering a tree with epiphytes at 3 to 7 m from a tree with lianas was slightly higher than expected, by chance only when the host species and dbh were taken into account (Figure 7). By contrast, the likelihood of a tree hosting lianas did not change in response to an epiphyte-hosting tree at any distance (Figure 7).

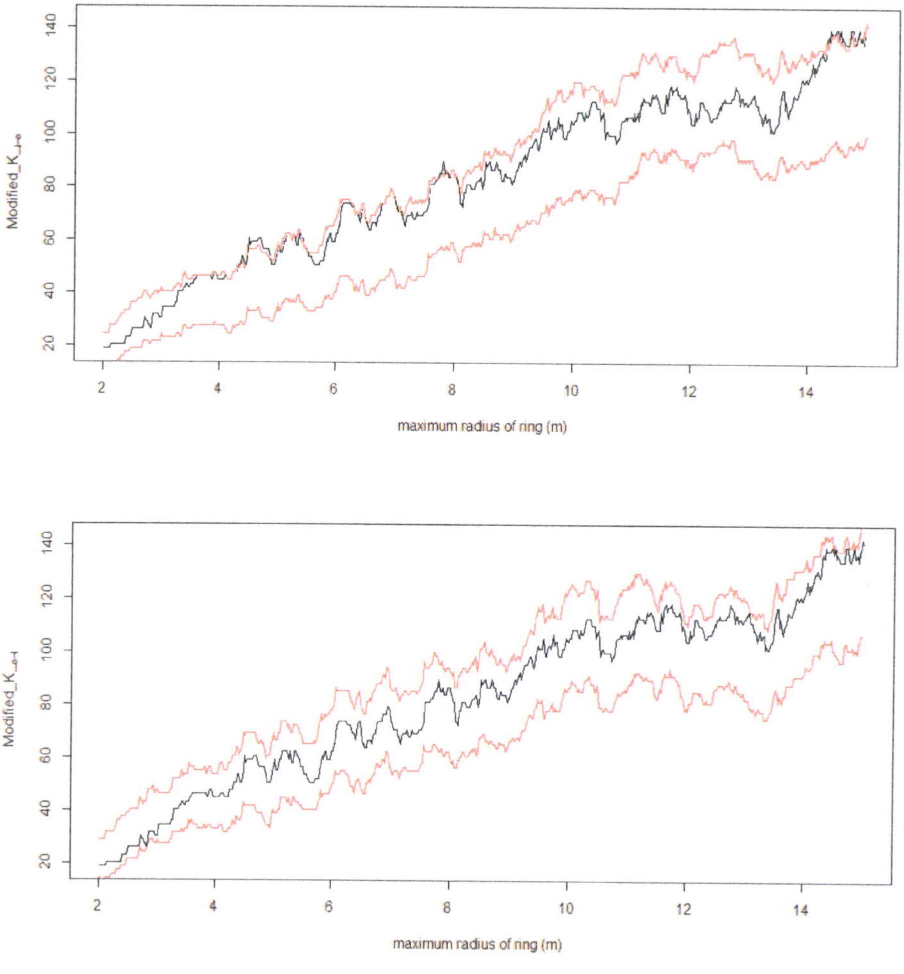

Figure 7. Spatial arrangement of trees with lianas and epiphytes. The curves in black indicate the pair of epiphytes and lianas at increasing distances. The confidence intervals in red indicate the central 95% values estimated from 10,000 permutations. In each of these permutations, we randomly distributed epiphytes (**above**) and lianas (**below**) considering the likelihood of a tree being colonized given its species identity and dbh.

4. Discussion

In this study, we evaluated the co-occurrence of lianas and epiphytes in the subtropical montane forests of northwestern Argentina as an initial step to investigate their potential interactions. To date, lianas and epiphytes have been studied separately, making this the first study to analyze both groups together in this region. We combined two databases of lianas and epiphytes surveyed in the same forest plots to explore patterns of occurrence at different ecological scales (tree zone, tree, and forest patch). The results showed that both lianas and epiphytes occur on the trunk and crown of trees, sharing, on average, 20% of the host trees within the same plots. However, the proportion of shared trees varied considerably between plots (3.3–52.5%), suggesting that local conditions may favor one group over the other [34,35].

The forest structure, composition, and successional status of forest patches may influence the varying prevalence of lianas and epiphytes. In particular, the forest patches Reserva, NogalCebil, Frontino, and CuestaVieja showed a higher proportion of trees with epiphytes compared to those hosting lianas. These forest patches are characterized by the dominance of one or two species of trees, such as *Anadenanthera colubrina* in "Reserva", *Juglans australis* and *Parapiptadenia excelsa* in "NogalCebil", *Ocotea porphyria* in "Frontino", and *Bauhinia forficata* in "CuestaVieja", which possess characteristics favorable for epiphyte colonization [26,34,45]. In addition, these patches have a homogeneous structure, with a single cohort of tall trees that are difficult for many species of lianas to reach. In contrast, forests with lower canopy heights and the presence of fallen trees tend to harbor more lianas, as these conditions enhance light availability and facilitate climbing [7,8]. For instance, the "Mora" forest has a substantial number of fallen trees [28,46,47], resulting in a prevalence of trees with lianas over those with epiphytes. However, these remain merely hypotheses that should be evaluated in the future and take into account the factors that determine whether a forest supports more epiphytes than lianas, or vice versa.

Tree diameter was an important factor influencing the occurrence of lianas and epiphytes. In smaller trees, co-occurrence was low, as they were generally colonized by either lianas or epiphytes, but not both. Nearly 50% of trees with a diameter greater than or equal to 33.8 cm were simultaneously colonized by lianas and epiphytes. In addition, tree diameter was positively associated with liana and epiphyte species richness, liana basal area, and epiphyte cover. This pattern could be explained by the size and age of the trees, as they offer a larger surface area and have been exposed to colonization by lianas and epiphytes for a longer time [9,26,27]. These findings align with results from other forests, showing that both epiphyte [17,19,48] and liana [49–51] prevalence increases among larger trees.

A higher species richness of lianas and epiphytes, as well as a greater liana basal area, was found in mature forests compared to successional forests. This pattern was expected for epiphytes but not for lianas, which tend to be more prevalent in successional stands [10,11,13,14,18]. These results underscore the conservation importance of mature forests, as they harbor the largest trees that support a high diversity of lianas and epiphytes. Moreover, mature forests are more stratified, allowing them to host liana species with diverse climbing modalities adapted to different tree sizes. However, this does not diminish the importance of successional forests for the conservation of these plants. It is essential to recognize that successional forests are in active recovery and, if protected from significant human impacts, could eventually achieve a diversity of lianas and epiphytes comparable to that of mature forests [10,18].

The presence of lianas and epiphytes was greater in large tree species that reach the canopy (e.g., *Ocotea porphyria*, *Terminalia triflora*, *Parapiptadenia excelsa*, and *Tipuana tipu*). On these tree species, epiphytes were found alongside liana species that had specialized climbing mechanisms to reach the canopy. For example, epiphyte species were commonly found in trees hosting lianas with tendrils and adventitious roots (e.g., *Dolichandra unguis-cati* and *Cissus striata*), as well as those with spines (*Celtis iguanaea* and *Acacia tucumanensis*) [36]. On the other hand, liana species that merely lean on tree branches (e.g., *Quechualia fulta* and *Chamissoa altissima*) were more frequent in smaller trees (e.g., *Eugenia uniflora* and

Piper tucumanum), which are typically not colonized by epiphytes [9]. Therefore, epiphytes could be spatially segregated from certain liana species due to the biology of lianas (i.e., specifically their ability to climb trees of certain sizes) rather than competition.

Other studies suggested that lianas could have a negative influence on epiphytes, as lianas cover host trees with a large number of leaves, leaving little space available for epiphytes [20,21]. It has also been suggested that in successional forests, abundant lianas may outcompete epiphytes by increasing shade and occupying colonization sites [52]. In this study, specifically in a successional forest highly dominated by lianas, we found a slight effect of the distance from liana-hosting trees on the likelihood of finding an epiphyte. However, caution is necessary when interpreting this as an effect of lianas on epiphytes, as the effect was not very strong. Furthermore, the non-replicated design and small plot size (0.36 ha) may have limited our ability to detect a conclusive pattern regarding the interactions between lianas and epiphytes.

In this study, the occurrence patterns of lianas and epiphytes were analyzed as an initial approach to understanding the interactions between these two groups of plants. The results indicate that, in this region, both lianas and epiphytes tend to co-occur primarily in mature forests and on larger-diameter trees. Furthermore, it was observed that epiphytes do not coexist uniformly with all liana species, but predominantly with those that have specialized climbing mechanisms. To further explore the nature of these interactions, future research should consider the patterns identified in this study. For instance, studies investigating the type of interaction between lianas and epiphytes in this region should focus on: (1) forests with high cover or abundance of both life forms, (2) tree species commonly hosting both lianas and epiphytes, and (3) specific interactions between epiphytes and certain liana species. In addition, it would be necessary to use a common variable for both lianas and epiphytes, such as canopy or bark occupancy. Building on these aspects, we can further investigate whether lianas and epiphytes compete for the same habitat or facilitate each other's presence.

Supplementary Materials: The following supporting information can be downloaded at: https://www.mdpi.com/article/10.3390/plants14010140/s1, Supplementary Materials S1. BGLMM results.

Author Contributions: S.J.C.: conceptualization, data curation, formal analysis, investigation, methodology, project administration, and writing. E.A.: conceptualization, formal analysis, investigation, methodology, and writing. T.N.R.: conceptualization, formal analysis, investigation, methodology, and writing. All authors have read and agreed to the published version of the manuscript.

Funding: Funding for this study was provided by the Programa de Bosques Andinos-CONDESAN.

Data Availability Statement: Data are available at this link: http://hdl.handle.net/11336/250932 (accessed on 28 November 2024).

Acknowledgments: We thank José Tisone and Iván Jerez for their field assistance, and Parque Sierra de San Javier for granting the permits to conduct this study.

Conflicts of Interest: The authors declare no conflicts of interest.

References

1. Parthasarathy, N. *Biodiversity of Lianas, Sustainable Development and Biodiversity*; Springer: Cham, Switzerland, 2015.
2. Zotz, G. *Plants on Plants—The Biology of Vascular Epiphytes*; Springer: Cham, Switzerland, 2016; 282p.
3. Schnitzer, S.A.; Bongers, F. The ecology of lianas and their role in forests. *Trends Ecol. Evol.* **2002**, *17*, 223–230. [CrossRef]
4. Woods, C.L. Primary ecological succession in vascular epiphytes: The species accumulation model. *Biotropica* **2017**, *49*, 452–460. [CrossRef]
5. Putz, F.E. The natural history of lianas on Barro Colorado Island, Panama. *Ecology* **1984**, *65*, 1713–1724. [CrossRef]

6. Putz, F.E.; Mooney, H.A. *The Biology of Vines*; Cambridge University Press: Cambridge, UK, 1991.
7. Schnitzer, S.A.; Dalling, J.W.; Carson, W.P. The impact of lianas on tree regeneration in tropical forest canopy gaps: Evidence for an alternative pathway of gap-phase regeneration. *J. Ecol.* **2000**, *88*, 655–666. [CrossRef]
8. Schnitzer, S.A.; Mangan, S.A.; Hubbell, S.P. The lianas of Barro Colorado Island, Panama. In *Ecology of Lianas*; Schnitzer, S.A., Bongers, F., Burnham, R.J., Putz, F.E., Eds.; Wiley Blackwell: Oxford, UK, 2015; pp. 76–90.
9. Ceballos, S.J.; Chacoff, N.P.; Malizia, A. Interaction network of vascular epiphytes and trees in a subtropical forest. *Acta Oecol.* **2016**, *77*, 152–159. [CrossRef]
10. Letcher, S.G. Patterns of liana succession in tropical forests. In *Ecology of Lianas*; Schnitzer, S.A., Bongers, F., Burnham, R.J., Putz, F.E., Eds.; John Wiley& Sons: Chichester, UK, 2015; pp. 116–130.
11. Dewalt, S.J.; Schnitzer, S.A.; Denslow, J.S. Density and diversity of lianas along a chronosequence in a central Panamanian lowland forest. *J. Trop. Ecol.* **2000**, *16*, 1–19. [CrossRef]
12. Letcher, S.G.; Chazdon, R.L. Lianas and self-supporting plants during tropical forest succession. *For. Ecol. Manag.* **2009**, *257*, 2150–2156. [CrossRef]
13. Barry, K.E.; Schnitzer, S.A.; van Breugel, M.; Hall, J.S. Rapid liana colonization along a secondary forest chronosequence. *Biotropica* **2015**, *47*, 672–680. [CrossRef]
14. Barthlott, W.; Schmit-Neuerburg, V.; Nieder, J.; Engwald, S. Diversity and abundance of vascular epiphytes: A comparison of secondary vegetation and primary montane rain forest in the Venezuelan Andes. *Plant Ecol.* **2001**, *152*, 145–156. [CrossRef]
15. Cascante-Marín, A.; Wolf, J.H.D.; Oostermeijer, J.G.B.; den Nijs, J.C.M.; Sanahuja, O.; Durán-Apuy, A. Epiphytic bromeliad communities in secondary and mature forest in a tropical premontane area. *Basic Appl. Ecol.* **2006**, *7*, 520–532. [CrossRef]
16. Shoo, L.P.; Freebody, K.; Kanowski, J.; Catterall, C.P. Slow recovery of tropical old-field rainforest regrowth and the value and limitations of active restoration. *Conserv. Biol.* **2015**, *30*, 121–132. [CrossRef] [PubMed]
17. Burns, K.C. Meta-community structure of vascular epiphytes in a temperate rainforest. *Botany* **2008**, *86*, 1252–1259. [CrossRef]
18. Woods, C.L.; DeWalt, S.J. The conservation value of secondary forests for vascular epiphytes in Central Panama. *Biotropica* **2013**, *45*, 119–127. [CrossRef]
19. Woods, C.L.; Cardelús, C.L.; DeWalt, S.J. Microhabitat associations of vascular epiphytes in a wet tropical forest canopy. *J. Ecol.* **2015**, *103*, 421–430. [CrossRef]
20. Magrach, A.; Rodríguez-Pérez, J.; Campbell, M.; Laurance, W.F. Edge effects shape the spatial distribution of lianas and epiphytic ferns in Australian tropical rain forest fragments. *Appl. Veg. Sci.* **2014**, *17*, 754–764. [CrossRef]
21. Bruy, D.; Ibanez, T.; Munzinger, J.; Isnard, S. Abundance, richness and composition of lianas in forest communities along an elevation gradient in New Caledonia. *Plant Ecolog. Divers.* **2017**, *10*, 469–481. [CrossRef]
22. Kelly, D.L. Epiphytes and climbers of a Jamaican rain forest: Vertical distribution, life forms and life histories. *J. Biogeogr.* **1985**, *12*, 223–241. [CrossRef]
23. DeWalt, S.J.; Schnitzer, S.A.; Alves, L.F.; Bongers, F.; Burnham, R.J.; Cai, Z.; Carson, W.P.; Chave, J.; Chuyong, G.B.; Costa, F.R.C.; et al. Biogeographical patterns of liana abundance and diversity. In *Ecology of Lianas*; Schnitzer, S.A., Bongers, F., Burnham, R.J., Putz, F.E., Eds.; John Wiley& Sons: Chichester, UK, 2015; pp. 131–146.
24. Tanaka, H.O.; Itioka, T. Ants inhabiting myrmecophytic ferns regulate the distribution of lianas on emergent trees in a Bornean tropical rainforest. *Biol. Lett.* **2011**, *7*, 706–709. [CrossRef]
25. Malizia, A.; Campanello, P.; Villagra, M.; Ceballos, S. Geographical, Taxonomical and Ecological Aspects of Lianas in Subtropical Forests of Argentina. In *Biodiversity of Lianas, Sustainable Development and Biodiversity*; Parthasarathy, N., Ed.; Springer: Cham, Switzerland, 2015; pp. 17–41. [CrossRef]
26. Ceballos, S.J. Vascular epiphytes in Argentinian Yungas: Distribution, diversity, and ecology. *Bot. Rev.* **2022**, *89*, 91–113. [CrossRef]
27. Malizia, A.; Grau, H.R. Liana-host tree associations in a subtropical montane forest of north western Argentina. *J. Trop. Ecol.* **2006**, *22*, 331–339. [CrossRef]
28. Grau, H.R.; Paolini, L.; Malizia, A.; Carilla, J. Distribución, estructura y dinámica de los bosques de la Sierra de San Javier. In *Ecología de Una Transición Natural Urbana: El Gran San Miguel de Tucumán y la Sierra de San Javier*; Grau, H.R., Ed.; EDUNT: Tucumán, Argentina, 2010; pp. 38–48.
29. Grau, H.R.; Arturi, M.F.; Brown, A.D.; Aceñolaza, P.G. Floristic and structural patterns along a chronosequence of secondary forest succession in Argentinean subtropical montane forests. *For. Ecol. Manag.* **1997**, *95*, 161–171. [CrossRef]
30. Brown, A.D.; Grau, H.R.; Malizia, L.R.; Grau, A. Argentina. In *Bosques Nublados del Neotrópico*; Kappelle, M., Brown, A.D., Eds.; Instituto Nacional de Biodiversidad: San José, Puerto Rico, 2001; pp. 623–659.
31. Hunzinger, H. Hydrology of montane forests in the Sierra de San Javier, Tucuman, Argentina. *Mt. Res. Dev.* **1997**, *17*, 299–308. [CrossRef]
32. Cabrera, A.; Willink, A. *Biogeografía de América Latina*, 2nd ed.; OEA: Washington, DC, USA, 1980.

33. Grau, H.R. Scale-dependent relationships between treefalls and species richness in a Neotropical montane forest. *Ecology* **2002**, *83*, 2591–2601. [CrossRef]
34. Ceballos, S.J. Vascular epiphyte communities in secondary and mature forests of a subtropical montane area. *Acta Oecol.* **2020**, *105*, 103571. [CrossRef]
35. Ceballos, S.J.; Malizia, A.; Chacoff, N. Alternative pathways of liana communities in the forests of northwestern Argentina. *Biotropica* **2020**, *52*, 533–540. [CrossRef]
36. Malizia, A.; Grau, H.R. Landscape context and microenvironment influences on liana communities within treefall gaps. *J. Veg. Sci.* **2008**, *19*, 597–604. [CrossRef]
37. Ceballos, S.J.; Malizia, A. Liana density declined and basal area increased over 12 y in a subtropical montane forest in Argentina. *J. Trop. Ecol.* **2017**, *33*, 241–248. [CrossRef]
38. Gerwing, J.J.; Schnitzer, S.A.; Burnham, R.J.; Bongers, F.; Chave, J.; DeWalt, S.J.; Ewango, C.E.; Foster, R.; Kenfack, D.; Martínez-Ramos, M.; et al. A standard protocol for liana censuses. *Biotropica* **2006**, *38*, 256–261. [CrossRef]
39. Flores-Palacios, A.; García-Franco, J.G. Sampling methods for vascular epiphytes: Their effectiveness in recording species richness and frequency. *Selbyana* **2001**, *22*, 181–191. Available online: https://www.jstor.org/stable/41760095 (accessed on 14 September 2024).
40. Acebey, A.; Krömer, T.; Maass, B.L.; Kessler, M. Ecoregional distribution of potentially useful species of Araceae and Bromeliaceae as non-timber forest products in Bolivia. *Biodivers. Conserv.* **2010**, *19*, 2553–2564. [CrossRef]
41. Johansson, D. Ecology of vascular epiphytes in West African rain forest. *Acta Phytogeogr. Suecica* **1974**, *59*, 1–136.
42. Dormann, C.F.; Fründ, J.; Blüthgen, N.; Gruber, B. Indices, graphs and null models: Analyzing bipartite ecological networks. *Open J. Ecol.* **2009**, *2*, 7–24. [CrossRef]
43. Baddeley, A.; Turner, R. Spatstat: An R package for analyzing spatial point patterns. *J. Stat. Softw.* **2005**, *12*, 1–42. [CrossRef]
44. R Core Team. *R: A Language and Environment for Statistical Computing*; R Foundation for Statistical Computing: Vienna, Austria, 2021. Available online: https://www.r-project.org (accessed on 14 September 2024).
45. Malizia, A. Host tree preference of vascular epiphytes and climbers in a subtropical montane cloud forest of Northwest Argentina. *Selbyana* **2003**, *24*, 196–205. Available online: https://www.jstor.org/stable/41760133 (accessed on 14 September 2024).
46. Grau, H.R.; Aragón, R. Árboles invasores de la Sierra de San Javier, Tucumán, Argentina. In *Ecología de Árboles Exóticos en las Yungas Argentinas*; Grau, H.R., Ed.; LIEY: Tucumán, Argentina, 2000; pp. 5–20.
47. Ceballos, S.J.; Blundo, C.; Malizia, A.; Osinaga Acosta, O.; Carilla, J. Dynamics of tree mortality in subtropical montane forests of Northwestern Argentina. *For. Ecol. Manag.* **2021**, *497*, 119528. [CrossRef]
48. Hietz, P.; Hietz-Seifert, U. Structure and ecology of epiphyte communities of a cloud forest in central Veracruz, Mexico. *J. Veg. Sci.* **1995**, *6*, 719–728. [CrossRef]
49. Chittibabu, C.V.; Parthasarathy, N. Liana diversity and host relationships in a tropical evergreen forest in the Indian Eastern Ghats. *Ecol. Res.* **2001**, *16*, 519–529. [CrossRef]
50. Nabe-Nielsen, J. Diversity and distribution of lianas in a neotropical rain forest, Yasuní National Park, Ecuador. *J. Trop. Ecol.* **2001**, *17*, 1–19. [CrossRef]
51. Pérez-Salicrup, D.R.; De Meijere, W. Number of lianas per tree and number of trees climbed by lianas at Los Tuxtlas, Mexico. *Biotropica* **2005**, *37*, 153–156. [CrossRef]
52. Martin, P.H.; Sherman, R.E.; Fahey, T.J. Forty years of tropical forest recovery from agriculture: Structure and floristics of secondary and old-growth riparian forests in the Dominican Republic. *Biotropica* **2004**, *36*, 297–317. [CrossRef]

plants

MDPI

Article

Diverging Elevational Patterns of Tree vs. Epiphyte Species Density, Beta Diversity, and Biomass in a Tropical Dry Forest

Florian A. Werner [1,*,†] and Jürgen Homeier [2,3,*]

1 Functional Ecology, Institute of Biology and Environmental Sciences, University of Oldenburg, Carl-von-Ossietzkystraße 9-11, 26111 Oldenburg, Germany
2 Faculty of Resource Management, HAWK University of Applied Sciences and Arts, Daimlerstraße 2, 37075 Göttingen, Germany
3 Plant Ecology, Georg-August University of Göttingen, 37073 Göttingen, Germany
* Correspondence: florianwerner@yahoo.com (F.A.W.); juergen.homeier@hawk.de (J.H.)
† Current address: International Climate Initiative (IKI) Support Project, Deutsche Gesellschaft für Internationale Zusammenarbeit (GIZ), Friedrich-Ebert-Allee 32–36, 53113 Bonn, Germany.

Abstract: There is evidence to suggest that vascular epiphytes experience low competition for resources (light, water, and nutrients) compared to terrestrial plants. We tested the hypothesis that low resource competition may lead to higher nestedness among vascular epiphyte assemblages compared to trees. We studied the species composition and biomass of epiphytes and trees along an elevation gradient in a tropical dry forest in SW Ecuador. Both life-forms were inventoried on 25 plots of 400 m^2 across five elevation levels (550–1250 m). Tree species density and total species richness increased with elevation, whereas basal area and biomass did not show significant trends. Epiphyte species density and richness both increased strongly with elevation, in parallel to biomass. Plot-level compositional changes were similarly strong for both life-forms. We attribute elevational increases in the species richness of trees and epiphytes to increasing humidity, i.e., more mesic growth conditions. We attribute the more pronounced elevational increase in epiphyte biomass, species density, and richness—the latter coupled with a higher degree of nestedness—to the greater moisture dependency of epiphytes and relatively low direct competition for resources. Our study provides a first comparison of elevational trends in epiphyte and tree diversity and biomass for a tropical dry forest.

Keywords: alpha diversity; beta diversity; biomass; biotic interactions; competition; determinants of nestedness; Ecuador; species richness; species turnover; Tumbes–Piura dry forest

1. Introduction

Rates of tropical forest loss remain high. Understanding elevational trends in biodiversity and their synergies and trade-offs with environmental services, especially carbon storage, is essential for prioritising biodiversity conservation efforts [1,2]. Moreover, climate change impacts are projected to increase strongly, likely to match land-use change as the main threat to biodiversity in the course of the coming decades [3]. Steep elevational gradients constitute the best natural study models available for predicting how vegetation and species assemblages may change due to global climate change [4].

Meanwhile, tropical dry forests remain poorly studied, in part due to the paucity of intact gradients suitable and available for study. Throughout the tropics, dry forest environments have long been preferred for agricultural use and have undergone intense settlement, conversion, and fragmentation [5]. Man-lit fires escape more easily in arid environments, further reducing and degrading natural vegetation. Today, tropical dry forest environments are among the most threatened, degraded and pressured [6]. As a consequence, few mountains in arid or semiarid environments continue to support contiguous gradients of natural vegetation sufficiently intact for ecological studies [7–9].

Lowland tropical dry forests tend to be markedly less speciose in trees than moist and wet forests, although this varies greatly among regions and with spatial scale [10]. This trend is mirrored by biomass [11]. Elevational patterns also diverge: in moist and wet tropical environments, tree diversity and biomass usually decrease with elevation, e.g., [12,13]. In contrast, in semi-humid and semi-arid environments, tree diversity can show mid-elevation peaks. This pattern has been attributed to increasing air humidity, decreasing vapour pressure deficit (VPD), and hence lower drought stress towards higher elevations, creating more mesic conditions suitable for a wider variety of taxa [14].

Due to their disconnect from soil water, epiphytes are considered particularly sensitive to measures of desiccation stress such as humidity and VPD [15–17]. While vascular epiphytes have developed a wealth of adaptations to avoid desiccation, desiccation tolerance is rare. Hence, few epiphytes are adapted to withstand prolonged periods of drought [18], especially during establishment [19,20]. This sensitivity of vascular epiphytes to water availability is closely reflected by geographical patterns of species diversity. For example, in coastal Ecuador, [21] found that the number of tree species in plots of 0.1 ha decreased moderately from wet forest (114 species) to dry forest (48), while the number of vascular epiphytes decreased dramatically from 127 to just 3. Regarding epiphytic biomass, the highest values have likewise been found in wet mountain environments [22]. However, our understanding of biomass patterns is confined by diverging methodologies, numerous site characteristics, a pronounced bias of studies towards cloud forest sites with exuberant epiphytic vegetation, and the paucity of studies from xeric environments and elevational gradients [22]. Moreover, the representativeness of individual studies is questionable. For example, the highest total epiphytic biomass density on record, 44 t ha^{-1} [23], widely cited in the literature, is based on extrapolations from one tree representing each of three vegetation strata.

There is evidence to suggest that competition does not strongly shape vascular epiphyte assemblages [24–26], and competition may hence be less critical than in terrestrial plants. In all but the most marginal of circumstances, the species composition of terrestrial plant communities is defined by fierce competition for resources. In humid environments, these are primarily light and nutrients [27]. In subhumid to arid environments, water availability plays an increasingly important role [28]. We argue that vascular epiphytes as a life-form are fundamentally distinct in this regard. Light is plentiful in tree crowns in most environments. Although water availability is critical, it is essentially not competed for among epiphytes due to their lack of access to soil water. Any amount not intercepted and adsorbed or stored by epiphytes—the bulk of a good rainfall event—flows and drips to the ground, becoming unavailable for epiphytes. Thus, water is an ephemeral and limiting resource, albeit not strongly competed for, even in humid environments. Quite correspondingly, nutrients are yielded from atmospheric deposition and decomposing litter, absorbed from rainfall and stemflow rather than from the soil, and hence—like water—not strongly competed for [29,30]. As a consequence of these limitations, almost nowhere is there a scarcity of host bark suitable for colonisation, which is further ensured by the challenges of colonising naked bark as well as by host growth and patch disturbance (e.g., dislodgement of epiphyte mats) [25]. Rather than constituting competition, the presence of epiphytes creates niches and microsites for additional species, facilitating successful colonisation by more mesic species and hence higher species density and richness.

Numerous studies have addressed succession in epiphyte assemblages along the twig–branch trajectory, which presents neat chronosequences in any host tree, e.g., [10,31–33]. A characteristic of the successional process is a high degree of facilitation. Xerophilous lichens tend to pioneer the colonisation of exposed twigs, retaining water for and spores of the first bryophytes, which subsequently overgrow, gradually expand, thicken, and diversify to form increasingly moisture-retaining substrates that facilitate the establishment of many vascular species [33,34]. However, it is also vascular species themselves that facilitate other vascular epiphytes. Tank bromeliads largely owe their remarkable success to their ability to store substantial quantities of rainfall over prolonged periods. The spongy base of bromeliad tanks creates suitable microhabitats for more mesic taxa such as orchids or ferns, offering both shade and water, which continues to leak from the tank-forming leaf bases long after precipitation events. Another prime example of facilitating epiphytes is nest ferns, which store water in accumulated litter and litter-derived humus rather than tanks [35]. Countless other vascular epiphytes offer microsites to more ombrophilous taxa in a less sophisticated fashion, by providing shade and increasing rainfall and litter retention through their roots' structuring of smooth host bark and the formation, retention, and accumulation of humus [29,36].

The apparent low competition among vascular epiphytes makes it intriguing to compare their biodiversity patterns to those of terrestrial plant life-forms. The change in species composition and richness along environmental gradients or under climate change tends to be pronounced. Such compositional change or beta diversity results from either one or, more commonly, the combination of two additive mechanisms: (a) replacement, typically by more competitive species, and (b) nestedness, where species drop out without being replaced [37]. These two mechanisms and the processes underlying them are antithetical. In order to understand how and why species assemblages change, we must discern these mechanisms and quantify their contributions to compositional change [38]. Such understanding is also a critical basis for conservation planning [39].

Our study site is characterised by a strong increase in humidity with elevation. We therefore expected pronounced differences in the composition and structure of plant communities. We hypothesised that (1) epiphytes compared to trees will increase more strongly in biomass (1a) and species density (1b) with elevation, due to their high sensitivity to humidity.

We further hypothesised that (2) epiphytes, being less strongly driven by resource competition, will display a greater degree of nestedness in their floristic variability among sampling plots than trees, which compete fiercely for resources.

2. Results

2.1. Environment

Soil properties related to nutrient availability, such as pH value, C/N ratio, or plant-available phosphorus (P_{resin}), tended to deteriorate slightly towards higher elevations (Table 1). Tree surface cover of both epiphytic lichens and bryophytes increased strongly at higher elevations (Table 1).

Table 1. Environmental characteristics (mean ± standard deviation for every 5 plots per elevational stratum).

	550 m			800 m			1050 m			1150 m			1250 m		
	Mean	**±**	**SD**	**Mean**	**±**	**SD**	**Mean**	**±**	**SD**	**Mean**	**±**	**SD**	**Mean**	**±**	**SD**
Elevation (m a.s.l.)	550.6	±	17.5	810.6	±	27.5	1057.4	±	18.1	1152.0	±	13.6	1248.6	±	31.5
Slope inclination (°)	30.6	±	4.8	29.8	±	8.1	22.0	±	3.8	26.5	±	6.3	29.2	±	4.2
Soil pH (KCl)	6.5	±	0.1	6.6	±	0.1	5.9	±	0.1	5.7	±	0.2	5.9	±	0.3
Soil N_{total} (%)	0.2	±	0.0	0.3	±	0.0	0.3	±	0.1	0.3	±	0.1	0.6	±	0.1
Soil C/N	10.1	±	1.0	10.7	±	0.7	11.6	±	0.5	11.2	±	0.5	11.0	±	0.3
Soil P_{resin} (μmol g^{-1})	0.6	±	1.2	2.5	±	0.9	0.5	±	0.2	1.4	±	0.7	1.9	±	1.0
Soil cation exchange capacity (μmol g^{-1})	23.1	±	29.6	58.0	±	42.1	32.4	±	22.6	176.1	±	202.5	296.5	±	206.3
Soil base saturation (%)	99.7	±	0.3	95.0	±	11.3	89.0	±	22.5	88.6	±	11.5	88.8	±	6.8
Lichen cover understorey (%)	1.2	±	0.4	7.0	±	3.7	26.6	±	15.7	37.8	±	12.4	24.0	±	9.5
Lichen cover canopy (%)	2.2	±	2.2	3.0	±	2.7	29.4	±	11.5	46.4	±	7.8	33.6	±	13.3
Bryophyte cover understorey (%)	1.2	±	0.4	1.0	±	0.0	7.4	±	5.9	9.6	±	2.7	41.4	±	6.8
Bryophyte cover canopy (%)	1.0	±	0.0	1.0	±	0.0	3.6	±	2.9	4.8	±	1.5	18.6	±	4.0

2.2. Biomass

Aboveground biomass (AGB) of trees did not change significantly with elevation. Tree AGB was not correlated with elevation nor with measures of soil fertility (C/N, P, and pH; Table S1). Vascular epiphyte biomass, in contrast, increased strongly above 800 m (Figure 1), mirroring the increase in the surface cover of non-vascular epiphytes (lichens and bryophytes) (Table 1). This increase in biomass was essentially a consequence of increasing epiphyte abundance (number of stands) since the mean stand biomass varied little across elevational belts (Table 2).

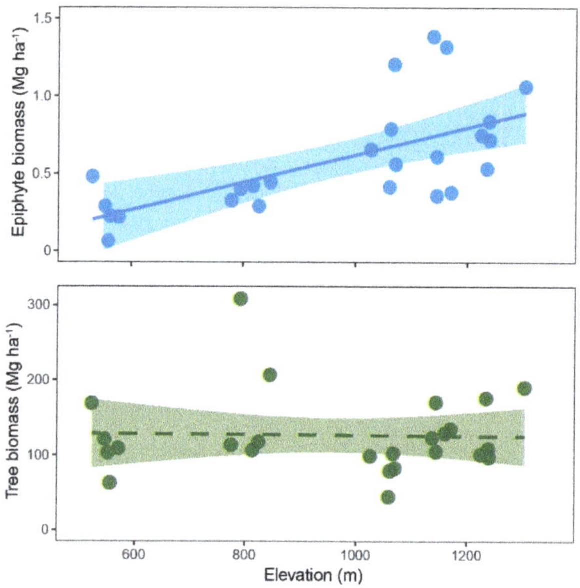

Figure 1. Aboveground biomass of vascular epiphytes (**top**) and trees (**bottom**) along the elevational gradient. Inserted are linear trendlines with 95% confidence intervals.

Table 2. Elevational trends in abundance, biomass, and diversity of trees and vascular epiphytes (mean ± standard deviation for 5 plots per elevational stratum).

	550 m			800 m			1050 m			1150 m			1250 m		
	Mean	**±**	**SD**	**Mean**	**±**	**SD**	**Mean**	**±**	**SD**	**Mean**	**±**	**SD**	**Mean**	**±**	**SD**
Trees															
Individuals	22.2	±	3.8	24.4	±	8.5	43.8	±	5.1	33.0	±	8.7	48.8	±	9.1
Basal area (cm^2 plot^{-1})	8902	±	2773	17,453	±	9400	10,695	±	1398	9838	±	2735	10,081	±	2595
AGB (Mg ha^{-1}) [a]	113	±	38	171	±	87	82	±	23	133	±	24	135	±	45
Species density raw	7.4	±	1.5	10.4	±	3.2	14.0	±	1.6	11.2	±	1.5	16.0	±	3.7
Species density rarefied (n = 13)	5.7	±	0.8	7.6	±	1.1	7.8	±	0.8	7.2	±	0.7	7.9	±	1.5
Species richness observed	17			21			28			22			38		
Species richness estimated															
Chao 1	20.3	±	4.1	28.0	±	7.1	29.3	±	1.7	24.5	±	3.2	39.3	±	1.6
Chao 2	31.4	±	13.1	25.5	±	4.4	30.9	±	2.8	23.7	±	2.1	41.1	±	2.7
Jackknife 1	24.2	±	1.5	27.4	±	3.5	35.2	±	1.5	26.8	±	2.3	47.6	±	2.7
Jackknife 2	29.2			30.4			36.6			27.7			47.6		
Michaelis–Menten (mean)	24.2			27.6			37.0			29.1			59.0		

Table 2. *Cont.*

	550 m			800 m			1050 m			1150 m			1250 m		
	Mean	±	SD	Mean	±	SD	Mean	±	SD	Mean	±	SD	Mean	±	SD
Epiphytes															
Stands (no.)	176	±	56	253	±	25	570	±	339	700	±	178	604	±	287
Plant dry weight (g) [b]	47.7	±	19.1	52.0	±	12.6	51.4	±	14.6	38.3	±	15.6	52.1	±	25.3
Biomass (Mg ha^{-1} slope-corrected)	0.26	±	0.15	0.38	±	0.06	0.73	±	0.30	0.81	±	0.50	0.78	±	0.19
Species density raw	8.4	±	0.9	8.4	±	1.5	14.4	±	2.9	24.4	±	4.0	27.6	±	3.1
Species density rarefied (n = 125)	8.2	±	0.9	7.9	±	1.3	11.2	±	1.3	15.9	±	1.7	19.8	±	2.4
Species richness observed	12			14			24			45			50		
Species richness estimated															
Chao 1	12.0	±	0.2	17.0	±	4.2	25.5	±	2.6	45.8	±	1.3	59.3	±	8.9
Chao 2	12.3	±	0.8	26.0	±	11.0	24.9	±	1.3	48.1	±	2.8	53.8	±	3.2
Jackknife 1	13.6	±	1.0	18.8	±	1.5	28.0	±	1.3	53.8	±	3.4	59.6	±	2.4
Jackknife 2	13.9			22.4			27.4			54.6			61.0		
Michaelis–Menten (mean)	13.3			15.8			28.6			57.0			62.0		

[a] slope-corrected (see Methods section for details). [b] calculated as total biomass × number of stands^{-1}.

2.3. Floristics

In total, we recorded 1128 stems from 60 tree species and 11,518 vascular epiphyte stands from 64 species.

Most speciose tree families were Fabaceae (12 species), Capparaceae (3), Moraceae (3), Polygonaceae (3), and Sapindaceae (3). The most abundant tree species were *Handroanthus chrysanthus* (Bignoniaceae, 74 stems), *Eriotheca ruizii* (Malvaceae, 68), *Ipomoea wolcottiana* subsp. *calodendron* (Convolvulaceae, 58), and *Erythrina velutina* (Fabaceae, 55). All four species are deciduous (Table S2).

Most speciose families of epiphytes were Orchidaceae (27 species) and Bromeliaceae (18), followed by Polypodiaceae (4). Across elevations, the most abundant epiphyte species were bromeliads: *Guzmania monostachia* (1600 individuals), *Tillandsia trichoglochinoides* (1170), *Vriesea spinosa* (1023), and *Tillandsia flagellata* (1018). Epiphyte assemblages at 550–800 m were composed almost entirely of Bromeliaceae, accompanied by few Cactaceae and Orchidaceae, and a stray individual of *Ficus*, whereas other angiosperm and fern taxa were only present at higher elevations (Table S3).

Our study yielded several noteworthy species records. Four epiphyte species are likely new to science ("*sp. nov.*" in Table S3). One tree species has only recently been described scientifically and is endemic to Loja province (*Pradosia aureae* [40]), while another tree species (*Schaefferia serrata*) was formerly only known from Peru.

2.4. Species Density and Richness

Species density was more strongly correlated with elevation in epiphytes (Spearman's $\rho = 0.88$, $p < 0.001$) than in trees ($\rho = 0.73$, $p < 0.001$) (Table S1). While tree species density and total species richness showed a moderate linear increase with elevation, epiphyte species density and richness showed a strong increase above 800 m (Table 2). In both trees and epiphytes, this pattern persisted after controlling for variation in abundance (Figure 2).

The early levelling-off of species accumulation curves (Figure 3) suggests that the size and number of plots sampled were adequate to address our site's species richness, as also confirmed by species richness estimation (Table 2).

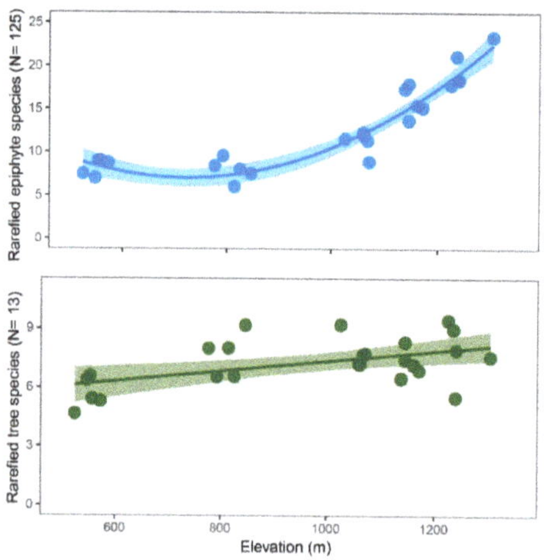

Figure 2. Species density (species per plot) of epiphytes (**top**) and trees (**bottom**) vs. elevation. Species numbers are calculated through individual-based rarefaction at n = 13 (trees) and n = 125 (epiphytes), respectively. Inserted are second-degree polynomial (epiphytes) and linear (trees) trendlines with 95% confidence intervals. Raw species densities are plotted in Supplementary Figure S1.

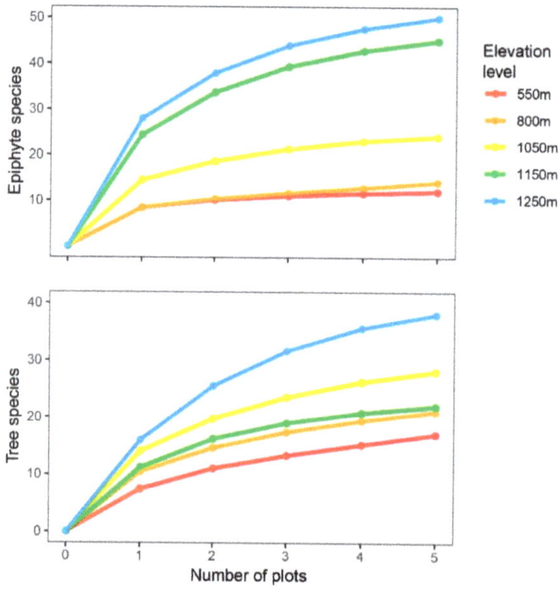

Figure 3. Species accumulation curves across elevational levels (strata) for epiphytes (**top**) and trees (**bottom** panel) as yielded via sample-based rarefaction (5000 runs).

2.5. Interactions between Trees and Epiphytes

Epiphyte stand number, biomass, and raw and rarefied species density showed significant ($p < 0.01$) and strong correlations (Spearman's $\rho \geq 0.69$) with elevation, lichen, and bryophyte cover. All of these vascular epiphyte measures also showed significant correlations with tree individual number (ρ = 0.63–0.67) and raw tree species density (ρ = 0.53–0.68) but not with rarefied tree species density (ρ = 0.63–0.67). No vascular epiphyte measures showed any significant correlation with tree basal area or AGB (Table S1).

2.6. Composition and Turnover

Floristic composition was aligned closely with elevation for both epiphytes and trees (Figure 4).

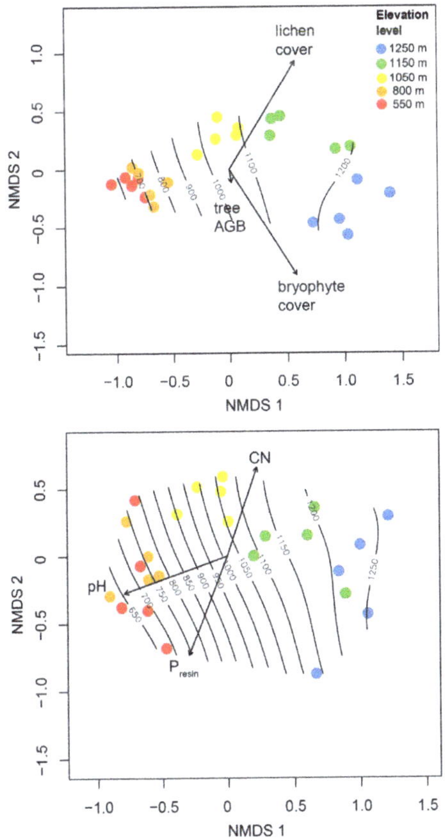

Figure 4. NMDS ordinations of epiphyte (**top**) and tree assemblages (**bottom**), with fitted environmental vectors and regression surfaces. The length and direction of an arrow indicate the strength and sign of the linear correlation of an environmental variable with ordination scores. The contour lines show smooth trends in the relationship between elevation and plot scores. Vectors for epiphytes: tree AGB (not slope-corrected), mean stem lichen cover, and mean bryophyte cover (covers were taken as the average between understorey and canopy values); trees: soil pH, soil C/N, and soil P_{resin}.

Trees exhibited significantly higher levels of compositional change (Sørensen dissimilarity, β_{sor}) at plot level relative to epiphytes (paired *t*-test $t = -12.2$, $p < 0.001$), and the same applied for its species turnover component (βsim; $t = -14.8$, $p < 0.001$). In contrast, the nestedness component (β_{sne}) was significantly higher in epiphytes ($t = 8.5$, $p < 0.001$). All three indices (β_{sor}, β_{sim}, and β_{sne}) increased more strongly with elevational distance in epiphytes than in trees (Figure 5).

Epiphyte assemblages were also significantly more strongly nested in terms of relative nestedness (β_{sne}/β_{sor}; paired *t*-test $t = 9.0$, $p < 0.001$) (Figure 6).

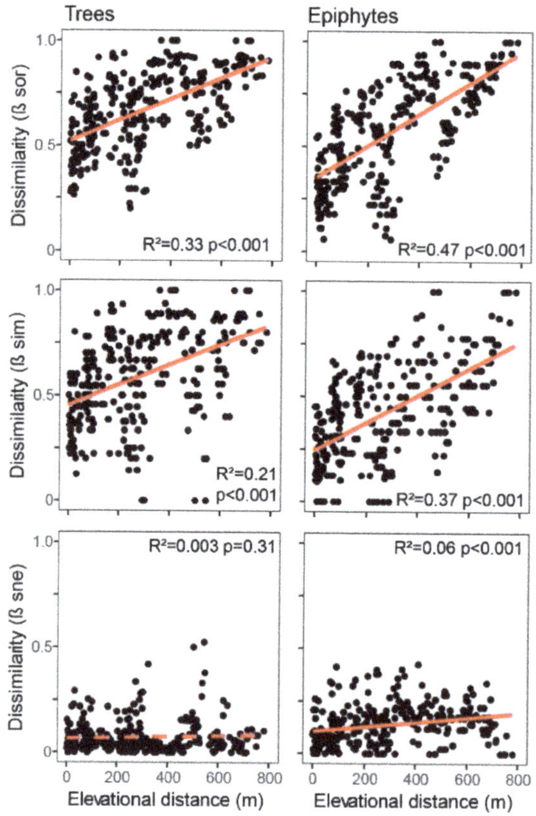

Figure 5. Relationship between compositional dissimilarity and plot elevational distance among trees (**left** panels) and epiphytes (**right** panels). Shown are Sørensen dissimilarity (β_{sor}; **top**) and its components turnover (β_{sim}; **centre**) and nestedness (β_{sne}; **bottom** panels). Linear regression lines with their corresponding R^2 and *p*-values are inserted in the panels. Differences between trees and epiphytes are significant for all three indices (paired *t*-test, df = 299): $t = -12.2$, $p < 0.001$ (β_{sor}); $t = -14.8$, $p < 0.001$ (β_{sim}); $t = 8.5$, $p < 0.001$(β_{sne}).

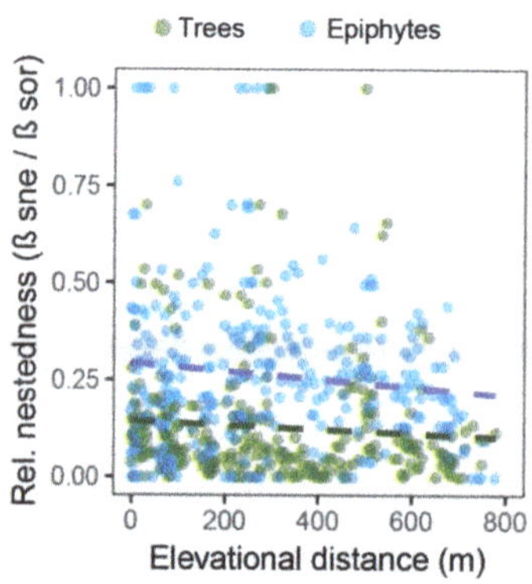

Figure 6. Relative nestedness (β_{sne}/β_{sor}) among trees (green circles) and epiphytes (blue circles). Dashed lines are linear trend lines. Relative nestedness is significantly higher in epiphytes than in trees (paired *t*-test: $t = 9.0$, df = 299, $p < 0.001$).

3. Discussion

3.1. Biomass

Variability in AGB was fairly high, reflecting the rather small plot size for tree biomass assessments. Mean tree AGB did not exhibit a change with elevation despite variability in soil nutrient availability (Tables 1 and 2). In general, dry forest soils are relatively fertile [28,41]. Compared to soils from Ecuadorian moist or wet forests at similar elevations [12,42], the soil nutrient availability in the studied transect is moderate to high and seems to be less limiting for plant growth than water availability. Epiphyte biomass, in contrast, increased drastically, averaging ca. threefold at 1050 m and above, relative to biomass levels at 550–800 m (Table 2). Epiphytic biomass was correlated strongly with elevation (Spearman's $\rho = 0.69$; $p < 0.01$), supporting hypothesis 1a (stronger increase in epiphyte biomass with elevation). This elevational increase was paralleled by even more pronounced increases in lichen and bryophyte cover (Table 1).

Elevation does not affect biodiversity directly, but rather via its effects on climate [43]. VPD decreases from lowland towards the mid-elevation of tropical mountains and beyond, driving elevational changes in epiphyte assemblages [44–48]. Bryophytes are highly sensitive to humidity and VPD, and epiphytic bryophyte cover has been shown to be a good proxy for mean air humidity [49]. It is thus likely that high vascular epiphyte biomass above 800 m is likewise a result of higher levels of humidity.

In humid environments, lichen cover can decrease with increasing humidity [50]. This is partly due to increasing competition from bryophytes, especially in low-exposure microsites such as the lower canopy or understorey, e.g., [51]. Under arid conditions, in contrast, there is little evidence for such competition (see, e.g., [52,53]). Due to the markedly xeric conditions at our study site, bryophyte competition with lichens is likely to be relatively weak even at the highest elevations. Hence, the observed elevational increase in lichen cover may also result from increasing moisture availability.

3.2. Diversity and Composition

Species density and richness of both trees and epiphytes increased markedly with elevation. This increase was more pronounced in epiphytes, supporting hypothesis 1b (stronger increase in epiphyte species density with elevation). With respect to trees, this pattern is rather uncommon. In moist forests, the measures of tree diversity usually decline with elevation, e.g., [13,54]. Interestingly, a recent study from subhumid to humid forests in Mexico [14] found a corresponding increase in species density with elevation. We attribute this unusual pattern to a substantial decrease in hygric stress with elevation, which favours the persistence of mesic species, of which a large species pool exists in the region. Floristic composition was also aligned closely with elevation for both epiphytes and trees (Figure 4). Lichen and bryophyte covers showed a strong relationship with the floristic composition of vascular epiphytes. In contrast, this was not the case for the tree basal area or AGB.

The more drastic increase in epiphyte species richness with elevation appears to confirm their high sensitivity to VPD and drought [15,29,55]. At lower elevations, atmospheric bromeliads strongly dominated epiphyte assemblages, with more mesic taxa (e.g., aroids, ferns, Piperaceae, and most orchids) successively showing presence with increasing elevation. Unlike the lower elevations of our transect, higher elevations are subject to frequent fog, as also evidenced by an abundance of pendant moss near ridge crests (F.A.W. and J.H., pers. obs.; compare also [56,57]). Many epiphytes possess highly evolved structures (e.g., absorbing trichomes in bromeliads), which allow them to absorb condensed fog water. Therefore, water availability may effectively rise more strongly with elevation for epiphytes than for trees, which may go a long way in explaining the more dramatic increase in epiphyte species density and richness with elevation.

All epiphyte response metrics—stand number, biomass, raw and rarefied species density—showed close correlations with both lichen and bryophyte cover, suggesting a strong deterministic role of moisture availability. In contrast, none of the epiphyte response metrics showed even weak correlations with tree basal area or AGB—proxies of epiphyte substrate availability which, moreover, themselves tend to be well-correlated with stand maturity, age, and hence time for epiphyte establishment and accumulation. This lack of correlation is somewhat surprising. Typically, epiphytic biomass and species number are strongly tied to host tree size [22,58,59] and would thus be expected to correlate with tree basal area and biomass at plot level. Possibly, this causality was simply dwarfed by the strength of climate effects.

A mid-elevation peak in species numbers can be assumed to occur wherever diversity does not decline monotonously with elevation in field studies (e.g., [46,55,60,61]; but see also [62]) and database analyses [63,64]. However, in our study, species density and richness increased unabatedly, with no sign of levelling off towards the upper limit of our study gradient. Most likely, our gradient's upper limit was too low to approach peak diversity. In the tropical Andes, vascular epiphyte species richness commonly peaks around 2000 m [64], whereas tree species richness (e.g., [13,54,65]) and tree AGB (e.g., [12]) usually decrease with elevation.

Discussing the differences in diversity patterns of trees vs. epiphytes is challenging due to a lack of references for comparison. Several studies have addressed the diversity of both trees and vascular epiphytes at plot level (e.g., [66–70]). However, to our knowledge, only the authors of [14] have studied both trees and epiphytes in forest plots along an elevational gradient. Their gradient of study was situated in (semi-)humid tropical E-Mexico, stretching from 0 to 3500 m a.s.l. For old-growth forests, they found that the tree species density in plots of 400 m^2 increased slightly from sea level before peaking at 1000–1500 m. Vascular epiphytes, in comparison, showed a much more pronounced peak at 1500 m.

3.3. *Turnover and Nestedness*

We found turnover to exceed nestedness greatly in both trees and epiphytes (Figure 5). This is in line with the bulk of data sets analysed for beta diversity components, with turnover within plant assemblages typically faring over 5 x larger than nestedness [71]. However, both absolute and relative nestedness were significantly greater in epiphytes than in trees, supporting hypothesis 2 (greater nestedness in epiphytes). We attribute higher nestedness in epiphytes primarily to two factors: (a) lower resource competition among epiphytes (most notably for light, nutrients, and water) relative to trees and other terrestrial growth habits, coupled with (b) pronouncedly xeric conditions for epiphytes at our study site, particularly so at lower elevations. Patterns in epiphytic lichen and bryophyte cover at our site underpin the relevance of factor (a): tree surface cover of lichens and bryophytes increased significantly with elevation and was correlated positively with each other and with vascular epiphyte abundance and biomass (Table 1).

Strong nestedness is typical of marginal (extreme) environments. The evidence for the deterministic role of marginal growth conditions (factor b) is provided by the consistent trend in increasing nestedness with latitude towards the poles, which has been found across a variety of taxa [71]. Ref. [72] showed that land snail assemblages are nested along a gradient of habitat quality (especially soil pH). Some evidence for nestedness in marginal environments can even be found in the epiphyte literature. For a Costa Rican wet forest, [58] demonstrated how vascular epiphyte assemblages on saplings and mid-sized trees constitute nested subsets of assemblages on large trees with their more mature and diverse substrates (see also [73]). Ref. [74] found high levels of nestedness in epiphyte assemblages at the highest, frost-prone elevations (3000–3500 m a.s.l.) in tropical Mexico. In a perarid inter-Andean landscape (12 arid months), [18] found epiphyte assemblages of disturbed habitats (especially edge-exposed and isolated trees) strongly nested within the closed woodland assemblage. In this extremely xeric environment, a considerable fraction of bryophyte and vascular local flora was unable to colonise these open-canopy, highly xeric habitats, despite the close proximity of populations in closed woodlands. These species dropped out entirely, without being replaced by even more xerotolerant species.

Floristic similarity for both epiphytes and trees decreased as elevational distance between plots rose, as would be expected as a consequence of gradually diverging environmental conditions (e.g., [75]). Absolute nestedness increased with increasing elevational distance between pairs of plots. This pattern may reflect communities at low elevations being subsets of high-elevation communities to a degree, with the resulting increase in nestedness being outpaced by species turnover as would be expected along such a lengthy gradient in a landscape with a large species pool. We attribute this result to (a) lower resource competition among epiphytes (most notably for light, nutrients, and water) relative to trees and other terrestrial growth habits, coupled with (b) extremely xeric conditions for epiphytes at our study site, especially at lower elevations.

The relationship between absolute pairwise nestedness and elevational distance between plots may not be linear but rather hump-shaped, with nestedness returning to shrink with increasing elevational distance beyond a certain gradient length. This seems reasonable to expect because few species are sufficiently plastic to inhabit both extremes of very long gradients. In fact, somewhat lower levels of nestedness at our largest elevational distances (i.e., 700–800 m, Figure 5) may possibly indicate that a respective inflection point was already passed in our gradient of study.

4. Methods

4.1. Study Site

Field work was carried out from March to September 2010 in the private protected area *Reserva Natural Laipuna*, Loja Province, southwestern Ecuador. The reserve is situated on the northern rim of the Andes' Huancabamba depression, where the exceptionally low stature of the Andes permits a peculiar mosaic of humidity conditions and facilitates species exchange between eastern and western Andean slopes [76,77]. The reserve constitutes one of the few well-preserved elevational gradients of Tumbesian dry forest [78,79]. Tumbesian dry forests span a narrow strip of approximately 50,000 km^2 [80] located between the Pacific Ocean and the Andes, extending from the southwestern tip of Ecuador to northwestern Peru. These forests are found at elevations ranging from sea level to mostly below 1500 m a.s.l. The rainy season lasts from January to May, and the dry season lasts from June to December. At 600 m a.s.l., annual mean temperature is 23.7 °C and annual rainfall is approximately 540 mm, with high year-on-year variability [81,82]. Higher altitudes receive additional moisture input, partly from fog driven by westerly winds [83]. At 1450 m a.s.l., annual mean temperature is 16.1 °C and rainfall is approximately 1260 mm [84]. The forest is deciduous, with an increasing number and abundance of semi-deciduous and evergreen species at higher elevation. Canopy height averages 11–16 m regardless of elevation, with tallest tree individuals reaching up to 20 m. The reserve has been under protection since 2002. Selective logging and extensive goat and cattle foraging had affected the forest in earlier decades [79].

4.2. Forest Inventory

We stratified the ca. 800 m of elevational range accessible within the reserve into five elevational levels. Within each elevational stratum, we laid out five plots of 20 × 20 m (in total 25 plots, at 525–1304 m a.s.l.) (Figure 7). For simplicity, these elevational strata are referred to as 550, 800, 1050, 1150, and 1250 m a.s.l. in this paper. Plots were positioned randomly within mature sections of the reserve, avoiding major recent disturbances such as landslides or larger tree fall gaps.

All trees ≥5 cm dbh (diameter at breast height, 130 cm) were recorded and tagged, measuring tree height with a Vertex IV height meter and a T3 transponder (Haglöf, Langsele, Sweden).

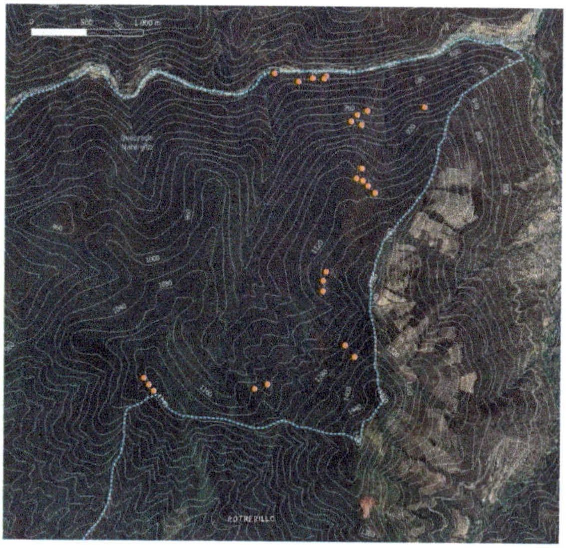

Figure 7. Map of the study area Reserva Natural Laipuna (reserve border: stippled blue line) and plot locations (orange dots).

Facultative and obligatory vascular holo-epiphytes and primary hemi-epiphytes (*sensu* [85]) were recorded for all woody host plants rooted within the plots. We counted the number of stands (an individual or indiscernibly dense groups of individuals of one species; [86]) rooted at >0.25 m above ground using binoculars while omitting seedlings and early juveniles. We modified the method of [60] for assessing epiphytic biomass: for each species and plot, we collected one specimen representative of a mean-sized stand using a telescope pruning pole. If such a stand could not be collected inside a plot, it was sought in the surroundings. In rare cases, it was approximated by dividing a larger stand to the right size or by assigning an estimated correction factor to a smaller stand. Primary hemi-epiphytes (*Ficus, Clusia*) with soil contact were excluded from biomass sampling and analysis.

Species identifications were made and specimens were deposited at the herbaria LOJA, QCA, and QCNE and at the institutional herbaria of selected taxonomic experts (see Acknowledgments).

The percentage of bryophyte and lichen cover of living woody substrate surfaces was estimated visually for each plot quadrant for understorey (conservatively defined as 0.25–2.25 m above ground) and canopy (>2.25 m), respectively. Vascular epiphyte cover was not recorded but probably did not exceed ca. 10% of woody plant surface in any of the plots (F.A.W., pers. obs.).

A topsoil sample representative of 0–10 cm depth was taken in the centre of each plot quadrant, lumped into one mixed sample per plot, and air-dried for lab analysis at the Department of Plant Ecology, University of Göttingen, Germany, following the analytical methods of [42].

4.3. Epiphyte and Tree Biomass Estimation

Epiphyte biomass samples were stored in paper bags and air-dried to constant weight in the lab at ca. 40 °C before determining dry weight on a digital scale. Biomass was then calculated for each plot by multiplying individual numbers by the dry weight of the sampled specimen. We consider this method to be very suitable for our study site characterised by rather low diversity coupled with regular occurrence of most epiphyte species.

We used the allometric equation developed by [87] specifically for tropical dry forests to estimate tree aboveground biomass (AGB):

$$AGB = 0.112 \times (\rho D^2 H)^{0.916}$$

where ρ is wood specific gravity (oven-dried wood mass over green volume), D is trunk dbh, and H is tree height in metres. Wood density data were taken from [88,89].

Since plots were laid out on 400 m^2 of land surface area we considered plot inclination for calculating biomass ha^{-1} and corrected the actual plot area by dividing plot length by cos α (with α being the inclination angle).

4.4. Data Analysis

Because plant abundance varied greatly among plots, we applied individual-based rarefaction to remove the effects of abundance on species density (species per plot) using PAST 4.03 [90]. Sample-based rarefaction and species richness estimations (bias-corrected Chao, Jackknife, and Michaelis–Menten mean) were calculated with EstimateS 8.2 [91]. Community composition was analysed using non-metric multidimensional scaling using Vegan 2.6-4 in R [92].

We followed [37] in separating beta diversity (Sørensen dissimilarity; β_{sor}) into its additive components, species turnover and nestedness-resultant turnover (termed nestedness henceforth). In contrast to other metrics for analysing beta diversity, the turnover component of the Baselga family of metrics is independent of differences in species richness [93]. We calculated Sørensen dissimilarity (β_{sor}) and its components species turnover (β_{sim}) and nestedness (β_{sne}) using betapart 1.6 in R [94] and compared the respective results for epiphytes and trees with paired *t*-tests.

Statistics were carried out using R 4.3.3 [95] unless stated otherwise.

5. Conclusions

Both trees and vascular epiphytes displayed increasing species density and richness with increasing elevation. Those of epiphytes, however, increased more strongly. Unlike trees, epiphytes also increased markedly in their projected biomass and as a function of increasing abundance. The plot-level compositional variability was similarly strong for both life-forms but with significantly higher nestedness in epiphytes than in trees.

We attribute the elevational increases in the species richness of trees and epiphytes to increasing humidity, i.e., more mesic growth conditions at higher elevations. The more pronounced elevational increase in epiphyte biomass, species density, and richness—the latter coupled with a higher degree of nestedness—we attribute to greater moisture dependency of epiphytes and low direct competition for resources (light, water, nutrients) among epiphytes.

To our knowledge, this study is novel in comparing epiphytes to terrestrial plants with regard to beta diversity components, i.e., the mechanisms of nestedness vs. species turnover, revealing a markedly distinct pattern for epiphytes. The notion that, unlike terrestrial plant life-forms, vascular epiphyte assemblages may not be driven by intense competition has long been implicit in epiphyte ecology. However, to our knowledge, it has not been tested in general terms, nor have its potential ecological implications received due attention (but see, e.g., [30,73]).

In ecosystems such as cloud forests, where growth conditions are ideal and plant density is very high, competition among vascular epiphytes may well be substantial [26]. In general, however, across life zones and with sparse exceptions, the vascular epiphyte cover on woody host surfaces is low [25], by and large remaining well below ca. 20–25% throughout the tropics (F.A.W., pers. obs.). This holds true even for many of the most speciose ecosystems, suggesting that comparably low competition among vascular epiphytes may be a phenomenon of general validity. If true, this circumstance has wide implications for epiphyte diversity and its underlying processes. For instance, low competitive pressure could help explain the volatility of epiphyte assemblages in terms of species density, as well as the remarkably high epiphyte α and γ diversity of some forest landscapes (e.g., [15,96,97]). It could further suggest that vascular epiphyte communities are more strongly driven by stochastic processes than the terrestrial life-forms of plants. Future studies will need to clarify if and how low levels of competition affect the diversity and abundance of vascular epiphytes.

To our knowledge, our study further provides a first comparison of elevational patterns in epiphyte and tree diversity and biomass for a tropical dry forest. It revealed considerable species richness and yielded several new species registers, underlining the poor status of floristic knowledge on and high relevance of tropical montane dry forests for global biodiversity conservation. Our results further highlight the importance of preserving contiguous elevational gradients for biodiversity conservation in the face of global climate change.

Supplementary Materials: The following supporting information can be downloaded at https://www.mdpi.com/article/10.3390/plants13182555/s1. Table S1. Correlations among the selected response and predictor variables. Table S2. Tree individuals (≥5 cm dbh) across the 25 study plots. Table S3. Vascular epiphyte abundance ('stands' *sensu* Sanford 1968) across the 25 study plots. Figure S1. Raw species density (species per plot) of epiphytes (top) and trees (bottom) vs. elevation.

Author Contributions: Conceptualisation, methodology, software, validation, formal analysis, investigation, resources, data curation, original draft preparation, review and editing, visualisation, and funding acquisition were carried out by J.H. and F.A.W. All authors have read and agreed to the published version of the manuscript.

Funding: Funding was provided by the Deutsche Forschungsgemeinschaft (DFG) via grants to J.H. (HO 3296/2-2 and 3296/6-2) and F.A.W. (WE 4551/1-2) within the framework of research units FOR 816 and FOR 2730.

Data Availability Statement: Epiphyte data are available through the EpiG database [98]; https://epigdatabase.weebly.com/). Tree data are available via the DFG-FOR816 Data Warehouse at http://www.tropicalmountainforest.org.

Acknowledgments: We thank Naturaleza & Cultura International (NCI) for their support with logistics. Field work was carried out during an ecological field course and a student excursion of the University of Göttingen, both logistically supported by NCI. Thanks are particularly due to Eduardo Cueva for his help with field work coordination and plant identifications. We would further like to thank the following taxonomic experts for their invaluable support with the identifications of specimens from this study site and others: Tom Croat (Araceae), Stig Dalstrøm (Oncidiinae), Carlyle Luer and Lisa Thoerle (Pleurothallidinae), Calaway H. Dodson (non-pleurothallidinid orchids), Eric Hágsater (*Epidendrum*), Harry Luther (Bromeliaceae), Jens Madsen (Cactaceae), José Manuel Manzanares (Bromeliaceae), Guido Matthieu (*Peperomia*), Alan R. Smith (Polypodiaceae), and Walter Till (Bromeliaceae).

Conflicts of Interest: The authors declare no conflicts of interest.

References

1. Fjeldså, J. How broad-scale studies of patterns and processes can serve to guide conservation planning in Africa. *Conserv. Biol.* **2007**, *21*, 659–667. [CrossRef] [PubMed]
2. Thomas, C.D.; Anderson, B.J.; Moilanen, A.; Eigenbrod, F.; Heinemeyer, A.; Quaife, T.; Roy, D.B.; Gillings, S.; Armsworth, P.R.; Gaston, K.J. Reconciling biodiversity and carbon conservation. *Ecol. Lett.* **2013**, *16*, 39–47. [CrossRef] [PubMed]
3. Marco, M.; Harwood, T.; Hoskins, A.; Ware, C.; Hill, S.; Ferrier, S. Projecting impacts of global climate and land-use scenarios on plant biodiversity using compositional-turnover modelling. *Glob. Chang. Biol.* **2019**, *25*, 2763–2778. [CrossRef]
4. Bründl, A.; Sallé, L.; Lejeune, L.; Sorato, E.; Thiney, A.; Chaine, A.; Russell, A. Elevational gradients as a model for understanding associations among temperature, breeding phenology and success. *Front. Ecol. Evol.* **2020**, *8*, 563377. [CrossRef]
5. DRYFLOR. Plant diversity patterns in neotropical dry forests and their conservation implications. *Science* **2016**, *353*, 1383–1387. [CrossRef] [PubMed]
6. Buchadas, A.; Jung, M.; Bustamante, M.; Fernández-Llamazares, Á.; Garnett, S.; Nanni, A.; Ribeiro, N.; Meyfroidt, P.; Kuemmerle, T. Tropical dry woodland loss occurs disproportionately in areas of highest conservation value. *Glob. Chang. Biol.* **2023**, *29*, 4880–4897. [CrossRef]

7. Aguirre Mendoza, Z.; Geada-Lopez, G. Estado de conservación de los bosques secos de la provincia de Loja, Ecuador. *Arnaldoa* **2017**, *24*, 207–228.
8. Rivas, C.A.; Guerrero-Casado, J., Navarro-Cerillo, R.M. Deforestation and fragmentation trends of seasonal dry tropical forest in Ecuador: Impact on conservation. *For. Ecosyst.* **2021**, *8*, 46. [CrossRef]
9. Buchadas, A.; Baumann, M.; Meyfroidt, P.; Kuemmerle, T. Uncovering major types of deforestation frontiers across the world's tropical dry woodlands. *Nat. Sustain.* **2022**, *5*, 619–627. [CrossRef]
10. Amici, A.; Nadkarni, N.; Williams, C.; Gotsch, S. Differences in epiphyte biomass and community composition along landscape and within-crown spatial scales. *Biotropica* **2019**, *52*, 46–58. [CrossRef]
11. Jaramillo, V.J.; Martínez-Yrízar, A.; Sanford, R.L. Primary productivity and biogeochemistry of seasonally dry tropical forests. In *Seasonally Dry Tropical Forests: Ecology and Conservation*; Bullock, S.H., Mooney, H.A., Eds.; Stanford University: Stanford, CA, USA, 2011; pp. 109–128.
12. Homeier, J.; Leuschner, C. Factors controlling the productivity of tropical Andean forests: Climate and soil are more important than tree diversity. *Biogeosciences* **2021**, *18*, 1525–1541. [CrossRef]
13. Malizia, A.; Blundo, C.; Carilla, J.; Osinaga Acosta, O.; Cuesta, F.; Duque, A.; Young, K.R. Elevation and latitude drives structure and tree species composition in Andean forests: Results from a large-scale plot network. *PLoS ONE* **2020**, *15*, e0231553. [CrossRef] [PubMed]
14. Gómez-Díaz, J.A.; Bautista-Bello, A.P.; Carvajal-Hernández, C.I.; Guzmán-Jacob, V.; Monge-González, M.L.; Krömer, T. Diversity Patterns of Vascular Plant Groups Along Gradients of Elevation and Disturbance in Mexico. In *Neotropical Gradients and Their Analysis*; Myster, R., Ed.; Springer: Heidelberg, Germany, 2023; pp. 391–418.
15. Kreft, H.; Köster, N.; Küper, W.; Nieder, J.; Barthlott, W. Diversity and biogeography of vascular epiphytes in Western Amazonia, Yasuní, Ecuador. *J. Biogeogr.* **2004**, *31*, 1463–1476. [CrossRef]
16. Werner, F.A. Reduced growth and survival of vascular epiphytes on isolated remnant trees in a recent tropical montane forest clear-cut. *Bas. Appl. Ecol.* **2009**, *12*, 172–181. [CrossRef]
17. Gotsch, S.; Davidson, K.; Murray, J.; Duarte, V.; Draguljic, D. Vapor pressure deficit predicts epiphyte abundance across an elevational gradient in a tropical montane region. *Am. J. Bot.* **2017**, *104*, 1790–1801. [CrossRef]
18. Werner, F.A.; Gradstein, S.R. Diversity of epiphytes along a gradient of disturbance in an interandean dry forest, northern Ecuador. *J. Veg. Sci.* **2009**, *20*, 59–68. [CrossRef]
19. Zotz, G.; Hietz, P.; Schmidt, G. Small plants large plants. *J. Experim. Bot.* **2001**, *52*, 2051–2056. [CrossRef]
20. Werner, F.A.; Gradstein, S.R. Establishment of vascular epiphytes on isolated and enclosed forest trees in an Andean landscape, Ecuador. *Biodiv. Cons.* **2008**, *17*, 3195–3207. [CrossRef]
21. Gentry, A.H.; Dodson, C.H. Contribution of nontrees to species richness of a tropical rain forest. *Biotropica* **1987**, *19*, 149–156. [CrossRef]
22. Werner, F.A.; Homeier, J.; Oesker, M.; Boy, J. Epiphytic biomass of a tropical Andean forest varies with topography. *J. Trop. Ecol.* **2012**, *28*, 23–31. [CrossRef]
23. Hofstede, R.G.M.; Wolf, J.H.D.; Benzing, D.H. Epiphytic biomass and nutrient status of a Colombian upper montane rain forest. *Selbyana* **1993**, *14*, 37–45.
24. Benzing, D.H. *Vascular Epiphytes: General biology and Related Biota*; Cambridge University Press: New York, NY, USA, 1990.
25. Burns, K.C.; Zotz, G. A hierarchical framework for investigating epiphyte assemblages: Networks, meta-communities, and scale. *Ecology* **2010**, *91*, 377–385. [CrossRef] [PubMed]
26. Spicer, M.E.; Woods, C.L. A case for studying biotic interactions in epiphyte ecology and evolution. *Perspect. Plant Ecol. Evol. Syst.* **2022**, *54*, 125658. [CrossRef]
27. Matsuo, T.; Martínez-Ramos, M.; Onoda, Y.; Bongers, F.; Lohbeck, M.; Poorter, L. Light competition drives species replacement during secondary tropical forest succession. *Oecologia* **2024**, *205*, 1–11. [CrossRef] [PubMed]
28. Gei, M.G.; Powers, J.S. Nutrient cycling in tropical dry forests. In *Tropical Dry Forests in the Americas: Ecology, Conservation, and Management*; Sanchez-Azofeifa, G.A., Powers, J.S., Fernandes, G.W., Quesada, M., Eds.; CRC Press: Boca Raton, FL, USA, 2014; pp. 141–155.
29. Mendieta-Leiva, G.; Porada, P.; Bader, M.Y. Interactions of epiphytes with precipitation partitioning. In *Precipitation Partitioning by Vegetation: A Global Synthesis*; Aubrey, D.P., Ed.; Springer: Heidelberg, Germany, 2020; pp. 133–146.
30. Victoriano-Romero, E.; Valencia-Díaz, S.; García-Franco, J.; Mehltreter, K.; Toledo-Hernández, V.; Flores-Palacios, A. Interactions between epiphytes during canopy soil formation: An experiment in a lower montane cloud forest of southeastern Mexico. *Plant Biol.* **2023**, *25*, 468–477. [CrossRef]
31. Dudgeon, W. Successions of epiphytes in the *Quercus incana* forests at Landour, Western Himalayas. *J. Ind. Bot. Soc.* **1923**, *3*, 270–272.
32. Johansson, D. Ecology of Vascular Epiphytes in West African Rain Forest. Ph.D. Thesis, Uppsala University, Uppsala, Sweden, 1974.
33. Wolf, J.H.D. Non-vascular epiphyte diversity patterns in the canopy of an upper montane rain forest (2550–3670 m), Central Cordillera, Colombia. *Selbyana* **1995**, *16*, 185–195.
34. Ingram, S.W.; Nadkarni, N.M. Composition and distribution of epiphytic organic matter in a Neotropical cloud forest, Costa Rica. *Biotropica* **1993**, *25*, 370–383. [CrossRef]

35. Taylor, A.; Burns, K. Plant composition patterns inside an endemic birds' nest fern (*Asplenium goudeyi*) on Lord Howe Island. *J. Trop. Ecol.* **2015**, *31*, 413–421. [CrossRef]
36. Nadkarni, N.M. Epiphyte biomass and nutrient capital of a Neotropical elfin forest. *Biotropica* **1984**, *16*, 249–256. [CrossRef]
37. Baselga, A. Partitioning the turnover and nestedness components of beta diversity. *Global Ecol. Biogeogr.* **2010**, *19*, 134–143. [CrossRef]
38. Si, X.; Baselga, A.; Ding, P. Revealing beta-diversity patterns of breeding bird and lizard communities on inundated land-bridge islands by separating the turnover and nestedness components. *PLoS ONE* **2015**, *10*, e0127692. [CrossRef] [PubMed]
39. Bergamin, R.; Bastazini, V.; Vélez-Martin, E.; Debastiani, V.; Zanini, K.; Loyola, R.; Müller, S. Linking beta diversity patterns to protected areas: Lessons from the Brazilian Atlantic Rainforest. *Biodiv. Cons.* **2017**, *26*, 1557–1568. [CrossRef]
40. Palacios, W. *Pradosia aureae*. *Cinchonia* **2024**, *19*, 296–302.
41. Pennington, R.T.; Lehmann, C.E.; Rowland, L.M. Tropical savannas and dry forests. *Curr. Biol.* **2018**, *28*, R541–R545. [CrossRef]
42. Unger, M.; Leuschner, C.; Homeier, J. Variability of indices of macronutrient availability in soils at different spatial scales along an elevation transect in tropical moist forests (NE Ecuador). *Plant Soil* **2010**, *336*, 443–458. [CrossRef]
43. Rahbek, C.; Borregaard, M.K.; Colwell, R.K.; Dalsgaard, B.; Holt, B.G.; Morueta-Holme, N.; Nogues-Bravo, D.; Whittaker, R.J.; Fjeldså, J. Humboldt's enigma: What causes global patterns of mountain biodiversity? *Science* **2019**, *365*, 1108–1113. [CrossRef]
44. Krömer, T.; Acebey, A.; Kluge, J.; Kessler, M. Effects of altitude and climate in determining elevational plant species richness patterns: A case study from Los Tuxtlas, Mexico. *Flora* **2013**, *208*, 197–210. [CrossRef]
45. Sanger, J.C.; Kirkpatrick, J.B. Moss and vascular epiphyte distributions over host tree and elevation gradients in Australian subtropical rainforest. *Austral. J. Bot.* **2015**, *63*, 696–704. [CrossRef]
46. Ding, Y.; Liu, G.; Zang, R.; Lu, X.; Huang, J. Distribution of vascular epiphytes along a tropical elevational gradient: Disentangling abiotic and biotic determinants. *Nature* **2016**, *6*, 19706. [CrossRef]
47. Rodríguez Quiel, C.; Zotz, G. Vascular epiphytes assemblages along an elevational gradient on isolated trees in Southwest Panama. *Diversity* **2021**, *13*, 49. [CrossRef]
48. Berrios, H.K.; Coronado, I.; Marsico, T.D. High species richness and turnover of vascular epiphytes is associated with water availability along the elevation gradient of Volcán Maderas, Nicaragua. *Ecol. Evol.* **2022**, *12*, e9501. [CrossRef]
49. Karger, D.N.; Kluge, J.; Abrahamczyk, S.; Salazar, L.; Homeier, J.; Lehnert, M.; Kessler, M. Bryophyte cover of trees as proxy for air humidity in the tropics. *Ecol. Indic.* **2012**, *20*, 277–281. [CrossRef]
50. Batke, S.P.; Murphy, B.R.; Hill, N.; Kelly, D.L. Can air humidity and temperature regimes within cloud forest canopies be predicted from bryophyte and lichen cover? *Ecol. Indic.* **2015**, *56*, 1–5. [CrossRef]
51. Pharo, E.J.; Vitt, D.H. Local variation in bryophyte and macro-lichen cover and diversity in montane forests of western Canada. *Bryologist* **2000**, *103*, 455–466. [CrossRef]
52. Werner, F.A.; Gradstein, S.R. Spatial distribution and abundance of epiphytes along a gradient of human disturbance in an interandean dry valley, Ecuador. *Selbyana* **2010**, *30*, 208–215.
53. Miranda-Gonzalez, R.; McCune, B. The weight of the crust: Biomass of crustose lichens in tropical dry forest represents more than half of foliar biomass. *Biotropica* **2020**, *52*, 1298–1308. [CrossRef]
54. Homeier, J.; Salazar, L.; Chinchero, M.A.; Bossen, S.; Trogisch, S.; Unger, M.; Leuschner, C. Diversidad de plantas vasculares en relación con factores edafológicos y climáticos en los bosques de la Reserva Biosfera Sumaco y sus alrededores. In *Gente, Bosque y Biodiversidad: El rol del Bosque Sobre la Biodiversidad y las Poblaciones Rurales*; Torres, B., Vargas, J.C., Arteaga, Y., Torres, A., Lozano, P., Eds.; Universidad Estatal Amazónica, Programa Economía de Recursos Naturales y Desarrollo Empresarial: Puyo, Ecuador, 2017; pp. 203–218.
55. Krömer, T.; Kessler, M.; Gradstein, S.R.; Acebey, A.R. Diversity patterns of vascular epiphytes along an elevational gradient in the Andes. *J. Biogeogr.* **2005**, *32*, 1799–1809. [CrossRef]
56. Gehrig-Downie, C.; Obregón, A.; Bendix, J.; Gradstein, S. Epiphyte biomass and canopy microclimate in the tropical lowland cloud forest of French Guiana. *Biotropica* **2011**, *43*, 591–596. [CrossRef]
57. Pardow, A.; Gehrig-Downie, C.; Gradstein, S.; Lakatos, M. Functional diversity of epiphytes in two tropical lowland rainforests, French Guiana: Using bryophyte life-forms to detect areas of high biodiversity. *Biodiv. Cons.* **2012**, *21*, 3637–3655. [CrossRef]
58. Woods, C. Primary ecological succession in vascular epiphytes: The species accumulation model. *Biotropica* **2017**, *49*, 452–460. [CrossRef]
59. Komada, N.; Itioka, T.; Nakanishi, A.; Tagane, S.; Shimizu-Kaya, U.; Nakagawa, M.; Meleng, P.; Pungga, R.; Kanzaki, M. Effects of host tree size on the species richness and abundance of epiphyte assemblages in a Bornean lowland tropical forest. *Tropics* **2022**, *30*, 53–61. [CrossRef]
60. Hietz, P.; Hietz-Seifert, U. Structure and ecology of epiphyte communities of a cloud forest in central Veracruz, Mexico. *J. Veg. Sci.* **1995**, *6*, 719–728. [CrossRef]
61. Cardelús, C.L.; Colwell, R.K.; Watkins, J.E. Vascular epiphyte distribution patterns: Explaining the mid-elevation richness peak. *J. Ecol.* **2006**, *94*, 144–156. [CrossRef]
62. Jiménez-López, D.A.; Martínez-Camilo, R.; Martínez-Meléndez, N.; Kessler, M. Diversity of epiphyte ferns along an elevational gradient in El Triunfo Biosphere Reserve, southern Mexico. *Plant Ecol. Evol.* **2023**, *153*, 12–21. [CrossRef]
63. Jørgensen, P.M.; León-Yánez, S. (Eds.) Catalogue of the vascular plants of Ecuador. In *Monographs in Systematic Botany from the Missouri Botanical Garden*; Missouri Botanical Garden Press: St. Louis, MO, USA, 1999; Volume 75, pp. 1–1182.

64. Küper, W.; Kreft, H.; Nieder, J.; Köster, N.; Barthlott, W. Large-scale diversity patterns of vascular epiphytes in Neotropical montane rain forests. *J. Biogeogr.* **2004**, *31*, 1477–1487. [CrossRef]
65. Homeier, J.; Breckle, S.W.; Günter, S.; Rollenbeck, R.T.; Leuschner, C. Tree diversity, forest structure and productivity along altitudinal and topographical gradients in a species-rich Ecuadorian montane rain forest. *Biotropica* **2010**, *42*, 140–148. [CrossRef]
66. Whitmore, T.C.; Peralta, R.; Brown, K. Total species count in a Costa Rican tropical rain forest. *J. Trop. Ecol.* **1985**, *1*, 375–378. [CrossRef]
67. Kelly, D.L.; Tanner, E.V.J.; Lughadha, E.N.; Kapos, V. Floristics and biogeography of a rain forest in the Venezuelan Andes. *J. Biogeogr.* **1994**, *21*, 421–440. [CrossRef]
68. ter Steege, H.; Ek, R.; van Andel, T. A comparison of diversity patterns of tree and non-tree groups. In *Plant Diversity in Guyana. With Recommendation for a Protected Areas Strategy;* Series 18; ter Steege, H., Ed.; Tropenbos International: Ede, The Netherlands, 2000; Chapter 9.
69. Balslev, H.; Valencia, R.; Paz y Miño, G.; Christensen, H.; Nielsen, I. *Forest Biodiversity in North, Central and South America and the Carribean: Research and Monitoring;* Man and the Biosphere Series; Dallmeier, F., Comiskey, J.A., Eds.; UNESCO and The Parthenon Publishing Group: Carnforth, UK, 1998; pp. 591–600.
70. Linares-Palomino, R.; Cardona, V.; Hennig, E.I.; Hensen, I.; Hoffmann, D.; Lendzion, J.; Soto, D.; Herzog, S.K.; Kessler, M. Non-woody life-form contribution to vascular plant species richness in a tropical American forest. *Plant Ecol.* **2009**, *201*, 87–99. [CrossRef]
71. Soininen, J.; Heino, J.; Wang, J. A meta-analysis of nestedness and turnover components of beta diversity across organisms and ecosystems. *Glob. Ecol. Biogeogr.* **2017**, *27*, 96–109. [CrossRef]
72. Hylander, K.; Nilsson, C.; Jonsson, B.; Göthner, T. Difference in habitat quality explain nestedness in a land snail meta-community. *Oikos* **2005**, *108*, 351–361. [CrossRef]
73. Mendieta-Leiva, G.; Buckley, H.L.; Zotz, G. Directional changes over time in the species composition of tropical vascular epiphyte assemblages. *J. Ecol.* **2021**, *110*, 553–568. [CrossRef]
74. Guzmán-Jacob, V.; Zotz, G.; Craven, D.; Taylor, A.; Krömer, T.; Monge, M.; Kreft, H. Effects of forest-use intensity on vascular epiphyte diversity along an elevational gradient. *Div. Distr.* **2019**, *26*, 4–15. [CrossRef]
75. Nascimbene, J.; Spitale, D. Patterns of beta-diversity along elevational gradients inform epiphyte conservation in alpine forests under a climate change scenario. *Biol. Cons.* **2017**, *216*, 26–32. [CrossRef]
76. Weigend, M. Observations on the biogeography of the Amotape-Huancabamba zone in northern Peru. *Bot. Rev.* **2002**, *68*, 38–54. [CrossRef]
77. Richter, M.; Diertl, K.-H.; Emck, P.; Peters, T.; Beck, E. Reasons for an outstanding plant diversity in the tropical Andes of Southern Ecuador. *Landsc. Online* **2009**, *12*, 1–35. [CrossRef]
78. Manchego, C.E.; Hildebrandt, P.; Cueva, J.; Espinosa, C.I.; Stimm, B.; Günter, S. Climate change versus deforestation: Implications for tree species distribution in the dry forests of southern Ecuador. *PLoS ONE* **2017**, *12*, e0190092. [CrossRef]
79. Wurz, A.; Bendix, J.; Homeier, J.; Matt, F.; Paladines, P.; Serrano, F.; Farwig, N. A hidden gem in the Tumbesian dry forest in southern Ecuador: Estacíon Científica Laipuna. *Ecotropica* **2023**, *25*, 202301.
80. Best, B.; Kessler, M. *Biodiversity and Conservation in Tumbesian Ecuador and Peru;* BirdLife International: Cambridge, UK, 1995.
81. Pucha Cofrep, D.; Peters, T.; Bräuning, A. Wet season precipitation during the past century reconstructed from tree-rings of a tropical dry forest in Southern Ecuador. *Glob. Planet. Chang.* **2015**, *133*, 65–78. [CrossRef]
82. Spannl, S.; Volland, F.; Pucha, D.; Peters, T.; Cueva, E.; Bräuning, A. Climate variability, tree increment patterns and ENSO-related carbon sequestration reduction of the tropical dry forest species *Loxopterygium huasango* of Southern Ecuador. *Trees* **2016**, *30*, 1245–1258. [CrossRef]
83. Butz, P.; Hölscher, D.; Cueva, E.; Graefe, S. Tree water use patterns as influenced by phenology in a dry forest of Southern Ecuador. *Front. Plant Sci.* **2018**, *9*, 945. [CrossRef] [PubMed]
84. Peters, T.; Richter, M. Climate Station Data Reserva Laipuna Mountain Peak. 2011. From DFG-FOR816dw. Available online: http://www.tropicalmountainforest.org/data_pre.do?citid=963 (accessed on 14 August 2024).
85. Nieder, J.; Prosperi, J.; Michaloud, G. Epiphytes and their contribution to canopy diversity. *Plant Ecol.* **2001**, *153*, 51–63. [CrossRef]
86. Sanford, W.W. Distribution of epiphytic orchids in semi-deciduous tropical forest in southern Nigeria. *J. Ecol.* **1968**, *56*, 697–705. [CrossRef]
87. Chave, J.; Andalo, C.; Brown, S.; Cairns, M.A.; Chambers, J.Q.; Eamus, D.; Fölster, H.; Fromard, F.; Higuchi, N.; Kira, T.; et al. Tree allometry and improved estimation of carbon stocks and balance in tropical forests. *Oecologia* **2005**, *145*, 87–99. [CrossRef]
88. Chave, J.; Muller-Landau, H.C.; Baker, T.R.; Easdale, T.A.; Steege, H.T.; Webb, C.O. Regional and phylogenetic variation of wood density across 2456 neotropical tree species. *Ecol. Appl.* **2006**, *16*, 2356–2367. [CrossRef]
89. Gonzalez, K.C. Variación Altitudinal del Contenido de Carbono Aéreo de Bosques Secos en la Reserva Natural Laipuna al sur de Ecuador. Master's Thesis, Universidad Nacional de Loja, Loja, Ecuador, 29 November 2023.
90. Hammer, Ø.; Harper, D.A.T.; Ryan, P.D. PAST: Paleontological statistics software package for education and data analysis. *Palaeontol. Electron.* **2001**, *4*, 1–9. Available online: http://palaeo-electronica.org/2001_1/past/issue1_01.htm (accessed on 14 August 2024).
91. Colwell, R.K. EstimateS: Statistical Estimation of Species Richness and Shared Species from Samples. 2006. Available online: http://purl.oclc.org/estimates (accessed on 14 August 2024).

92. Oksanen, J.; Simpson, G.; Blanchet, F.; Kindt, R.; Legendre, P.; Minchin, P.; O'Hara, R.; Solymos, P.; Stevens, M.; Szoecs, E.; et al. Vegan: Community Ecology Package_. R package version 2.6-4, 2022. Available online: https://CRAN.R-project.org/package=vegan (accessed on 14 August 2024).
93. Baselga, A.; Leprieur, F. Comparing methods to separate components of beta diversity. *Meth. Ecol. Evol.* **2015**, *6*, 1069–1079. [CrossRef]
94. Baselga, A.; Orme, D.; Villeger, S.; De Bortoli, J.; Leprieur, F.; Logez, M.; Martinez-Santalla, S.; Martin-Devasa, R.; Gomez-Rodriguez, C.; Crujeiras, R. Betapart: Partitioning Beta Diversity into Turnover and Nestedness Components. R package version 1.6. 2023. Available online: https://CRAN.R-project.org/package=betapart (accessed on 14 August 2024).
95. R Core Team. *R: A Language and Environment for Statistical Computing*; R Foundation for Statistical Computing: Vienna, Austria, 2024. Available online: https://www.R-project.org/ (accessed on 14 August 2024).
96. Werner, F.A.; Homeier, J.; Gradstein, S.R. Ecology of vascular epiphytes on isolated remnant trees in the montane forest belt of southern Ecuador. *Ecotropica* **2005**, *11*, 21–40.
97. Köster, N.; Friedrich, K.; Nieder, J.; Barthlott, W. Conservation of epiphyte diversity in an Andean landscape transformed by human land use. *Conserv. Biol.* **2009**, *23*, 911–919. [CrossRef]
98. Mendieta-Leiva, G.; Ramos, F.N.; Elias, J.P.; Zotz, G.; Acuña-Tarazona, M.; Alvim, F.S.; Barbosa, D.E.F.; Basílio, G.A.; Batke, S.P.; Benavides, A.M.; et al. EpIG-DB: A database of vascular epiphyte assemblages in the Neotropics. *J. Veg. Sci.* **2020**, *31*, 518–528. [CrossRef]

MDPI AG
Grosspeteranlage 5
4052 Basel
Switzerland
Tel.: +41 61 683 77 34

Plants Editorial Office
E-mail: plants@mdpi.com
www.mdpi.com/journal/plants

www.ingramcontent.com/pod-product-compliance
Lightning Source LLC
Chambersburg PA
CBHW061554210926
51850CB00009B/2097